Bacterial Resistance to Antibiotics – From Molecules to Man

Bacterial Resistance to Antibiotics – From Molecules to Man

Edited by

Boyan B. Bonev
School of Life Sciences
University of Nottingham
Queen's Medical Centre
Nottingham, UK

Nicholas M. Brown
Consultant Medical Microbiologist
Cambridge University Hospitals NHS Foundation Trust
Cambridge Biomedical Campus
Cambridge, UK

Registered Offices
John Wiley & Sons, Inc., 111 River Street, Hoboken, NJ 07030, USA
John Wiley & Sons Ltd, The Atrium, Southern Gate, Chichester, West Sussex, PO19 8SQ, UK

Editorial Office
The Atrium, Southern Gate, Chichester, West Sussex, PO19 8SQ, UK

For details of our global editorial offices, customer services, and more information about Wiley products visit us at www.wiley.com.

Wiley also publishes its books in a variety of electronic formats and by print-on-demand. Some content that appears in standard print versions of this book may not be available in other formats.

Library of Congress Cataloging-in-Publication Data

Names: Bonev, Boyan B., 1966– editor. | Brown, Nicholas M., 1962– editor.
Title: Bacterial resistance to antibiotics – from molecules to man / edited by Boyan B. Bonev,
    Nicholas M. Brown.
Description: Hoboken, NJ : Wiley, 2019. | Includes bibliographical references and index. |
Identifiers: LCCN 2019003167 (print) | LCCN 2019003905 (ebook) | ISBN 9781119558200 (Adobe PDF) |
    ISBN 9781119558224 (ePub) | ISBN 9781119940777 (pbk.)
Subjects: | MESH: Drug Resistance, Bacterial
Classification: LCC RM267 (ebook) | LCC RM267 (print) | NLM QW 45 | DDC 615.7/922–dc23
LC record available at https://lccn.loc.gov/2019003167

Cover Design: Wiley
Cover Images: © science photo/Shutterstock, © nobeastsofierce/Shutterstock

Set in 10/12pt Warnock by SPi Global, Pondicherry, India

Printed in the UK by Bell & Bain Ltd, Glasgow

10 9 8 7 6 5 4 3 2 1

Contents

List of Contributors

Kristin J. Adolfsen
Department of Chemical and Biological
Engineering, Princeton University,
Princeton, NJ, USA

Vassiliy N. Bavro
School of Biological Sciences, University
of Essex, Colchester, UK

Alison J. Baylay
Institute of Microbiology and
Infection, University of Birmingham,
Birmingham, UK

Boyan B. Bonev
School of Life Sciences, University of
Nottingham, Queen's Medical Centre,
Nottingham, UK

Nicholas M. Brown
Cambridge University Hospitals NHS
Foundation Trust, Cambridge Biomedical
Campus, Cambridge, UK

Mark P. Brynildsen
Department of Chemical and Biological
Engineering, Princeton University,
Princeton, NJ, USA

Clemente Capasso
Istituto di Bioscienze e Biorisorse – CNR,
Napoli, Italy

Vincent Cattoir
Inserm Unit U1230, University of Rennes
1, Rennes, France
Department of Clinical Microbiology,
Rennes University Hospital, Rennes,
France
National Reference Center for
Antimicrobial Resistance (Lab
Enterococci), Rennes, France

Karl Drlica
New Jersey Medical School, Rutgers
Biomedical and Health Sciences, Public
Health Research Institute, Newark,
NJ, USA

Hiroshi Hiasa
Department of Pharmacology,
University of Minnesota Medical School,
Minneapolis, MN, USA

Robert Kerns
Division of Medicinal and Natural
Products Chemistry, University of Iowa,
Iowa City, IA, USA

François Lebreton
Departments of Ophthalmology, Microbiology and Immunobiology, Harvard Medical School, Massachusetts Eye and Ear Infirmary, Boston, MA, USA

Muhammad Malik
New Jersey Medical School, Rutgers Biomedical and Health Sciences, Public Health Research Institute, Newark, NJ, USA

Robert L. Marshall
Institute of Microbiology and Infection, University of Birmingham, Birmingham, UK

Wendy W.K. Mok
Department of Chemical and Biological Engineering, Princeton University, Princeton, NJ, USA

Arkady Mustaev
New Jersey Medical School, Rutgers Biomedical and Health Sciences, Public Health Research Institute, Newark, NJ, USA

Alex J. O'Neill
University of Leeds, Leeds, UK

Laura J.V. Piddock
Institute of Microbiology and Infection, University of Birmingham, Birmingham, UK

Marilyn C. Roberts
Department of Environmental and Occupational Health Sciences, University of Washington, Seattle, WA, USA

Liam K.R. Sharkey
Institute of Infection and Immunity, University of Melbourne, Melbourne, Australia

Claudiu T. Supuran
Dipartimento di Scienze Farmaceutiche, Università degli Studi di Firenze, Polo Scientifico, Florence, Italy

Mark A. Webber
Quadram Institute, Norwich, UK

Ada Yonath
The Helen and Milton A. Kimmelman Center for Biomolecular Structure and Assembly, Weizmann Institute, Rehovot, Israel

Ying Zhang
Department of Molecular Microbiology and Immunology, Bloomberg School of Public Health, Johns Hopkins University, Baltimore, MD, USA

Xilin Zhao
New Jersey Medical School, Rutgers Biomedical and Health Sciences, Public Health Research Institute, Newark, NJ, USA
State Key Laboratory of Molecular Vaccinology and Molecular Diagnostics, School of Public Health, Xiamen University, Xiamen, China

Preface

The use of antibiotics alongside aseptic medical procedures, improved hygiene, and vaccination has revolutionized health care and has resulted in a major decline in morbidity and mortality from acute bacterial infections. Wound and post-surgical complications associated with localized or systemic bacterial infections are now less common and survival rates are high in vulnerable patient populations, such as neonates, and patients undergoing cancer chemotherapy or solid organ transplantation. Infections that were rampant in the past, such as cholera, diphtheria, typhoid fever, plague, and syphilis are now generally manageable, and their frequency has been reduced by many orders of magnitude.

The initial adoption and adaptation of antibiotic compounds for bacterial management has proven successful and resulted in societal and political developments. In the second half of the twentieth century, the widespread use of antibiotic prophylaxis and the use of antibiotics in animals became commonplace and we now appreciate that this seems to have swayed the balance in favor of bacterial adaptations and the subsequent emergence and spread of antibiotic resistance. In the late 1960s, the problem of bacterial infection was considered "solved" and scientific efforts were directed toward other areas, such as cancer. This led to a decline in antibiotic development and we currently have a dearth of concept drugs undergoing development. There are broadly three main reasons for this decline: research – the lack of new leads; regulatory – complex and costly approval routes for antibiotic development; and financial – the high cost of development paralleled by a poor return on investment due to low use of the final product and short treatment regimens.

Antibiotics are comparatively low molecular weight compounds with selective inhibitory activity, which are derived from or synthetically guided by natural microbial products that suppress competing bacteria to secure access to nutrients and to dominate the ecological niche. In turn, natural evolutionary adaptations in competing bacteria subjected to such antibiotic pressures have driven the development of resistance as a survival mechanism and the selection of bacterial populations that have an enhanced tolerance to antibiotics. The evolution of new antimicrobials and bacterial adaptations to such compounds is an ongoing natural process, presently influenced by human intervention through the extensive use of antibiotics.

Adaptations under sustained antibiotic pressure are beginning shift the balance in favor of bacterial pathogens. As a result, bacterial populations refractive to antibiotic management present serious emergent threats including, but not restricted to, carbapenem-resistant and extended-spectrum Beta lactamase-producing *Enterobacteriaceae*, multiply-resistant *Neisseria gonorrhoeae*, resistant enteric pathogens (such as *Salmonella typhi* and *Campylobacter* sp.), methicillin-resistant *Staphylococcus aureus (MRSA)* and vancomycin-resistant enterococci (VRE). These adaptive changes are communicated between bacterial populations and crossing interspecies boundaries with ease. Combined with a globalized societal and economic dynamics, the increased prevalence of resistant bacterial pathogens is rapidly becoming one of the biggest threats to health that humans face now and in the coming decades.

We have compiled this book with the intention that it serves as an introduction to antibiotic action and resistance for advanced undergraduates, graduate students, and clinicians, as well as a reference. Recognizing the complexity of bacterial resistance to antibiotics and its global impact, as well as the need of multidisciplinary and worldwide effort in tackling resistance, we have brought together contributions from experts from diverse fields and complementary expertise that shares a common interest in antimicrobial chemotherapy and in tackling the resistance of bacteria to antibiotics.

The Editors are UK based and met at a British Society for Antimicrobial Chemotherapy (BSAC) initiative to reboot the Antimicrobial Drug Development and Design pipeline. Boyan B. Bonev is a physical biochemist and structural biologist at the University of Nottingham, UK, with interests in antibiotic mechanisms, resistance, molecular structure, and drug design. Nicholas M. Brown is a consultant microbiologist at Addenbrooke's Hospital in Cambridge, UK, and a former president of BSAC with a career-long interest in the use of antibiotics and antibiotic resistance. We thank David Turner, a microbiology consultant at Nottingham, who was co-Editor during the inception of the project. And, thanks to Nick for taking over in his stead.

The book includes contributions from worldwide leaders with complementary expertise and common interest in antimicrobial mechanisms and resistance. Ada Yonath is a structural biologist, Director of the Centre for Biomolecular Structure and Assembly of the Weizmann Institute of Science in Rehovot, Israel, who shares the 2009 Nobel Prize in Chemistry for solving the structure of the ribosome, an important molecular target for antibiotics. Molecular mechanisms of antibiotic action and resistance are introduced by Laura Piddock, who is a molecular microbiologist at the University of Birmingham, Fellow of the American Academy of Microbiology and a founding Fellow of the European Society of Clinical Microbiology and Infectious Diseases and recently joined the Global Antibiotic Research and Development Partnership (GARDP) as Head of Scientific Affairs; and, by Alex O'Neill, a molecular microbiologist at the University of Leeds working on antibiotic resistance and drug discovery. Glycopeptide resistance is discussed by Vincent Cattoir, a microbiologist and Director of the Centre National de Référence de la Résistance aux Antibiotiques, Rennes, France; resistance to tetracycline antibiotics is introduced by Marilyn Roberts, a microbiologist from the University of Washington, USA; Karl Drlica, an expert in microbiology, biochemistry, and molecular genetics from Rutgers Public Health Research Institute, Newark, USA, offers a detailed description of fluoroquinolones mechanisms and resistance; resistance to dihydrofolate reductase inhibitors is introduced by Claudiu Supuran from the University of Florence, Italy. Tuberculosis is a special challenge and an important case of resistance to anti-tuberculosis drugs is made by Ying Zhang, a molecular microbiologist and immunologist from Johns Hopkins Bloomberg School of Public Health, Baltimore, USA, with a special interest in antibiotic resistance in mycobacteria. Multidrug resistance, a common mechanism in many bacterial adaptations to antibiotics, is described by Vassiliy

Bavro, a structural biologist from the University of Essex, Colchester, UK. Looking into alternative approaches beyond classic antimicrobial chemotherapy, anti-virulence and other alternative or complementary strategies are discussed by Mark Brynildsen, an expert in host–pathogen interactions, quorum sensing, and bacterial persistence from the Princeton School of Chemical and Biological Engineering, Princeton, USA, who also introduces aminoglycoside resistance.

The scale and magnitude of the antimicrobial resistance problem is created and exacerbated by a globalized and complex society. Managing bacterial infections affects all and tackling antibiotic resistance is a problem of equal measure for molecular scientists, pharmacologists, clinicians, veterinarians, and mathematicians, as it is for individuals, communities, sociologists, economists, and politicians. Ensuring successful management of bacterial infections requires a thorough understanding of bacterial organization and life at the molecular, systemic, and ecological level, of bacterial/host interactions and dynamics, as well as of the flexible and adaptive arsenal of measures used to control and regulate our immediate microbial environment. A global effort in antimicrobial stewardship and surveillance, public awareness, and detailed advice for clinicians will be needed to maintain the high levels of success in the management of bacterial infections along with joint input from governments, regulators, manufacturers, academics, economists, and clinicians.

Boyan B. Bonev and Nicholas M. Brown

Foreword

Could a bright outlook for antibiotics usage emerge from the colossal health issue?

Ada Yonath

The Helen and Milton A. Kimmelman Center for Biomolecular Structure and Assembly, Weizmann Institute, Rehovot, Israel

One of the major problems in modern medicine is the increasing resistance of pathogenic bacteria to antibiotics. Since the production of the first pharmaceutically active antibiotics around the mid twentieth century, which revolutionized the treatment of infectious diseases and led to an unforeseen decrease in mortality and an increase in life expectancy, the clinical usage of the currently available antibiotics has suffered from a number of severe problems. These fallouts include (i) the development of resistance to one or several antibiotics (namely, multidrug resistance), caused by pathogens capability of undergoing modifications and mutations that minimize or remove the contacts between the antibiotics and their targets, (ii) the unintentional damage of the microbiome owing to the preference for using broad-spectrum antibiotics alongside the structural similarities of the antibiotics' binding sites among diverse bacteria, and (iii) the contamination of the environment caused by significant amounts of antibiotic metabolites that enter it. This crucial environmental issue results from the chemical nature of the molecular scaffolds of most currently used antibiotics, which are composed of organic metabolites that cannot be fully digested by humans or animals. These nondigestible, rather toxic compounds are also nonbiodegradable and contaminate the environment. Furthermore, following release into agricultural irrigation systems, these compounds are increasingly being consumed by humans and animals and thereby spreading antibiotic resistance.

Currently, almost all clinically useful antibiotic therapeutics are derived from natural compounds produced by microorganisms for inhibiting the growth of competing bacteria so they can defend themselves. Many of the natural antibiotics that are medically useful have undergone subsequent chemical modifications to improve their effectiveness. In addition to the natural and semisynthetic substances, very few fully synthetic drugs are in use.

Similarly, the various resistance mechanisms are basic natural processes for the survival of microorganisms, regardless of their exposure to modern clinical treatment and/or nutrition, thus suggesting that microbes have long evolved the capability to fight toxins, including antibiotics. Resistance to antibiotics is generally acquired by molecular mechanisms, some of which, such as activation of cellular efflux pumps, are common to almost all antibiotics. In conjunction, many bacteria have developed specific molecular pathways that cause resistance. The prominent frequently used mechanisms of acquiring resistance to a single or several antibiotics (called multidrug resistance) include modifications of the antibiotic binding pockets by mutations; activation of key enzymatic processes, such as methylation; enzymatic inactivation of the antibiotic; removal of the antibiotic drug from its target by cellular components; or disruption of the interactions between cellular components that play key roles in key life processes. **Indeed, it seems that combating resistance to antibiotics is unlikely, since bacteria "want" to live and because bacteria are extremely "clever" in terms of survival!**

The increasing development of multidrug-resistant bacterial strains, together with the minimal (negligible) number of new antibiotic drugs that are presently undergoing development and/or clinical trials by the major pharmaceutical companies, is becoming a colossal health threat. Thus, the World Health Organization stated that it seems that we will soon revert back to the pre-antibiotic era, during which diseases caused by parasites or by simple (*e.g.* pneumonia, wounds) or severe infections (such as tuberculosis), were almost untreatable and resulted in frequent deaths. The World Bank estimated that up to 3.8% of the global economy will be lost by 2050 because of resistance to antibiotics and several funding agencies, such as the National Institutes of Health, European Research Council, and the Group of Eight, came up with grants for researching antibiotics resistance. On the other hand, most pharmaceutical companies have stopped developing new antibiotics, owing to the expected development of resistance and to the huge mismatch between the investment needed and the low profit anticipated.

Is there a way out from this depressing and frightening situation? Will longevity decrease to the pre-mid-twentieth century level soon?

Not necessarily. An encouraging initial outcome that may indicate partial winning was recently obtained from investigating the process of protein biosynthesis, which plays a key role in life and is performed by ribosomes in all living cells. In fact, owing to its key role in life, about half of the existing antimicrobial drugs hamper this process. The clinical use of antibiotics is facilitated by the minute differences between the prokaryotic and eukaryotic (human and animal) ribosomes that enable their selectivity toward prokaryotic ribosomes.

Analyses of high resolution molecular structures and sequences of ribosomes from nonpathogenic and multidrug resistant pathogenic bacteria showed that the clinically used antibiotics bind exclusively to ribosomal active sites, mostly located at the ribosome's core. These analyses also revealed unique structural motifs crucial to protein biosynthesis and specific to each pathogen, which are located mostly on the ribosome periphery and are not involved in the primary ribosomal activity hence, currently no pathogen contains genes for their modification.

Promising preliminary results in designing inhibitors exploiting these unique motifs indicated that these sites may provide specific novel antibiotic binding sites with the potential for minimized resistance alongside preserving the microbiome that is occasionally unintentionally damaged by the broad-spectrum antibiotics used clinically. Applying a multifaceted approach could lead to optimization of the novel antibiotics for maximum potency, minimal toxicity (namely high selectivity, obtained by the location of the potential binding sites on the ribosomal periphery, which is the maximal evolving region), and appropriated degradability. Thus, in addition to the medical advantages, these new antibiotics should reduce the ecological burden caused by the non-degradable cores of many of the currently available antibiotics.

Although there will clearly be resistance to the next-generation of novel antibiotics, a much slower resistance development is predicted as principles for the design of further antibiotics have been identified. This approach represents a revolution in the future arsenal of antibiotics as it combines conventional and nonconventional critical aspects that may relieve, to some extent, the current problematic medical situation.

1

Molecular Mechanisms of Antibiotic Resistance – Part I

Alison J. Baylay[1], Laura J.V. Piddock[1], and Mark A. Webber[2]

[1] Institute of Microbiology and Infection, University of Birmingham, Birmingham, UK
[2] Quadram Institute, Norwich, UK

1.1   Introduction

Since the 1940s, pathogens resistant to antibiotics have emerged and spread around the globe, such that antibacterial resistance is one of the greatest challenges to human health in the twenty-first century. The mechanisms underpinning this evolution of resistance are complex and include the mutations in genes that either encode the targets of antibiotics or factors that control production of proteins that influence bacterial susceptibility to antibiotics, as well as the transfer of genes between strains and species including nonpathogenic bacteria. In this chapter, we give an overview of the molecular mechanisms by which bacteria can survive exposure to some of the most clinically important antibiotics currently available.

To exert its antimicrobial effect, a drug must reach and successfully bind to its target. Bacteria have evolved multiple and different antibiotic resistance mechanisms. These can be broadly grouped into four categories (summarized in Figure 1.1):

- Reducing the concentration of drug able to reach its target, either by preventing its entry or actively removing it from the cell,
- Inactivating or modifying the drug before it reaches the target, either extracellularly or intracellularly,
- Changing the target so that the drug can no longer bind,
- Acquiring an alternative route to carry out the cellular process blocked by the drug.

1.2   Molecular Mechanisms of Resistance

1.2.1   Reducing the Intracellular Concentration of a Drug

1.2.1.1   Increased Efflux

All bacterial genomes encode multiple efflux pumps, which extrude a variety of compounds from the cell. Efflux pumps are ancestrally ancient proteins and their original function is often unknown, but some are known to export naturally occurring molecules

Bacterial Resistance to Antibiotics – From Molecules to Man, First Edition.
Edited by Boyan B. Bonev and Nicholas M. Brown.
© 2020 John Wiley & Sons Ltd. Published 2020 by John Wiley & Sons Ltd.

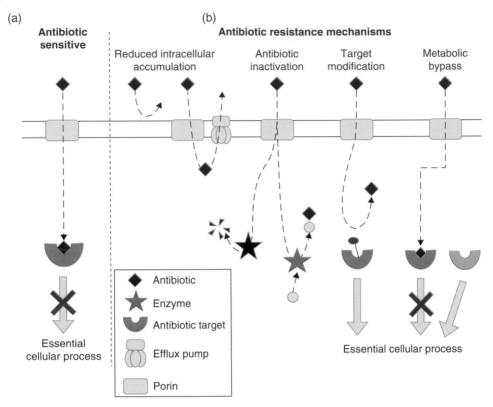

Figure 1.1 **Overview of antibiotic resistance mechanisms.** (a) In general, antibiotics function by binding to a cellular target such that an important biochemical process is blocked. (b) Bacteria may resist the action of antibiotic by a variety of mechanisms, which are summarized here. These include reduction of the intracellular concentration of the antibiotic by increased efflux or reduced permeability, inactivation of the antibiotic by hydrolysis or modification, modification of the target to prevent antibiotic binding, and metabolic bypass of the cellular process blocked by the antibiotic.

that are toxic to the cell [1]. In addition, functional efflux pumps have been shown to be important for other cellular processes, such as virulence and biofilm formation in particular [2–5].

Efflux pumps play a major role in determining the intrinsic level of susceptibility of a bacterial species to a particular drug but can also cause further clinically important antibiotic resistance when they are over-expressed. This can occur *via* mutations in local or global regulators [6–8], or by acquisition of insertion sequence (IS) elements that act as strong promoters upstream of efflux pump genes [9, 10]. Alternatively, new pump genes can be acquired on mobile genetic elements, for example, the *mef* and *msr* genes that encode macrolide transporters in Gram-positive bacteria [11, 12].

While some efflux pumps have a narrow specificity, such as Tet pumps that confer high level resistance to tetracyclines [13], others known as multidrug efflux systems

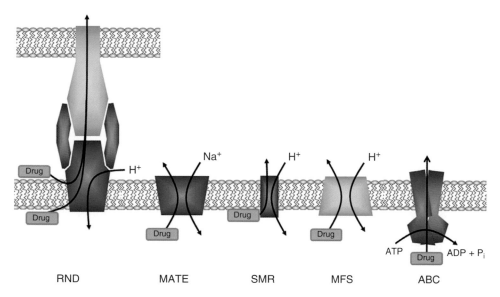

RND   MATE   SMR   MFS   ABC

Figure 1.2 Multidrug efflux systems. There are five known classes of multidrug efflux systems, summarized here. RND, resistance-nodulation-division family; MATE, multidrug and toxic compound extrusion family; SMR, small multidrug resistance family; MFS, major facilitator superfamily; ABC, ATP-binding cassette superfamily.

export a wide range of substrates, often including multiple antibiotics [1]. There are five known families of multidrug efflux pump (Figure 1.2):

- Efflux pumps of the resistance-nodulation-division (RND) family are tripartite transporters found in Gram-negative bacteria, which consist of an inner membrane pump, an outer membrane channel and a periplasmic adaptor protein that connects the two channels. Substrate export is powered by the proton motive force. The best studied example is the AcrAB–TolC pump which was initially discovered in *Escherichia coli*, but close homologs are widely distributed among Gram-negative bacteria [14].
- Major facilitator superfamily (MFS) pumps are the largest group of solute transporters and are responsible for most efflux-mediated resistance in Gram-positive bacteria [15, 16], although they are also found in Gram-negative bacteria [17]. They consist of a single polypeptide chain with 12 or 14 membrane spanning domains, with substrate efflux powered by the proton motive force. As an example, several members of this family cause clinically relevant resistance in *Staphylococcus aureus*. NorA confers resistance to fluoroquinolone antibiotics, QacA exports cationic lipophilic drugs, including biocides such as benzalkonium chloride, and LmrS exports a variety of agents such as lincomycin, linezolid, chloramphenicol and trimethoprim [18–20].
- Small multidrug resistance (SMR) transporters are, as the name suggests, small, having 110–120 amino acid proteins with four membrane spanning domains [15]. They form functional transporters by oligomerizing in the membrane, where they

transport substrates using the proton motive force [21]. Examples include QacC from *S. aureus* and EmrA from *E. coli*, both of which transport toxic organic cations such as methyl viologen [22, 23].

- Multidrug and toxic compound extrusion (MATE) efflux pumps are commonly found in Gram-negative bacteria. Unlike RND pumps, they are formed from a single polypeptide chain with 12 membrane spanning domains. They obtain power for efflux using the proton motive force or sodium antiport mechanisms. Examples include VcrM from *Vibrio cholerae*, MepA from *S. aureus* and PmpM from *Pseudomonas aeruginosa*, which transport a variety of substrates including fluoroquinolones and benzalkonium chloride [24–26].
- Some ATP-binding cassette (ABC) transporters confer antibiotic resistance, such as PatAB from *Streptococcus pneumoniae*, which transports fluoroquinolones, and MacAB from *E. coli* which exports macrolides [27–30]. ABC transporters are a very widespread family of transporters found in all three kingdoms of life. They consist of four subunits: two membrane spanning domains and two ATP binding domains. The family of ABC transporters involved in substrate export are usually formed from homo- or heterodimers of two half-transporters, each consisting of one membrane spanning domain and one nucleotide binding domain [31, 32]. Unlike the other classes of multidrug efflux pumps, the ABC pumps are primary transporters, meaning that transport is coupled to ATP hydrolysis instead of ion transport.

The molecular mechanisms of transport differ between the families of transporters and most are not well understood. In general, MDR efflux systems bind multiple substrates, transduce potential energy to power transport, and traffic the substrates in a unidirectional manner.

1.2.1.2 Reduced Entry (Permeability)

The intracellular concentration of an antibiotic can be reduced by preventing its entry into the cell. This mechanism is particularly relevant in Gram-negative bacteria as the outer membrane forms an efficient permeability barrier. Many antibiotics diffuse across the outer membrane *via* porin proteins, which form relatively large, non-selective channels allowing solutes to move across the membrane. The major porins in *E. coli* are OmpC and OmpF. Outer membrane permeability can be reduced, resulting in decreased permeability to antibiotics, by two mechanisms:

- reducing expression of porins or replacing them with other porins that form smaller channels,
- mutation of porin genes in ways which alter the permeability of the porin channel.

Pseudomonas aeruginosa is a good example of the effect of reduction of outer membrane permeability on antibiotic resistance. This bacterium is intrinsically resistant to many agents as it does not express many general diffusion porins, and instead produces smaller, dedicated porins to allow acquisition of nutrients [33, 34]. OprF, a homolog of OmpF, is expressed at high levels but mostly exists in a closed confirmation, while the open conformation is present at low levels [35]. Other changes in outer membrane protein expression have also been observed, for example, reduction in OprD expression causes resistance to the carbapenem imipenem [36]. Additionally, the

MexAB-OprM efflux pump is co-regulated with OprD, leading to further imipenem resistance [37].

In *E. coli* and *Salmonella spp.* reduction in OmpF and OmpC levels can occur, often in conjunction with de-repression of the AcrAB-TolC efflux pump, which causes resistance to multiple antibiotics [38–40]. In *Klebsiella pneumoniae*, replacement of the major porins OmpK35 and OmpK36 with an alternative porin with a narrower channel, OmpK37, has a similar effect to porin loss [41].

Mutations that change the structure of porins, reducing their permeability to β-lactam antibiotics have been found in mutation hotspots such as the L3 loop, which forms the constriction zone of OmpC/OmpF-like porins [42].

Gram-positive bacteria tend to be less intrinsically tolerant to antibiotics than Gram-negative bacteria as they do not possess an outer membrane so are less able to control their permeability. However, reduced permeability to some drugs has been documented. For example, vancomycin intermediate *S. aureus* (VISA) produce a thickened cell wall. As vancomycin functions by binding peptidoglycan precursors and preventing cross-linking, this thickened cell wall sequesters the vancomycin, increasing the concentration required to permeate through the cell wall and weaken its structure sufficiently to cause cell lysis [43].

1.2.2 Antibiotic Inactivation

Antibiotic resistance mediated by degradation or inactivation of an antibiotic before it reaches its target is achieved by either hydrolysis of a key structural feature of the antibiotic, or modification of the antibiotic structure by transfer of a chemical group.

1.2.2.1 Degradation by Hydrolysis

The main advantage of hydrolysis as a strategy for antibiotic inactivation is that the enzymatic reaction only requires water as a co-substrate, meaning that degradative enzymes can function outside the cell. This means that enzymes can be extracellularly secreted and destroy antibiotics before they reach the bacterium.

The classic examples of antibiotic degradation by hydrolysis are the β-lactamases, which inactivate β-lactam antibiotics such as penicillin by hydrolyzing the key β-lactam ring.

The β-lactamase enzymes can be separated into two groups based on their mechanism of hydrolysis. The majority of β-lactamases carry out hydrolysis by nucleophilic attack of the β-lactam ring by a key active site serine residue. These enzymes can be further classified into three groups according to the systems of Bush or Ambler [44, 45]. The remaining β-lactamases, forming class 3 under the Bush classification system and class B under the Ambler classification, are the metallo-β-lactamases, which catalyze hydrolysis by activation of a water molecule *via* a coordinated zinc ion [46].

Of particular current clinical concern is the prevalence of enzymes capable of hydrolyzing cephalosporins and carbapenems, the classes of β-lactam antibiotics developed to combat the problem of increasing resistance to the penicillins and early cephalosporins. Extended-spectrum β-lactamases (ESBLs) that can hydrolyse third and fourth generation cephalosporins were identified in clinical isolates of *K. pneumoniae* and *Serratia marcescens* in the mid-1980s [47–49]. These ESBLs, variants of the TEM-1/2 and

SHV-1 penicillinases, contain point mutations in the active site that extend the substrate spectrum of the enzyme [48, 50]. Since then, several additional classes of non-TEM and non-SHV ESBL have been discovered, such as the widespread CTX-M (reviewed in [51]) and OXA families [51–57].

Carbapenems are often used to treat infections caused by ESBL-producing organisms as they are resistant to hydrolysis by most β-lactamase enzymes. Several families of carbapenemase enzymes have been identified which are able to hydrolyze carbapenems in addition to other β-lactams. These consist of serine active site enzymes belonging to groups A and D of the Ambler classification, and also metallo-β-lactamases belonging to group B. The most prevalent group A carbapenemases are the KPC family, which were initially detected in *K. pneumoniae* [58], but are often plasmid-encoded so are widespread in various species of *Enterobacteriaciae* [59–64]. The class D carbapenemases primarily consist of variants of the OXA type ESBLs that also have weak carbapenem hydrolyzing activity [65]. Worldwide spread of carbapenem resistance caused by Group B enzymes (metallo-β-lactamases) in *P. aeruginosa* and the *Enterobacteriaceae* is predominantly caused by enzymes belonging to the VIM and IMP classes (reviewed in [66, 67]). Both families are found on class I integrons which facilitate their spread between bacteria. A particularly worrying example of this is the acquisition of the New Delhi metallo-β-lactamase 1 (NDM-1) enzyme, which is able to inactivate all β-lactam antibiotics except aztreonam, by a class I integron [68]. This element was originally found carried on a plasmid in *K. pneumoniae* that also carried multiple other resistance determinants, rendering the *K. pneumoniae* isolate resistant to all antibiotics except colistin and ciprofloxacin [68]. Spread of NDM-1 onto other plasmids, mediated by the class I integron, has led to an increasing number of cases of disease caused by NDM-1-producing Enterobacteriaceae worldwide [69].

Macrolide esterase enzymes provide a further example of antibiotic degradation by hydrolysis. Macrolides are cyclic molecules and the ring structure is closed by an ester bond catalyzed by the thioesterase molecule of the polyketide synthetase [70]. It is this ester bond that is targeted by macrolide esterases. The first of these (*ereA*) was found in *E. coli* [71]. Macrolide esterases are not as common as other, ribosome modification macrolide resistance mechanisms (discussed below) but, where present, they result in very high levels of macrolide resistance, as is typical for enzyme-based resistance [72]. They have been found to be disseminated on a class 2 integron [73], and have been found in *Providencia stuartii*, *S aureus* and *Pseudomonas spp.*, as well as *E. coli* [74–76].

1.2.2.2 Modification by Transfer of a Chemical Group

The structure of antibiotics can be modified by addition of a variety of chemical groups to vulnerable hydroxyls and amines. Addition of these chemical side chains prevents efficient binding of the drugs to their targets. Groups transferred include acyl, nucleotidyl and phosphate groups and, less commonly, ribosyl, glycosyl or thiol groups. The variety of groups that can be transferred makes the group transfer enzymes the most diverse and largest family of antibiotic resistance enzymes known [77].

The aminoglycoside antibiotics provide a good example of the effects of group transfer resistance mechanisms. Aminoglycosides are a diverse class of molecules characterized by an aminocylitol nucleus linked to various amino sugar groups by glycosidic bonds. They function by occupying the A-site of the ribosome and preventing binding of

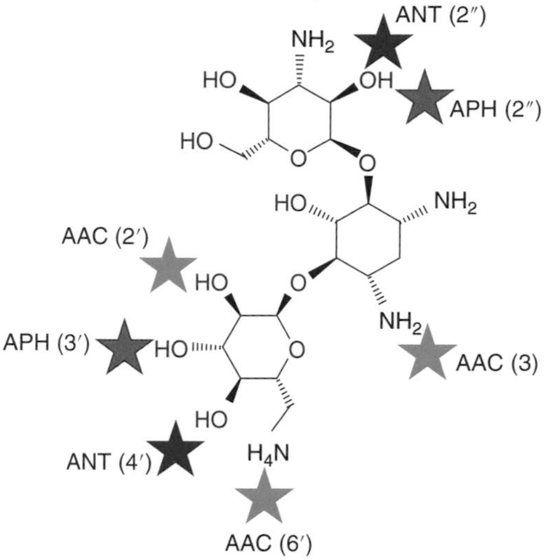

Figure 1.3 **Aminoglycoside modifying enzymes.** Aminoglycosides are large molecules with several hydroxyl and amine groups that are vulnerable to modification by aminoglycoside modifying enzymes. To illustrate this, the possible modification sites of kanamycin are indicated here, along with the class of aminoglycoside modifying enzyme that can recognize each site. AAC, aminoglycoside acetyltransferase; ANT, aminoglycoside nucleotidyltransferase; APH, aminoglycoside phosphotransferase.

aminoacyl-tRNA, which then disrupts protein synthesis [78, 79]. This process relies on specific interactions between key functional groups in the aminoglycoside molecule and residues in the ribosome A-site. This binding interaction can be easily disrupted by chemical modification of vulnerable hydroxyl and amine groups found both on the aminocylitol nucleus and the sugar moieties. A wide range of aminoglycoside resistance enzymes therefore exist that can catalyze transfer of chemical groups to several different reactive centers within the molecule. Figure 1.3 shows the possible sites of modification of kanamycin as an example [80].

There are three types of aminoglycoside modifying enzyme: N-acetyltransferases, which catalyze transfer of an acetyl group from acetyl-CoA to an amine group; O-phosphotransferases, which catalyze phosphorylation of hydroxyl groups; and O-nucleotidyltransferases, which transfer adenine to hydroxyl groups using ATP as a co-substrate. These enzymes are further classified, firstly by their stereospecificity, and then by the particular resistance profile they confer. There are four known classes of aminoglycoside acetyltransferase, five classes of aminoglycoside phosphotransferase and seven classes of aminoglycoside nucleotidyltransferase, which are differentiated by the position of the hydroxyl or amine group they modify [80]. New variants with different stereochemistry are generated by mutation of the enzyme active site, and many of these enzymes are encoded on mobile genetic elements allowing them to spread between different bacteria [77, 81].

These group transfer resistance mechanisms are not restricted to the aminoglycosides. For example, chloramphenicol resistance can also be conferred by acyltransferases. Chloramphenicol acetyltransferases are trimeric enzymes with two major types, A and B [82]. Again, this is a very diverse family of enzymes, with at least 16 known subfamilies of class A enzymes distributed throughout both Gram-positive and Gram-negative bacteria, and five subfamilies of class B enzymes mostly found in Gram-negative species. A few macrolide kinases (MPHs) are also known, for example in *E. coli* [72, 83, 84] and *S. aureus* [85]. As for macrolide esterases, macrolide kinases are rare compared to ribosome modifying mechanisms but, where present, they confer very high levels of resistance [86]. Lin proteins adenylate lincosamides and clindamycin. Three such enzymes have been described in the Gram-positive bacteria *S. haemolyticus*, *S. aureus* and *Enterococcus faecium* [87–89].

Acylation, phosphorylation, and adenylation are the most common types of group transfer mechanism conferring antibiotic resistance. However, other modifications do occur. O-Glycosylation of antibiotic molecules is used as a self-protection mechanism by antibiotic-producing species such as *Streptomyces lividans* [90, 91], but has not been yet found as an antibiotic resistance mechanism in human pathogens. O-Ribosylation, using NAD as an ADP-ribosyl donor, is used as a mechanism of rifampin resistance by *Mycobacterium smegmatis* [92]. Genes encoding a similar system have also been found on class I integrons in *Acinetobacter spp.* [93]. Finally, thiol transfer is used as a mechanism of resistance to fosfomycin in some bacteria, as it inactivates the antibiotic by opening a key epoxide ring. The FosA fosfomycin resistance metalloenzyme, which catalyzes thiol transfer using glutathione as a co-substrate, has been found encoded on several Gram-negative plasmids and on the *P. aeruginosa* chromosome. An equivalent enzyme, FosB, has been found in Gram-positive bacteria, encoded on staphylococcal plasmids and the *Bacillus subtilis* chromosome [94–96]. FosB uses cysteine as a co-substrate as Gram-positive bacteria do not produce glutathione [97].

1.2.3   Changes in Antibiotic Target

1.2.3.1   Molecular Modification of Target
Antibiotic targets can be modified either by mutation of the coding sequence or post-transcriptionally by addition of chemical groups. Alternatively, bacteria can acquire dedicated resistance proteins that protect key intracellular targets. Finally, perhaps uniquely, some bacteria have acquired mechanisms of vancomycin resistance that involve changing the fundamental chemical composition of the cell wall to prevent antibiotic binding.

1.2.3.1.1   *Mutational*
The development of resistance to an antibiotic through mutation of the gene encoding the target protein is common, particularly for drugs where enzymatic mechanisms of resistance do not exist.

Fluoroquinolones are broad-spectrum bacteriocidal antibiotics that function by inhibiting the type II topoisomerases DNA gyrase and topoisomerase IV. DNA gyrase is a tetrameric enzyme formed from two GyrA and two GyrB subunits, which catalyzes the negative supercoiling of DNA. Topoisomerase II is also a tetrameric enzyme consisting of two ParC and two ParE subunits, which are responsible for decatenation of daughter chromosomes following DNA replication. Fluoroquinolone antibiotics inhibit the religation of double-stranded breaks in the DNA created by these enzymes by stabilizing the enzyme-DNA complex, which results in formation of lethal double-stranded breaks in the DNA (reviewed in [98]). The main mechanism for the development of high-level fluoroquinolone resistance in most bacteria is mutation within the quinolone-resistance determining regions in the *gyrA* subunit of DNA gyrase and/or the *parC* subunit of topoisomerase II [99, 100]. These mutations alter the fluoroquinolone binding site, while still allowing the enzyme to function.

1.2.3.1.2 Post-Translational Modification of Target

Modifications preventing antibiotic binding to an intracellular target can also be added post-translation, often by methylation of the target. Resistance to antibiotics such as macrolides, lincosamides and streptogramin, which all target the 50S ribosomal subunit, can be conferred by methylation of a particular adenine residue in the 23S ribosomal RNA (A2058 in *E. coli*; [101]). The methylation is catalyzed by the Erm (erythromycin resistance methylase) protein. There are more than thirty known classes of Erm protein, and the *erm* gene is often found on transferrable elements, making ribosome methylation the predominant mechanism of macrolide resistance.

1.2.3.1.3 Target Protection

As discussed earlier, fluoroquinolone resistance is most commonly caused by chromosomal topoisomerase mutations. However, several plasmid-mediated mechanisms of quinolone resistance have been discovered in bacteria. One of these mechanisms is conferred by the Qnr proteins. These proteins belong to the pentapeptide repeat protein family, which are characterized by containing several tandem A(D/N)LXX repeats [102]. The first of these, QnrA, was found encoded on a plasmid from a clinical isolate of *K. pneumoniae* [103, 104]. QnrA binds to DNA gyrase and topoisomerase IV early in their catalytic cycle, reducing DNA binding, and it has been suggested that this reduces the number of topoisomerase holoenzyme-DNA targets available for fluoroquinolones to inhibit [105, 106].

A target protection mechanism conferring tetracycline resistance has also been observed. Tetracyclines inhibit binding of aminoacyl tRNA to the A-site of the ribosome [107]. Tetracycline resistance can be mediated by acquisition of ribosome protection proteins (RPPs), which catalyze release of tetracycline molecules from the ribosome, allowing protein synthesis to proceed. There are 11 known types of RPP [13], but the best studied are Tet(O) from *Campylobacter jejuni* and Tet(M) from *Streptococcus spp.* [108]. RPPs are GTPases in the translation factor superfamily [109] and are similar in structure to ribosomal elongation factors EF-G and EF-Tu [110]. In the presence of GTP they dislodge tetracycline from the ribosome, increasing the concentration of tetracycline required for inhibition [111, 112].

1.2.3.1.4 Other Mechanisms of Target Modification

Some Gram-positive bacteria have developed resistance to glycopeptide antibiotics such as vancomycin by an unusual antibiotic target modification strategy.

Unlike the β-lactams, which also inhibit cell wall biosynthesis, vancomycin functions by binding to the D-Ala-D-Ala C-terminus of peptidoglycan pentapeptide precursors, preventing their successful incorporation into the peptidoglycan. This means that the antibiotic does not interact directly with any of the cell wall biosynthesis enzymes, so resistance cannot be conferred simply by amino acid substitutions in a target protein. Instead, vancomycin resistance is conferred by the presence of alternative biosynthetic genes that produce peptidoglycan precursors with altered C-terminal ends. Six vancomycin resistance systems have been identified to date, but all of these result in production of either D-Ala-D-Lac (VanA, VanB and VanD) or D-Ala-D-Ser (VanC, VanE and VanG) pentapeptide termini [113]. The most common type of vancomycin resistance mechanism is VanA, which was first detected in *E. faecium* [114]. This is encoded by a nine-gene operon which is transferred on a Tn*1546* -like transposon, and it is particularly worrying from a clinical perspective as it causes high level vancomycin resistance and, to date, is the only vancomycin resistance system that has spread into *S. aureus*, although this remains rare to date [113]. Most of the vancomycin resistance operons are spread by mobile genetic elements and therefore are a system of acquired resistance [114]. However, VanC-type resistance, which is encoded by a three gene operon that catalyzes formation of D-Ala-D-Ser termini, confers intrinsic, low-level vancomycin resistance in *Enterococcus* species [115, 116].

1.2.4 Metabolic Bypass and Titration

The resistance mechanisms described earlier all center around preventing the drug from reaching and binding to its cellular target. An alternative mechanism employed by some bacteria is to find a way to carry out the cellular process normally blocked by the antibiotic, despite binding of the drug to the target. The inhibited step of a metabolic reaction can be bypassed by acquisition of an alternative enzyme that can carry out the required reaction but is not inhibited by the antibiotic. Alternatively, the target enzyme can be overproduced, such that the concentration of drug required for complete inhibition is increased.

A high-profile example of a metabolic bypass resistance mechanism is seen in β-lactam resistance in methicillin-resistant *S. aureus* (MRSA). The β-lactam antibiotics act by inhibiting the transpeptidase reaction catalyzed by peptidoglycan cross-linking enzymes known as penicillin binding proteins (PBPs) (reviewed in [117]). This leads to cell lysis as the bacterium grows and the unlinked cell wall is unable to counter the internal osmotic pressure. As discussed earlier, bacteria often develop resistance to β-lactams by acquiring genes encoding β-lactamase enzymes that inactivate the drug extracellularly. However, other bacteria develop β-lactam resistance by acquisition or evolution of alternative PBPs that can catalyze peptidoglycan cross-linking but are not inhibited by β-lactam antibiotics. MRSA have developed resistance to many β-lactam antibiotics by acquisition of the PBP2A enzyme encoded by the *mecA* gene carried on the SCC*mec* mobile genetic element that is characteristic of MRSA [118]. This enzyme consists of a penicillin-insensitive transpeptidase domain, which can work with the transglycosylase domain of the native PBP2 enzyme to maintain cell wall biosynthesis when the transpeptidase domain of the native enzyme is inhibited [119, 120].

Another example of metabolic bypass is seen in the resistance mechanisms of some bacteria to the sulfonamide class of antibiotics and trimethoprim, which inhibit different stages of the synthesis of the essential nutrient folate. Sulfonamides inhibit dihydropteroate synthase (DHPS), which converts para-aminobenzoic acid to dihydropteroate [121]. Trimethoprim disrupts a later step in the pathway by competitively inhibiting the bacterial dihydrofolate reductase (DHFR) enzyme, which converts dihydrofolate to tetrahydrofolate [122]. Trimethoprim and the sulfonamide sulfamethoxazole are often used together in a combination called co-trimoxazole as they have slightly different spectra of action among pathogenic bacteria [123]. Bacterial resistance to both agents occurs primarily through metabolic bypass mechanisms, where bacteria acquire a drug resistant version of the relevant biosynthesis enzyme that allows production of folate despite the presence of the drug. While chromosomal mutations causing drug insensitivity have been observed [124–127], more commonly drug resistant versions of an enzyme, which originated in intrinsically resistant species, are transferred on mobile genetic elements. There are twenty trimethoprim resistant *dhfr* variants, and two sulfonamide resistant *dhps* genes (*sulI* and *sulII*) that are known to be transferrable (reviewed in [128, 129]). Use of the co-trimoxazole combination therapy was expected to prevent development of resistance as it inhibits two stages of the folate synthesis pathway. However, this has not entirely prevented resistance as bacteria can acquire resistance to both agents, for example by co-transfer of trimethoprim and sulfonamide resistance genes on Tn*21* transposons [130].

Resistance to trimethoprim can also be conferred by overproduction of the antibiotic target. Chromosomal mutations in the promoter of *dhfr* have been observed in *Haemophilus influenzae*, which cause over-expression of the gene [131]. Overproduction of DHFR increases the number of enzymes available to carry out the reductase reaction, which increases the amount of trimethoprim required to achieve sufficient inhibition of reductase activity to have the desired antibacterial effect.

1.3   Acquisition and Transfer of Resistance

1.3.1   Intrinsic Resistance

Intrinsic resistance (otherwise known as inherent or innate resistance) can be defined as a relatively high tolerance to a specific antibiotic that is common to all members of a particular bacterial species. A bacterial species may be more tolerant to a drug than another species for various reasons. Differences in levels of uptake of drugs between bacterial species can be a general mechanism of tolerance to multiple antibiotic classes. *P. aeruginosa*, as mentioned earlier, naturally accumulates lower levels of antibiotics intracellularly than other Gram-negative bacteria due to its low membrane permeability and expression of multidrug efflux pumps [132]. Alternatively, specific classes of antibiotic may be unable to enter certain bacterial species for biochemical reasons. For example, anaerobic bacteria, such as *Bacteroides* and *Clostridium spp.*, are intrinsically resistant to most aminoglycoside antibiotics as the mechanism of aminoglycoside uptake is dependent on the presence of a functional electron transport chain [133].

In some species the antibiotic target may not be present or may be significantly altered in structure such that the drug is not active. For example, *P. aeruginosa* is intrinsically resistant to the biocide triclosan that attacks FabI, an enzyme involved in fatty acid biosynthesis, as this species possesses an alternative allele that is drug resistant [134].

Intrinsic antibiotic resistance can also be conferred by the presence of chromosomally encoded resistance mechanisms that are common to all members of a species. There are many examples of this. Some *Enterobacter* and *Citrobacter* species express a chromosomally encoded cephalosporinase, AmpC, which causes clinically relevant resistance to some β-lactams induced in the presence of the drug [135–137]. VanC-type vancomycin resistance is chromosomally encoded in *Enterococcus gallinarum*, *Enterococcus casseliflavus* and *Enterococcus flavescens*, as discussed above. In a non-clinical setting antibiotic-producing bacteria carry genes that confer self-resistance to the antibiotic that they produce [138], and soil dwelling bacteria often contain determinants causing resistance to antimicrobial compounds produced by other soil organisms such as fungi [139, 140]. Transferrable resistance genes that cause acquired resistance in normally susceptible bacteria can often be traced back to the capture of chromosomal genes from a species with intrinsic drug resistance by a mobile genetic element [141].

1.3.2   Acquired Resistance

1.3.2.1   Mutation

Due to their short generation times, large populations and haploid genome structure, bacteria can rapidly acquire antibiotic resistance through mutation. Mutational resistance to many antibiotics can be easily selected *in vitro* and has been recorded in most bacteria. It is particularly clinically relevant in bacterial species that do not participate in horizontal gene transfer, such as *Mycobacterium tuberculosis* [142].

Additionally, mutation is important for the development of clinical resistance to synthetic antibiotics for which there is little or no natural reservoir of resistance genes [137]. For example, mutation of topoisomerase genes is the predominant mechanism of resistance to fluoroquinolones [100], and mutation of 23S rRNA is a frequently reported resistance mechanism for the oxazolidinone antibiotic linezolid [143, 144].

1.3.2.2   Horizontal Gene Transfer

Another way by which bacteria can develop antibiotic resistance is by acquisition of an exogenous resistance gene from the environment or another bacterium in a process called horizontal gene transfer. Resistance elements can come from antibiotic-producing bacteria [138] or environmental species living in competition with fungi, plants, *etc.*, that produce antibacterial compounds [140].

There are three major mechanisms of gene transfer, which correspond to the three methods by which bacteria acquire exogenous DNA (summarized in Figure 1.4):

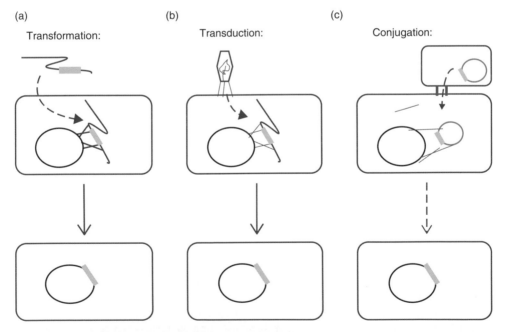

Figure 1.4 Mechanisms of horizontal gene transfer (HGT). There are three mechanisms by which bacteria can acquire resistance determinants by HGT. (a) Transformation: uptake of free DNA from the environment, followed by recombination with the bacterial chromosome. (b) Transduction: transfer of DNA from a donor to a recipient cell *via* a bacteriophage. (c) Conjugation: transfer of DNA from a donor to a recipient cell *via* direct cell–cell contact.

- **Transformation:** Transformation is the uptake of free DNA from the environment, and its subsequent recombination into the genome. This DNA can be in the form of plasmids, or chromosomal DNA from dying cells. About 60 species of bacteria are known to be naturally competent, meaning that they actively take up DNA from their environment [144]. It is especially prevalent among bacteria colonizing the upper respiratory tract such as *Streptococcus* spp. and *H. influenzae* [145]. Resistance to β-lactams in *S. pneumoniae* is typically caused by recombination of the gene encoding the major penicillin binding protein PBPX with homologous DNA from the closely related *Streptococcus mitis* acquired by transformation. This results in mosaic forms of *pbpX* resulting in an altered protein to which β-lactams cannot bind [146].
- **Transduction:** DNA can also be transferred between bacteria by the action of bacteriophages by two mechanisms [147]. DNA from a donor cell can be packaged into a bacteriophage particle along with the viral genome and can then be transferred to a new cell when the phage infects. Alternatively, bacteriophages can insert into the genome of an infected bacterium to form a lysogen instead of causing cell lysis. These inserted bacteriophage genomes are known as prophages. The contribution of gene transfer by bacteriophages to clinically relevant antibiotic resistance has not been fully explored, but isolation of phage containing antibiotic resistance genes from wastewater and activated sludge suggests that this may be a significant mechanism of resistance transfer in the environment [148–150].
- **Conjugation:** Conjugation is the direct transfer of DNA between bacteria, first described for the F plasmid in *E. coli* in 1946 [151]. It is a contact-dependent process that transfers mobilizable elements such as plasmids and conjugative transposons (discussed below) from a donor cell to a recipient cell, mediated by factors encoded by the transposable element (reviewed in [152]). The F-plasmids of *E. coli* are the best described examples of conjugative plasmids. The *tra* locus of these plasmids encodes an extracellular filament known as the F pilus, which makes contact with recipient cells to form a mating bridge allowing DNA transfer [153, 154]. Non-conjugative plasmids can be mobilized *in trans* when present in the same cell as a conjugative plasmid or transposon [155, 156].

For an acquired gene to confer antibiotic resistance on the recipient bacterium it must be incorporated into the chromosome or be carried on a plasmid so that it can be stably replicated and expressed. Some naturally competent organisms can incorporate DNA directly into their genomes by homologous recombination, but this requires regions of significant sequence similarity so is restricted to acquisition of DNA from closely related species. The majority of horizontally transferred resistance genes are therefore transferred on mobile genetic elements, which allow transfer between much more distantly related species.

1.3.3 Types of Mobile Genetic Element

- **Transposons:** Transposons are mobile DNA elements that have terminal regions that allow insertion into recipient DNA by site-specific recombination. The insertion is catalyzed by a transposase enzyme, usually encoded within the transposon itself. Transposons can exist on plasmids, or be integrated into the chromosome, and can also insert at sites adjacent to other transposons, flanking resistance genes and

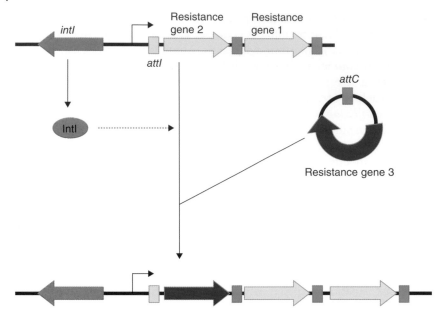

Figure 1.5 **Resistance gene capture by integrons.** Integrons are DNA elements that capture and promote expression of multiple genes. Gene capture occurs *via* recombination between an *attI* site on the integron and an *attC* site associated with the captured gene. Captured genes are expressed from a strong promoter associated with the integron.

making larger composite transposons. Conjugative transposons are a class of transposons that, like conjugative plasmids, encode the genetic machinery required for their own transfer. This usually consists of genes that catalyzes excision of the transposon from the donor genome, and a set of transfer genes that promote conjugation between the donor and recipient bacterial cells [157]. Non-conjugative transposons can be transferred intercellularly by transformation, transduction, or by carriage on plasmids.

- **Integrons:** Bacteria can collect multiple resistance genes at a single site through the action of integrons, which are DNA elements that capture genes through site-specific recombination and drive their expression [158]. Integrons consist of three essential components: an integrase gene, which codes for an enzyme catalyzing incorporation and excision of gene cassettes; an attachment (*att*) site to facilitate site-specific recombination; and a strong promoter which allows expression of incorporated genes (Figure 1.5). Integrons are not able to self-mobilize, but they are often found incorporated into transposons and IS elements, either chromosomally or on plasmids [159–165].

1.3.4 Fitness Costs of Antibiotic Resistance

The ultimate fate of any given strain of bacterium in a polymicrobial environment is dictated by its relative fitness, which can be defined as its ability to survive and reproduce to pass on its genetic material in competition with other strains. Well adapted strains will increase in proportion over time at the expense of less fit members of the community. In terms of antibiotic resistance, extreme selection and population changes occur when an antibiotic is applied to an environment and kills all the cells apart from pre-existing mutants that carry a target site point mutation, which can then expand and dominate the population. However, acquisition of antibiotic resistance can incur a cost; antibiotics target essential components of the cellular machinery and changes to the relevant target molecules that ablate drug binding can also impact function. Similarly, the acquisition of mobile genetic elements carrying new resistance genes can impact on the host in terms of energy requirements and ultimately the ability to grow.

The biological cost of resistance mutations can vary widely between mutations and between species. For example, in *gyrA* in *E. coli* and *Salmonella,* common mutations within the quinolone-resistance-determining region of *gyrA* often impose a small growth defect. However, in *Campylobacter,* analogous mutations have been shown to increase growth rate and promote competitive fitness of the mutants under some conditions [166]. Similarly, the impact of mobile elements acquisition varies. Some large plasmids have been shown to reduce significantly growth rates of host strains and impose a metabolic burden, while others, such as the recently characterized PCT plasmid that encodes CTX-M-mediated β-lactam resistance and has spread globally in *E. coli* from humans and animals, maintain a low copy number and have no detectable impact on fitness of various host strains [167]. Furthermore, plasmid *pA-CYC184* (encoding tetracycline resistance) was found to confer a fitness benefit to the host cell after a period of co-evolution [168].

Once a resistance mutation is selected or a resistance gene acquired, secondary or compensatory mutations can occur that alleviate the fitness cost of the primary resistance mechanism. These are selected for based on improving fitness but often also impact on resistance, as mutations that alleviate the impact of an initial resistance mutation often affect the gene in which the initial mutation occurred or partner genes contributing to the same process or metabolic pathway. As a result, these compensatory mutations may also allow further development of high level resistance. This has been observed for fluoroquinolone resistance in both *E. coli* and *S. pneumoniae.* Several quinolone resistance mutations in *parC* and *gyrA* were shown to cause a fitness defect alone but acquisition of a second topoisomerase mutation was able to alleviate the impact of the primary mutations as well as increasing fluoroquinolone resistance [169, 170].

The level of selective pressure is also important. High level exposure to drug dictates that only a small number of mutations may exist that can confer very high level resistance, such as a target site mutation within *gyrA* causing quinolone resistance, while at lower levels of drug there may be many genes in which mutations can, individually or in combination, impact on survival in the presence of the drug. Importantly, if drug levels are low enough to allow growth, even if at a reduced rate, populations can produce progeny which may acquire combinations of mutations or multiple compensatory mutations allowing gradual development of high level resistance with a minimal impact on fitness.

In summary, the fate of any particular resistant mutant is determined by a complex interplay between the strength of selective pressure favoring that mutant, mobility of any resistance genes, and the biological cost of carriage of the resistance allele/gene/vector on the host. Predicting how quickly resistance to a given agent will emerge and whether it will persist is extremely difficult with very different outcomes possible; for example, the extraordinarily rapid global spread of the CTX-M beta-lactamases in contrast to the extremely rare cases of vancomycin resistant *S. aureus*. Unfortunately, the latter example has been the exception rather than the rule and bacteria have proven adept at acquiring and then maintaining resistance genes in the absence of selection and, while prudent use of antibiotics will minimize selection, it is unlikely that "the clock can be reversed." Well-adapted antibacterial-resistant pathogens are likely to be hard to eradicate.

1.4 Summary and Conclusions

In this chapter we have illustrated some of the complexity of how antibiotic resistance is underpinned at a molecular level. Resistance often involves multiple genes, which may be mobile, and multidrug resistant pathogens are often able to express many antibiotic resistance genes in concert to provide a broad degree of protection to common antibiotics. The ecology of antibiotic resistance is also complicated, with an interplay between benefits in terms of acquisition of resistance genes and costs to fitness, although often these can be alleviated in a manner that promotes maintenance of the resistance phenotype.

Antibiotic resistance is not a recent phenomenon and bacteria have been evolving mechanisms of resistance to naturally produced antimicrobials for billions of years, allowing the efficient development of strategies to survive exposure to toxic xenobiotics. A thorough understanding of the mechanisms of resistance and the spread of resistance genes and strains is needed to help devise improved antibiotic usage strategies so that bacteria are targeted efficiently and selection of resistance is minimized in pathogenic organisms.

References

1 Piddock, L.J.V. (2006). Clinically relevant chromosomally encoded multidrug resistance efflux pumps in Bacteria. *Clinical Microbiology Reviews* 19 (2): 382–402.
2 Alvarez-Ortega, C., Olivares, J., Martinez, J. L. (2013). RND multidrug efflux pumps: what are they good for? *Frontiers in Microbiology* 4 (February), 7. https://doi.org/10.3389/fmicb.2013.00007
3 Baugh, S., Ekanayaka, A.S., Piddock, L.J.V., and Webber, M.A. (2012). Loss of or inhibition of all multidrug resistance efflux pumps of *Salmonella enterica* serovar Typhimurium results in impaired ability to form a biofilm. *Journal of Antimicrobial Chemotherapy* 67 (10): 2409–2417.
4 Piddock, L. (2006). Multidrug-resistance efflux pumps? Not just for resistance. *Nature Reviews Microbiology* 4 (8): 629–636.
5 Webber, M.A., Bailey, A.M., Blair, J.M.A. et al. (2009). The global consequence of disruption of the AcrAB-TolC efflux pump in *Salmonella enterica* includes reduced expression of SPI-1 and other attributes required to infect the host. *Journal of Bacteriology* 191 (13): 4276–4285.
6 Alekshun, M.N., Kim, Y.S., and Levy, S.B. (2000). Mutational analysis of MarR, the negative regulator of *marRAB* expression in *Escherichia coli*, suggests the presence of two regions required for DNA binding. *Molecular Microbiology* 35 (6): 1394–1404.

7 Baranova, N.N., Danchin, A., and Neyfakh, A.A. (1999). Mta, a global MerR-type regulator of the *Bacillus subtilis* multidrug-efflux transporters. *Molecular Microbiology* 31 (5): 1549–1559.

8 Webber, M.A., Talukder, A., and Piddock, L.J.V. (2005). Contribution of mutation at amino acid 45 of AcrR to *acrB* expression and ciprofloxacin resistance in clinical and veterinary *Escherichia coli* isolates. *Antimicrobial Agents and Chemotherapy* 49 (10): 4390–4392.

9 Boutoille, D., Corvec, S., Caroff, N. et al. (2004). Detection of an IS21 insertion sequence in the *mexR* gene of *Pseudomonas aeruginosa* increasing beta-lactam resistance. *FEMS Microbiology Letters* 230 (1): 143–146.

10 Rouquette-Loughlin, C.E., Balthazar, J.T., Hill, S.A., and Shafer, W.M. (2004). Modulation of the *mtrCDE*-encoded efflux pump gene complex of *Neisseria meningitidis* due to a Correia element insertion sequence. *Molecular Microbiology* 54 (3): 731–741.

11 Butaye, P., Cloeckaert, A., and Schwarz, S. (2003). Mobile genes coding for efflux-mediated antimicrobial resistance in gram-positive and gram-negative bacteria. *International Journal of Antimicrobial Agents* 22 (3): 205–210.

12 Luna, V.a., Coates, P., Eady, E.A. et al. (1999). A variety of gram-positive bacteria carry mobile *mef* genes. *Journal of Antimicrobial Chemotherapy* 44 (1): 19–25.

13 Thaker, M., Spanogiannopoulos, P., and Wright, G.D. (2010). The tetracycline resistome. *Cellular and Molecular Life Sciences* 67 (3): 419–431.

14 Tikhonova, E.B. and Zgurskaya, H.I. (2004). AcrA, AcrB, and TolC of *Escherichia coli* form a stable intermembrane multidrug efflux complex. *The Journal of Biological Chemistry* 279 (31): 32116–32124.

15 Paulsen, I.T., Brown, M.H., and Skurray, R.A. (1996). Proton-dependent multidrug efflux systems. *Microbiological Reviews* 60 (4): 575–608.

16 Poole, K. (2005). Efflux-mediated antimicrobial resistance. *Journal of Antimicrobial Chemotherapy* 56 (1): 20–51.

17 Saidijam, M., Benedetti, G., Ren, Q. et al. (2006). Microbial drug efflux proteins of the major facilitator superfamily. *Current Drug Targets* 7 (7): 793–811.

18 Brown, M.H. and Skurray, R.A. (2001). Staphylococcal multidrug efflux protein QacA. *Journal of Molecular Microbiology and Biotechnology* 3 (2): 163–170.

19 Floyd, J.L., Smith, K.P., Kumar, S.H. et al. (2010). LmrS is a multidrug efflux pump of the major facilitator superfamily from *Staphylococcus aureus*. *Antimicrobial Agents and Chemotherapy* 54 (12): 5406–5412.

20 Yoshida, H., Bogaki, M., Nakamura, S. et al. (1990). Nucleotide sequence and characterization of the *Staphylococcus aureus norA* gene, which confers resistance to quinolones. *Journal of Bacteriology* 172 (12): 6942–6949.

21 Yerushalmi, H., Lebendiker, M., and Schuldiner, S. (1996). Negative dominance studies demonstrate the oligomeric structure of EmrE, a multidrug antiporter from *Escherichia coli*. *The Journal of Biological Chemistry* 271 (49): 31044–31048.

22 Paulsen, I.T., Brown, M.H., Dunstan, S.J., and Skurray, R.A. (1995). Molecular characterization of the staphylococcal multidrug resistance export protein QacC. *Journal of Bacteriology* 177 (10): 2827–2833.

23 Yerushalmi, H., Lebendiker, M., and Schuldiner, S. (1995). EmrE, an *Escherichia coli* 12-kDa multidrug transporter, exchanges toxic cations and H+ and is soluble in organic solvents. *The Journal of Biological Chemistry* 270 (12): 6856–6863.

24 He, G.-X., Kuroda, T., Mima, T. et al. (2004). An H(+)-coupled multidrug efflux pump, PmpM, a member of the MATE family of transporters, from *Pseudomonas aeruginosa*. *Journal of Bacteriology* 186 (1): 262–265.

25 Huda, M.N., Chen, J., Morita, Y. et al. (2003). Gene cloning and characterization of VcrM, a Na$^+$-coupled multidrug efflux pump, from *Vibrio cholerae* non-O1. *Microbiology and Immunology* 47 (6): 419–427.

26 Kaatz, G.W., McAleese, F., and Seo, S.M. (2005). Multidrug resistance in *Staphylococcus aureus* due to overexpression of a novel multidrug and toxin extrusion (MATE) transport protein. *Antimicrobial Agents and Chemotherapy* 49 (5): 1857–1864.

27 Garvey, M.I., Baylay, A.J., Wong, R.L., and Piddock, L.J.V. (2011). Overexpression of *patA* and *patB*, which encode ABC transporters, is associated with fluoroquinolone resistance in clinical isolates of *Streptococcus pneumoniae*. *Antimicrobial Agents and Chemotherapy* 55 (1): 190–196.

28 Garvey, M.I. and Piddock, L.J.V. (2008). The efflux pump inhibitor reserpine selects multidrug-resistant *Streptococcus pneumoniae* strains that overexpress the ABC transporters PatA and PatB. *Antimicrobial Agents and Chemotherapy* 52 (5): 1677–1685.

29 Lin, H.T., Bavro, V.N., Barrera, N.P. et al. (2009). MacB ABC transporter is a dimer whose ATPase activity and macrolide-binding capacity are regulated by the membrane fusion protein MacA. *The Journal of Biological Chemistry* 284 (2): 1145–1154.

30 Marrer, E., Schad, K., Satoh, A. et al. (2006). Involvement of the putative ATP-dependent efflux proteins PatA and PatB in fluoroquinolone resistance of a multidrug-resistant mutant of *Streptococcus pneumoniae*. *Antimicrobial Agents and Chemotherapy* 50 (2): 685.

31 Dassa, E. and Bouige, P. (2001). The ABC of ABCs: a phylogenetic and functional classification of ABC systems in living organisms. *Research in Microbiology* 152 (3–4): 211–229.

32 Higgins, C. (2001). ABC transporters: physiology, structure and mechanisman overview. *Research in Microbiology* 152 (3–4): 205–210.

33 Livermore, D.M. (2001). Of *Pseudomonas*, porins, pumps and carbapenems. *Journal of Antimicrobial Chemotherapy* 47 (3): 247–250.

34 Nikaido, H. (2003). Molecular basis of bacterial outer membrane permeability revisited. *Microbiology and Molecular Biology Reviews* 67 (4): 593–656.

35 Sugawara, E., Nestorovich, E.M., Bezrukov, S.M., and Nikaido, H. (2006). *Pseudomonas aeruginosa* porin OprF exists in two different conformations. *The Journal of Biological Chemistry* 281 (24): 16220–16229.

36 Margaret, B.S., Drusano, G.L., and Standiford, H.C. (1989). Emergence of resistance to carbapenem antibiotics in *Pseudomonas aeruginosa*. *Journal of Antimicrobial Chemotherapy* 24 (Suppl A): 161–167.

37 Ochs, M.M., McCusker, M.P., Bains, M., and Hancock, R.E. (1999). Negative regulation of the *Pseudomonas aeruginosa* outer membrane porin OprD selective for imipenem and basic amino acids. *Antimicrobial Agents and Chemotherapy* 43 (5): 1085–1090.

38 Cohen, S.P., McMurry, L.M., and Levy, S.B. (1988). *marA* locus causes decreased expression of OmpF porin in multiple-antibiotic-resistant (Mar) mutants of *Escherichia coli*. *Journal of Bacteriology* 170 (12): 5416–5422.

39 Harder, K.J., Nikaido, H., and Matsuhashi, M. (1981). Mutants of *Escherichia coli* that are resistant to certain beta-lactam compounds lack the OmpF porin. *Antimicrobial Agents and Chemotherapy* 20 (4): 549–552.

40 Medeiros, A. (1987). Loss of OmpC porin in a strain of *Salmonella typhimurium* causes increased resistance to cephalosporins during therapy. *The Journal of Infectious Diseases* 156 (5): 751–757.

41 Doménech-Sánchez, A., Hernández-Allés, S., Martínez-Martínez, L. et al. (1999). Identification and characterization of a new porin gene of *Klebsiella pneumoniae*: its role in beta-lactam antibiotic resistance. *Journal of Bacteriology* 181 (9): 2726–2732.

42 Dé, E., Baslé, A., Jaquinod, M. et al. (2001). A new mechanism of antibiotic resistance in *Enterobacteriaceae* induced by a structural modification of the major porin. *Molecular Microbiology* 41 (1): 189–198.

43 Cui, L., Iwamoto, A., Lian, J.-Q. et al. (2006). Novel mechanism of antibiotic resistance originating in vancomycin-intermediate *Staphylococcus aureus*. *Antimicrobial Agents and Chemotherapy* 50 (2): 428–438.

44 Ambler, R.P. (1980). The structure of beta-lactamases. *Philosophical Transactions of the Royal Society of London. Series B, Biological Sciences* 289 (1036): 321–331.

45 Bush, L.M., Calmon, J., and Johnson, C.C. (1995). Newer penicillins and beta-lactamase inhibitors. *Infectious Disease Clinics of North America* 9 (3): 653–686.

46 Nordmann, P. and Poirel, L. (2002). Emerging carbapenemases in gram-negative aerobes. *Clinical Microbiology and Infection* 8 (6): 321–331.

47 Brun-Buisson, C., Legrand, P., Philippon, A. et al. (1987). Transferable enzymatic resistance to third-generation cephalosporins during nosocomial outbreak of multiresistant *Klebsiella pneumoniae*. *Lancet* 2 (8554): 302–306.

48 Knothe, H., Shah, P., Krcmery, V. et al. (1983). Transferable resistance to cefotaxime, cefoxitin, cefamandole and cefuroxime in clinical isolates of *Klebsiella pneumoniae* and *Serratia marcescens*. *Infection* 11 (6): 315–317.

49 Sirot, D., Sirot, J., Labia, R. et al. (1987). Transferable resistance to third-generation cephalosporins in clinical isolates of *Klebsiella pneumoniae*: identification of CTX-1, a novel beta-lactamase. *The Journal of Antimicrobial Chemotherapy* 20 (3): 323–334.

50 Sougakoff, W., Goussard, S., Gerbaud, G., and Courvalin, P. (1988). Plasmid-mediated resistance to third-generation cephalosporins caused by point mutations in TEM-type penicillinase genes. *Reviews of Infectious Diseases* 10 (4): 879–884.

51 Bonnet, R. (2004). Growing group of extended-spectrum beta-lactamases: the CTX-M enzymes. *Antimicrobial Agents and Chemotherapy* 48 (1): 1–14.

52 Danel, F., Hall, L.M., Duke, B. et al. (1999). OXA-17, a further extended-spectrum variant of OXA-10 beta-lactamase, isolated from *Pseudomonas aeruginosa*. *Antimicrobial Agents and Chemotherapy* 43 (6): 1362–1366.

53 Danel, F., Hall, L.M., Gur, D., and Livermore, D.M. (1998). OXA-16, a further extended-spectrum variant of OXA-10 beta-lactamase, from two *Pseudomonas aeruginosa* isolates. *Antimicrobial Agents and Chemotherapy* 42 (12): 3117–3122.

54 Hall, L.M., Livermore, D.M., Gur, D. et al. (1993). OXA-11, an extended-spectrum variant of OXA-10 (PSE-2) beta-lactamase from *Pseudomonas aeruginosa*. *Antimicrobial Agents and Chemotherapy* 37 (8): 1637–1644.

55 Philippon, L.N., Naas, T., Bouthors, A.T. et al. (1997). OXA-18, a class D clavulanic acid-inhibited extended-spectrum beta-lactamase from *Pseudomonas aeruginosa*. *Antimicrobial Agents and Chemotherapy* 41 (10): 2188–2195.

56 Poirel, L., Girlich, D., Naas, T., and Nordmann, P. (2001). OXA-28, an extended-spectrum variant of OXA-10 beta-lactamase from *Pseudomonas aeruginosa* and its plasmid- and integron-located gene. *Antimicrobial Agents and Chemotherapy* 45 (2): 447–453.

57 Toleman, M.A., Rolston, K., Jones, R.N., and Walsh, T.R. (2003). Molecular and biochemical characterization of OXA-45, an extended-spectrum class 2d' beta-lactamase in *Pseudomonas aeruginosa*. *Antimicrobial Agents and Chemotherapy* 47 (9): 2859–2863.

58 Yigit, H., Queenan, A.M., Anderson, G.J. et al. (2001). Novel carbapenem-hydrolyzing beta-lactamase, KPC-1, from a carbapenem-resistant strain of *Klebsiella pneumoniae*. *Antimicrobial Agents and Chemotherapy* 45 (4): 1151–1161.

59 Bratu, S., Tolaney, P., Karumudi, U. et al. (2005). Carbapenemase-producing *Klebsiella pneumoniae* in Brooklyn, NY: molecular epidemiology and *in vitro* activity of polymyxin B and other agents. *Journal of Antimicrobial Chemotherapy* 56 (1): 128–132.

60 Hossain, A., Ferraro, M.J., Pino, R.M. et al. (2004). Plasmid-mediated carbapenem-hydrolyzing enzyme KPC-2 in an *Enterobacter* sp. *Antimicrobial Agents and Chemotherapy* 48 (11): 4438–4440.

61 Miriagou, V., Tzelepi, E., Gianneli, D., and Tzouvelekis, L.S. (2003a). *Escherichia coli* with a self-transferable, multiresistant plasmid coding for metallo-beta-lactamase VIM-1. *Antimicrobial Agents and Chemotherapy* 47 (1): 395–397.

62 Miriagou, V., Tzouvelekis, L.S., Rossiter, S. et al. (2003b). Imipenem resistance in a *Salmonella* clinical strain due to plasmid-mediated class a carbapenemase KPC-2. *Antimicrobial Agents and Chemotherapy* 47 (4): 1297–1300.

63 Naas, T., Nordmann, P., Vedel, G., and Poyart, C. (2005). Plasmid-mediated carbapenem-hydrolyzing beta-lactamase KPC in a *Klebsiella pneumoniae* isolate from France. *Antimicrobial Agents and Chemotherapy* 49 (10): 4423–4424.

64 Smith Moland, E., Hanson, N.D., Herrera, V.L. et al. (2003). Plasmid-mediated, carbapenem-hydrolysing beta-lactamase, KPC-2, in *Klebsiella pneumoniae* isolates. *Journal of Antimicrobial Chemotherapy* 51 (3): 711–714.

65 Walther-Rasmussen, J. and Høiby, N. (2006). OXA-type carbapenemases. *The Journal of Antimicrobial Chemotherapy* 57 (3): 373–383.

66 Zhao, W.-H. and Hu, Z.-Q. (2011a). Epidemiology and genetics of VIM-type metallo-β-lactamases in gram-negative bacilli. *Future Microbiology* 6 (3): 317–333.

67 Zhao, W.-H. and Hu, Z.-Q. (2011b). IMP-type metallo-β-lactamases in gram-negative bacilli: distribution, phylogeny, and association with integrons. *Critical Reviews in Microbiology* 37 (3): 214–226.

68 Yong, D., Toleman, M.A., Giske, C.G. et al. (2009). Characterization of a new metallo-beta-lactamase gene, bla(NDM-1), and a novel erythromycin esterase gene carried on a unique genetic structure in *Klebsiella pneumoniae* sequence type 14 from India. *Antimicrobial Agents and Chemotherapy* 53 (12): 5046–5054.

69 Nordmann, P., Poirel, L., Walsh, T.R., and Livermore, D.M. (2011). The emerging NDM carbapenemases. *Trends in Microbiology* 19 (12): 588–595.

70 Donadio, S., Staver, M.J., McAlpine, J.B. et al. (1991). Modular organization of genes required for complex polyketide biosynthesis. *Science* 252 (5006): 675–679.

71 Barthélémy, P., Autissier, D., Gerbaud, G., and Courvalin, P. (1984). Enzymic hydrolysis of erythromycin by a strain of *Escherichia coli*. A new mechanism of resistance. *The Journal of Antibiotics* 37 (12): 1692–1696.

72 Nakamura, A., Miyakozawa, I., Nakazawa, K. et al. (2000). Detection and characterization of a macrolide 2′-phosphotransferase from a *Pseudomonas aeruginosa* clinical isolate. *Antimicrobial Agents and Chemotherapy* 44 (11): 3241–3242.

73 Biskri, L. and Mazel, D. (2003). Erythromycin esterase gene *ere*(A) is located in a functional gene cassette in an unusual class 2 integron. *Antimicrobial Agents and Chemotherapy* 47 (10): 3326–3331.

74 Kim, Y.-H., Cha, C.-J., and Cerniglia, C.E. (2002). Purification and characterization of an erythromycin esterase from an erythromycin-resistant *Pseudomonas* sp. *FEMS Microbiology Letters* 210 (2): 239–244.

75 Plante, I., Centr'on, D., and Roy, P.H. (2003). An integron cassette encoding erythromycin esterase, ere(A), from *Providencia stuartii. Journal of Antimicrobial Chemotherapy* 51 (4): 787–790.

76 Wondrack, L., Massa, M., Yang, B.V., and Sutcliffe, J. (1996). Clinical strain of *Staphylococcus aureus* inactivates and causes efflux of macrolides. *Antimicrobial Agents and Chemotherapy* 40 (4): 992–998.

77 Wright, G.D. (2005). Bacterial resistance to antibiotics: enzymatic degradation and modification. *Advanced Drug Delivery Reviews* 57 (10): 1451–1470.

78 Davis, B.B. (1988). The lethal action of aminoglycosides. *Journal of Antimicrobial Chemotherapy* 22 (1): 1–3.

79 Davis, B.D. (1987). Mechanism of bactericidal action of aminoglycosides. *Microbiological Reviews* 51 (3): 341–350.

80 Ramirez, M.S. and Tolmasky, M.E. (2010). Aminoglycoside modifying enzymes. *Drug Resistance Updates* 13 (6): 151–171.

81 Tolmasky, M.E. (2000). Bacterial resistance to aminoglycosides and beta-lactams: the Tn1331 transposon paradigm. *Frontiers in Bioscience* 5: D20–D29.

82 Schwarz, S., Kehrenberg, C., Doublet, B., and Cloeckaert, A. (2004). Molecular basis of bacterial resistance to chloramphenicol and florfenicol. *FEMS Microbiology Reviews* 28 (5): 519–542.

83 Noguchi, N., Emura, A., Matsuyama, H. et al. (1995). Nucleotide sequence and characterization of erythromycin resistance determinant that encodes macrolide 2′-phosphotransferase I in *Escherichia coli. Antimicrobial Agents and Chemotherapy* 39 (10): 2359–2363.

84 Noguchi, N., Katayama, J., and O'Hara, K. (1996). Cloning and nucleotide sequence of the *mphB* gene for macrolide 2′-phosphotransferase II in *Escherichia coli. FEMS Microbiology Letters* 144 (2–3): 197–202.

85 Matsuoka, M., Endou, K., Kobayashi, H. et al. (1998). A plasmid that encodes three genes for resistance to macrolide antibiotics in *Staphylococcus aureus. FEMS Microbiology Letters* 167 (2): 221–227.

86 Chesneau, O., Tsvetkova, K., and Courvalin, P. (2007). Resistance phenotypes conferred by macrolide phosphotransferases. *FEMS Microbiology Letters* 269 (2): 317–322.

87 Leclercq, R., Carlier, C., Duval, J., and Courvalin, P. (1985). Plasmid-mediated resistance to lincomycin by inactivation in *Staphylococcus haemolyticus. Antimicrobial Agents and Chemotherapy* 28 (3): 421–424.

88 Leclercq, R., Brisson-No¨el, A., Duval, J., and Courvalin, P. (1987). Phenotypic expression and genetic heterogeneity of lincosamide inactivation in *Staphylococcus* spp. *Antimicrobial Agents and Chemotherapy* 31 (12): 1887–1891.

89 Bozdogan, B., Berrezouga, L., Kuo, M.S. et al. (1999). A new resistance gene, *linB*, conferring resistance to lincosamides by nucleotidylation in *Enterococcus faecium* HM1025. *Antimicrobial Agents and Chemotherapy* 43 (4): 925–929.

90 Cundliffe, E. (1992). Glycosylation of macrolide antibiotics in extracts of *Streptomyces lividans*. *Antimicrobial Agents and Chemotherapy* 36 (2): 348–352.

91 Jenkins, G. and Cundliffe, E. (1991). Cloning and characterization of two genes from *Streptomyces lividans* that confer inducible resistance to lincomycin and macrolide antibiotics. *Gene* 108 (1): 55–62.

92 Quan, S., Venter, H., and Dabbs, E.R. (1997). Ribosylative inactivation of rifampin by *Mycobacterium smegmatis* is a principal contributor to its low susceptibility to this antibiotic. *Antimicrobial Agents and Chemotherapy* 41 (11): 2456–2460.

93 Houang, E.T.S., Chu, Y.-W., Lo, W.-S. et al. (2003). Epidemiology of rifampin ADP-ribosyltransferase (arr-2) and metallo-beta-lactamase (*bla*IMP-4) gene cassettes in class 1 integrons in *Acinetobacter* strains isolated from blood cultures in 1997 to 2000. *Antimicrobial Agents and Chemotherapy* 47 (4): 1382–1390.

94 Cao, M., Bernat, B.A., Wang, Z. et al. (2001). FosB, a cysteine-dependent fosfomycin resistance protein under the control of sigma(W), an extracytoplasmic-function sigma factor in *Bacillus subtilis*. *Journal of Bacteriology* 183 (7): 2380–2383.

95 Etienne, J., Gerbaud, G., Fleurette, J., and Courvalin, P. (1991). Characterization of staphylococcal plasmids hybridizing with the fosfomycin resistance gene *fosB*. *FEMS Microbiology Letters* 68 (1): 119–122.

96 Zilhao, R. and Courvalin, P. (1990). Nucleotide sequence of the *fosB* gene conferring fosfomycin resistance in *Staphylococcus epidermidis*. *FEMS Microbiology Letters* 56 (3): 267–272.

97 Fahey, R.C., Brown, W.C., Adams, W.B., and Worsham, M.B. (1978). Occurrence of glutathione in bacteria. *Journal of Bacteriology* 133 (3): 1126–1129.

98 Drlica, K. and Zhao, X. (1997). DNA gyrase, topoisomerase IV, and the 4-quinolones. *Microbiology and Molecular Biology Reviews: MMBR* 61 (3): 377–392.

99 Everett, M.J., Jin, Y.F., Ricci, V., and Piddock, L.J. (1996). Contributions of individual mechanisms to fluoroquinolone resistance in 36 *Escherichia coli* strains isolated from humans and animals. *Antimicrobial Agents and Chemotherapy* 40 (10): 2380–2386.

100 Ruiz, J. (2003). Mechanisms of resistance to quinolones: target alterations, decreased accumulation and DNA gyrase protection. *The Journal of Antimicrobial Chemotherapy* 51 (5): 1109–1117.

101 Weisblum, B. (1995). Erythromycin resistance by ribosome modification. *Antimicrobial Agents and Chemotherapy* 39 (3): 577–585.

102 Bateman, A., Murzin, A.G., and Teichmann, S.A. (1998). Structure and distribution of pentapeptide repeats in bacteria. *Protein Science* 7 (6): 1477–1480.

103 Martınez-Martınez, L., Pascual, A., and Jacoby, G.A. (1998). Quinolone resistance from a transferable plasmid. *Lancet* 351 (9105): 797–799.

104 Tran, J.H. and Jacoby, G.A. (2002). Mechanism of plasmid-mediated quinolone resistance. *Proceedings of the National Academy of Sciences of the United States of America* 99 (8): 5638–5642.

105 Tran, J.H., Jacoby, G.A., and Hooper, D.C. (2005a). Interaction of the plasmid-encoded quinolone resistance protein Qnr with *Escherichia coli* DNA gyrase. *Antimicrobial Agents and Chemotherapy* 49 (1): 118–125.

106 Tran, J.H., Jacoby, G.A., and Hooper, D.C. (2005b). Interaction of the plasmid-encoded quinolone resistance protein QnrA with *Escherichia coli* topoisomerase IV. *Antimicrobial Agents and Chemotherapy* 49 (7): 3050–3052.

107 Cundliffe, E. and McQuillen, K. (1967). Bacterial protein synthesis: the effects of antibiotics. *Journal of Molecular Biology* 30 (1): 137–146.

108 Burdett, V. (1991). Purification and characterization of Tet(M), a protein that renders ribosomes resistant to tetracycline. *The Journal of Biological Chemistry* 266 (5): 2872–2877.

109 Leipe, D.D., Wolf, Y.I., Koonin, E.V., and Aravind, L. (2002). Classification and evolution of P-loop GTPases and related ATPases. *Journal of Molecular Biology* 317 (1): 41–72.

110 Sanchez-Pescador, R., Brown, J.T., Roberts, M., and Urdea, M.S. (1988). Homology of the *TetM* with translational elongation factors: implications for potential modes of tetM-conferred tetracycline resistance. *Nucleic Acids Research* 16 (3): 1218.

111 Burdett, V. (1996). Tet(M)-promoted release of tetracycline from ribosomes is GTP dependent. *Journal of Bacteriology* 178 (11): 3246–3251.

112 Trieber, C.A., Burkhardt, N., Nierhaus, K.H., and Taylor, D.E. (1998). Ribosomal protection from tetracycline mediated by Tet(O): Tet(O) interaction with ribosomes is GTP-dependent. *Biological Chemistry* 379 (7): 847–855.

113 Courvalin, P. (2005). Genetics of glycopeptide resistance in gram-positive pathogens. *International Journal of Medical Microbiology* 294 (8): 479–486.

114 Arthur, M. and Courvalin, P. (1993). Genetics and mechanisms of glycopeptide resistance in enterococci. *Antimicrobial Agents and Chemotherapy* 37 (8): 1563–1571.

115 Leclercq, R., Dutka-Malen, S., Duval, J., and Courvalin, P. (1992). Vancomycin resistance gene *vanC* is specific to *Enterococcus gallinarum*. *Antimicrobial Agents and Chemotherapy* 36 (9): 2005–2008.

116 Navarro, F. and Courvalin, P. (1994). Analysis of genes encoding D-alanine-D-alanine ligase-related enzymes in *Enterococcus casseliflavus* and *Enterococcus flavescens*. *Antimicrobial Agents and Chemotherapy* 38 (8): 1788–1793.

117 Waxman, D.J. and Strominger, J.L. (1983). Penicillin-binding proteins and the mechanism of action of beta-lactam antibiotics. *Annual Review of Biochemistry* 52: 825–869.

118 Katayama, Y., Ito, T., and Hiramatsu, K. (2000). A new class of genetic element, *staphylococcus* cassette chromosome *mec*, encodes methicillin resistance in *Staphylococcus aureus*. *Antimicrobial Agents and Chemotherapy* 44 (6): 1549–1555.

119 Fuda, C., Suvorov, M., Vakulenko, S.B., and Mobashery, S. (2004). The basis for resistance to beta-lactam antibiotics by penicillin-binding protein 2a of methicillin-resistant *Staphylococcus aureus*. *The Journal of Biological Chemistry* 279 (39): 40802–40806.

120 Pinho, M.G., de Lencastre, H., and Tomasz, A. (2001). An acquired and a native penicillin-binding protein cooperate in building the cell wall of drug-resistant staphylococci. *Proceedings of the National Academy of Sciences of the United States of America* 98 (19): 10886–10891.

121 Brown, G.M. (1962). The biosynthesis of folic acid. II. Inhibition by sulfonamides. *The Journal of Biological Chemistry* 237: 536–540.

122 Hitchings, G.H. (1973). Mechanism of action of trimethoprim-sulfamethoxazole. I. *The Journal of Infectious Diseases* 128 (Suppl): 433–436.

123 Bushby, S.R. and Hitchings, G.H. (1968). Trimethoprim, a sulphonamide potentiator. *British Journal of Pharmacology and Chemotherapy* 33 (1): 72–90.

124 Dale, G.E., Broger, C., D'Arcy, A. et al. (1997). A single amino acid substitution in *Staphylococcus aureus* dihydrofolate reductase determines trimethoprim resistance. *Journal of Molecular Biology* 266 (1): 23–30.

125 Padayachee, T. and Klugman, K.P. (1999). Novel expansions of the gene encoding dihydropteroate synthase in trimethoprim-sulfamethoxazole-resistant *Streptococcus pneumoniae*. *Antimicrobial Agents and Chemotherapy* 43 (9): 2225–2230.

126 Pikis, A., Donkersloot, J.A., Rodriguez, W.J., and Keith, J.M. (1998). A conservative amino acid mutation in the chromosome-encoded dihydrofolate reductase confers trimethoprim resistance in *Streptococcus pneumoniae*. *The Journal of Infectious Diseases* 178 (3): 700–706.

127 Swedberg, G., Ringertz, S., and Sköld, O. (1998). Sulfonamide resistance in *Streptococcus pyogenes* is associated with differences in the amino acid sequence of its chromosomal dihydropteroate synthase. *Antimicrobial Agents and Chemotherapy* 42 (5): 1062–1067.

128 Huovinen, P. (2001). Resistance to trimethoprim-sulfamethoxazole. *Clinical Infectious Diseases* 32 (11): 1608–1614.

129 Huovinen, P., Sundström, L., Swedberg, G., and Sköld, O. (1995). Trimethoprim and sulfonamide resistance. *Antimicrobial Agents and Chemotherapy* 39 (2): 279–289.

130 Sundström, L., Rådström, P., Swedberg, G., and Sköld, O. (1988). Site-specific recombination promotes linkage between trimethoprim- and sulfonamide resistance genes. Sequence characterization of *dhfrV* and *sulI* and a recombination active locus of Tn21. *Molecular & General Genetics* 213 (2–3): 191–201.

131 de Groot, R., Sluijter, M., de Bruyn, A. et al. (1996). Genetic characterization of trimethoprim resistance in *Haemophilus influenzae*. *Antimicrobial Agents and Chemotherapy* 40 (9): 2131–2136.

132 Masuda, N., Sakagawa, E., Ohya, S. et al. (2000). Contribution of the MexX-MexY-OprM efflux system to intrinsic resistance in *Pseudomonas aeruginosa*. *Antimicrobial Agents and Chemotherapy* 44 (9): 2242–2246.

133 Bryan, L.E., Kowand, S.K., and Van Den Elzen, H.M. (1979). Mechanism of aminoglycoside antibiotic resistance in anaerobic bacteria: *Clostridium perfringens* and *Bacteroides fragilis*. *Antimicrobial Agents and Chemotherapy* 15 (1): 7–13.

134 Zhu, L., Lin, J., Ma, J. et al. (2010). Triclosan resistance of *Pseudomonas aeruginosa* PAO1 is due to FabV, a triclosan-resistant enoyl-acyl carrier protein reductase. *Antimicrobial Agents and Chemotherapy* 54 (2): 689–698.

135 Kaye, K.S., Cosgrove, S., Harris, A. et al. (2001). Risk factors for emergence of resistance to broad-spectrum cephalosporins among *Enterobacter* spp. *Antimicrobial Agents and Chemotherapy* 45 (9): 2628–2630.

136 Lindberg, F., Lindquist, S., and Normark, S. (1987). Inactivation of the *ampD* gene causes semiconstitutive overproduction of the inducible *Citrobacter freundii* beta-lactamase. *Journal of Bacteriology* 169 (5): 1923–1928.

137 Woodford, N., Reddy, S., Fagan, E.J. et al. (2007). Wide geographic spread of diverse acquired AmpC beta-lactamases among *Escherichia coli* and *Klebsiella* spp. in the UK and Ireland. *Journal of Antimicrobial Chemotherapy* 59 (1): 102–105.

138 Benveniste, R. and Davies, J. (1973). Aminoglycoside antibiotic-inactivating enzymes in actinomycetes similar to those present in clinical isolates of antibiotic-resistant bacteria. *Proceedings of the National Academy of Sciences of the United States of America* 70 (8): 2276–2280.

139 Dantas, G., Sommer, M.O.A., Oluwasegun, R.D., and Church, G.M. (2008). Bacteria subsisting on antibiotics. *Science* 320 (5872): 100–103.

140 D'Costa, V.M., McGrann, K.M., Hughes, D.W., and Wright, G.D. (2006). Sampling the antibiotic resistome. *Science* 311 (5759): 374–377.

141 Aminov, R.I. and Mackie, R.I. (2007). Evolution and ecology of antibiotic resistance genes. *FEMS Microbiology Letters* 271 (2): 147–161.

142 Behr, M.A. (2013). Evolution of *Mycobacterium tuberculosis*. *Advances in Experimental Medicine and Biology* 783: 81–91.

143 Auckland, C., Teare, L., Cooke, F. et al. (2002). Linezolid-resistant enterococci: report of the first isolates in the United Kingdom. *Journal of Antimicrobial Chemotherapy* 50 (5): 743–746.

144 Meka, V.G. and Gold, H.S. (2004). Antimicrobial resistance to linezolid. *Clinical Infectious Diseases* 39 (7): 1010–1015.

145 Johnsborg, O., Eldholm, V., and H°avarstein, L.S. (2007). Natural genetic transformation: prevalence, mechanisms and function. *Research in Microbiology* 158 (10): 767–778.

146 Sibold, C., Henrichsen, J., König, A. et al. (1994). Mosaic pbpX genes of major clones of penicillin-resistant *Streptococcus pneumoniae* have evolved from pbpX genes of a penicillin-sensitive *Streptococcus oralis*. *Molecular Microbiology* 12 (6): 1013–1023.

147 Weinbauer, M.G. (2004). Ecology of prokaryotic viruses. *FEMS Microbiology Reviews* 28 (2): 127–181.

148 Colomer-Lluch, M., Imamovic, L., Jofre, J., and Muniesa, M. (2011). Bacteriophages carrying antibiotic resistance genes in fecal waste from cattle, pigs, and poultry. *Antimicrobial Agents and Chemotherapy* 55 (10): 4908–4911.

149 Muniesa, M., García, A., Miró, E. et al. (2004). Bacteriophages and diffusion of beta-lactamase genes. *Emerging Infectious Diseases* 10 (6): 1134–1137.

150 Parsley, L.C., Consuegra, E.J., Kakirde, K.S. et al. (2010). Identification of diverse antimicrobial resistance determinants carried on bacterial, plasmid, or viral metagenomes from an activated sludge microbial assemblage. *Applied and Environmental Microbiology* 76 (11): 3753–3757.

151 Lederberg, J. and Tatum, E.L. (1946). Gene recombination in *Escherichia coli*. *Nature* 158 (4016): 558.

152 Silverman, P.M. (1997). Towards a structural biology of bacterial conjugation. *Molecular Microbiology* 23 (3): 423–429.

153 Grossman, T.H. and Silverman, P.M. (1989). Structure and function of conjugative pili: inducible synthesis of functional F pili by *Escherichia coli* K-12 containing a lac-tra operon fusion. *Journal of Bacteriology* 171 (2): 650–656.

154 Willetts, N. and Skurray, R. (1980). The conjugation system of F-like plasmids. *Annual Review of Genetics* 14: 41–76.

155 Udo, E.E., Love, H., and Grubb, W.B. (1992). Intra- and inter-species mobilisation of non-conjugative plasmids in staphylococci. *Journal of Medical Microbiology* 37 (3): 180–186.

156 Warren, G.J., Twigg, A.J., and Sherratt, D.J. (1978). ColE1 plasmid mobility and relaxation complex. *Nature* 274 (5668): 259–261.

157 Salyers, A.A., Shoemaker, N.B., Stevens, A.M., and Li, L.Y. (1995). Conjugative transposons: an unusual and diverse set of integrated gene transfer elements. *Microbiological Reviews* 59 (4): 579–590.

158 Hall, R.M. and Stokes, H.W. (1993). Integrons: novel DNA elements which capture genes by site-specific recombination. *Genetica* 90 (2–3): 115–132.

159 Carattoli, A. (2001). Importance of integrons in the diffusion of resistance. *Veterinary Research* 32 (3–4): 243–259.

160 Morabito, S., Tozzoli, R., Caprioli, A. et al. (2002). Detection and characterization of class 1 integrons in enterohemorrhagic *Escherichia coli*. *Microbial Drug Resistance* 8 (2): 85–91.

161 Naas, T., Mikami, Y., Imai, T. et al. (2001). Characterization of In53, a class 1 plasmid- and composite transposon-located integron of *Escherichia coli* which carries an unusual array of gene cassettes. *Journal of Bacteriology* 183 (1): 235–249.

162 Partridge, S.R., Brown, H.J., and Hall, R.M. (2002a). Characterization and movement of the class 1 integron known as Tn2521 and Tn1405. *Antimicrobial Agents and Chemotherapy* 46 (5): 1288–1294.

163 Partridge, S.R., Collis, C.M., and Hall, R.M. (2002b). Class 1 integron containing a new gene cassette, aadA10, associated with Tn1404 from R151. *Antimicrobial Agents and Chemotherapy* 46 (8): 2400–2408.

164 Partridge, S.R., Brown, H.J., Stokes, H.W., and Hall, R.M. (2001). Transposons Tn1696 and Tn21 and their integrons In4 and In2 have independent origins. *Antimicrobial Agents and Chemotherapy* 45 (4): 1263–1270.

165 Villa, L., Visca, P., Tosini, F. et al. (2002). Composite integron array generated by insertion of an ORF341-type integron within a Tn21-like element. *Microbial Drug Resistance* 8 (1): 1–8.

166 Han, J., Wang, Y., Sahin, O. et al. (2012). A fluoroquinolone resistance associated mutation in *gyrA* affects DNA supercoiling in *Campylobacter jejuni*. *Frontiers in Cellular and Infection Microbiology* 2 (March): 21.

167 Cottell, J.L., Webber, M.A., and Piddock, L.J.V. (2012). Persistence of transferable extended-spectrum-β-lactamase resistance in the absence of antibiotic pressure. *Antimicrobial Agents and Chemotherapy* 56 (9): 4703–4706.

168 Lenski, R.E., Simpson, S.C., and Nguyen, T.T. (1994). Genetic analysis of a plasmid-encoded, host genotype-specific enhancement of bacterial fitness. *Journal of Bacteriology* 176 (11): 3140–3147.

169 Marcusson, L.L., Frimodt-Møller, N., and Hughes, D. (2009). Interplay in the selection of fluoroquinolone resistance and bacterial fitness. *PLoS Pathogens* 5 (8): e1000541.

170 Rozen, D.E., McGee, L., Levin, B.R., and Klugman, K.P. (2007). Fitness costs of fluoroquinolone resistance in *Streptococcus pneumoniae*. *Antimicrobial Agents and Chemotherapy* 51 (2): 412–416.

2

Molecular Mechanisms of Antibiotic Resistance – Part II

Liam K.R. Sharkey[1] and Alex J. O'Neill[2]

[1] Institute of Infection and Immunity, University of Melbourne, Melbourne, Australia
[2] University of Leeds, Leeds, UK

2.1   Introduction

In the early years of the antibiotic era it was already apparent that bacteria could sometimes resist the growth-inhibitory effects of antibiotics, a situation that can compromise the treatment of bacterial infection in the patient. The considerable threat that this poses to effective antibacterial chemotherapy has prompted intensive research efforts toward understanding the phenomenon of antibiotic resistance. An important facet of these efforts has been to dissect the mechanisms underlying antibiotic resistance; in other words, to establish in molecular detail how bacteria evade the inhibitory action of antibiotics. Aside from yielding information that is of fundamental biological interest, these studies also serve an important practical purpose – they offer essential intelligence for those engaged in antibacterial drug discovery to inform the development of new approaches to circumvent or overcome antibiotic resistance.

The ability to resist the action of an antibiotic may result from inherent properties of a bacterium (intrinsic or natural resistance) or may evolve in a bacterial population that was previously antibiotic-susceptible (acquired resistance). In both cases, the means by which resistance is ultimately achieved are the same – by preventing or mitigating the effects of an antibiotic binding to its cellular target. However, this outcome can be realised in several different ways, and the mechanisms by which bacteria resist antibiotics can be grouped into four major classes that are summarized in Figure 2.1. The aim of this chapter is to describe these different mechanisms of resistance and to provide examples as they occur in medically-important bacteria.

Akin to the evolution of other traits in microorganisms, acquired antibiotic resistance can arise *via* two distinct routes: the endogenous route, in which evolution results from changes to the existing genetic material of a bacterium, and the exogenous route, which involves the acquisition of new genetic material from other microorganisms by horizontal gene transfer (HGT). The endogenous route to resistance usually involves point mutations in DNA that arise spontaneously and occur predominantly during the process of replication. These mutations alter the genetic code and can in turn lead to

Bacterial Resistance to Antibiotics – From Molecules to Man, First Edition.
Edited by Boyan B. Bonev and Nicholas M. Brown.
© 2020 John Wiley & Sons Ltd. Published 2020 by John Wiley & Sons Ltd.

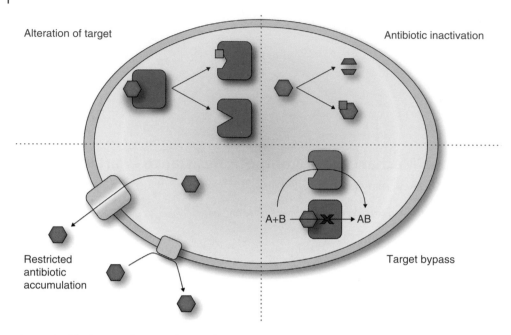

Figure 2.1 The four mechanistic classes of antibiotic resistance that occur in bacteria (antibiotic shown in purple). See text for further explanation.

changes in the organisms' phenotype, including changes in antibiotic susceptibility. The frequency with which such DNA mutations arise is typically low, approximately 1 in every 10^9 bases replicated. Nevertheless, the genetic diversity that this low rate of mutation can engender is substantial in the context of microorganisms that can replicate their entire genomes on a timescale measured in minutes and can readily attain very large population counts ($>10^9$ cells/ml), even in patients [1]. Thus, mutants exhibiting reduced susceptibility to an antibiotic as a consequence of spontaneous genetic changes may arise readily. In the absence of the corresponding antibiotic, these mutants usually exhibit a survival disadvantage compared to their antibiotic-susceptible counterparts and are likely to become outcompeted. However, in the presence of antibiotic selective pressure (for example, in a patient undergoing antibiotic treatment), such mutants will exhibit a substantial survival advantage, resulting in their selection and enrichment over the antibiotic-susceptible members of the population.

Exogenous antibiotic resistance involves the horizontal acquisition of antibiotic resistance determinants by previously antibiotic-susceptible bacteria. This process reflects the fact that microorganisms are in general genetically promiscuous and, at least in principle, have access to an enormous shared gene pool *via* HGT. The horizontal transfer of antibiotic resistance genes is facilitated by mobile genetic elements such as plasmids, transposons, and integrons, which can carry antibiotic resistance determinants to new hosts by the standard routes of HGT (conjugation, transduction, transformation). As with endogenous resistance, in environments where antibiotics are present, natural selection will favor those bacteria that have acquired the ability to resist the growth-inhibitory effects of antibiotics and are thereby able to outcompete their antibiotic-susceptible counterparts.

It is interesting to consider the origin of horizontally-acquired antibiotic resistance determinants and the reasons for their existence. A considerable proportion of transferable antibiotic resistance can be attributed to the enormous diversity of biological functions that bacteria possess; many "antibiotic resistance" genes apparently confer resistance as a by-product of their original evolved function, *i.e.*, antibiotic resistance is incidental to their native cellular role. In other cases, antibiotic resistance mechanisms have undoubtedly evolved as dedicated systems of bacterial protection against particular antibiotic classes. Most antibiotic classes in clinical use derive from natural compounds synthesized by microorganisms as part of their secondary metabolism. Consequently, it is unsurprising that self-protection mechanisms exist in the organisms that produce such compounds and that these resistance genes are in principle only a single HGT event from acquisition by medically-important bacteria.

2.1.1 Resistance as a Consequence of Target Alteration

2.1.1.1 Genetic Alteration of Antibiotic Target

An antibiotic recognizes and binds its cognate target through specific chemical contacts; any alteration in the target that prevents these contacts from being made will reduce the affinity of the interaction and can thereby cause resistance (Figure 2.1). In its simplest form, such an alteration can result from a point mutation in the gene encoding the target, leading to a single amino acid substitution in the protein. Mutational alteration of the target is a common endogenous route to resistance against antibiotics that inhibit a bacterial target encoded by a single gene copy (*e.g.*, fusidic acid, trimethoprim, mupirocin). By contrast, antibiotic targets encoded by multiple alleles (*e.g.*, the ribosome) are less prone to resistance development by this route, since the resistance mutations tend to be recessive and must present in several of the alleles for the resistance phenotype to become apparent [2]. Target site alteration by mutation represents a common source of acquired antibiotic resistance in all bacteria but its importance as a route to resistance is particularly evident in microorganisms that do not ordinarily engage in genetic exchange (*e.g.*, *Mycobacterium tuberculosis*).

Genetic alteration of the target represents the primary mechanism of acquired resistance to the antibiotic rifampicin (US: rifampin) in bacterial pathogens such as *Staphylococcus aureus* and *M. tuberculosis*. Rifampicin is an inhibitor of bacterial RNA transcription that acts by binding to the β subunit of the RNA polymerase multi-subunit enzyme complex and sterically blocking initiation of RNA synthesis. Although it is one of the most potent antibiotics in clinical use, capable of inhibiting the growth of susceptible bacteria at concentrations of only nanograms per milliliter, the useful antibacterial effect of the drug can become completely abrogated by single point mutations in the *rpoB* gene that encodes the β subunit [3] (Figure 2.2).

Mutation-based target alteration also represents an important source of resistance to the fluoroquinolones in both Gram-positive and Gram-negative bacteria. The fluoroquinolones are synthetic antibacterial agents that interfere with correct functioning of two essential topoisomerases (DNA gyrase and topoisomerase IV), enzymes that alter DNA topology in the bacterial cell [4]. As part of their functional cycle, both enzymes catalyze transient breakages in the sugar–phosphate backbone of DNA duplexes. Fluoroquinolones act by binding to, and stabilizing, these enzyme-cleaved DNA

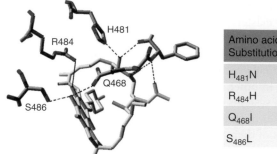

Amino acid Substitution	Loss of rifampicin activity
$H_{481}N$	250-fold
$R_{484}H$	32 000-fold
$Q_{468}I$	128 000-fold
$S_{486}L$	128 000-fold

Figure 2.2 Single amino acid substitutions in the β subunit of RNA polymerase can mediate high-level resistance to the antibiotic rifampicin. The figure on the left shows the key contacts that the rifampicin molecule (yellow) makes with amino acids lying within the β subunit. Shown on the right is the fold reduction in antibacterial activity of rifampicin against *Staphylococcus aureus* when one of these key contacts is lost by substitution of a single amino acid residue.

complexes, with the resulting accumulation of double-stranded DNA breaks leading to cessation of DNA replication and ultimately, to bacterial cell death [5]. A common route to high-level fluoroquinolone resistance involves the stepwise accumulation of point mutations in the genes encoding DNA gyrase and topioisomerase IV. In a similar fashion to that described for rifampicin, these point mutations lead to amino acid changes in the target proteins that seem to reduce their affinity for the drug. In Gram-negative organisms such as *Escherichia coli*, the primary target of the fluoroquinolones is usually DNA gyrase; consequently, most fluoroquinolone resistance mutations in such bacteria arise in the two genes (*gyrA* and *gyrB*) that encode the subunits of the DNA gyrase holoenzyme (GyrA and GyrB, respectively). Of these resistance mutations, most occur within *gyrA* and map to the quinolone resistance-determining region (QRDR), a region of the gene that encodes a positively charged portion of the GyrA protein that is involved in DNA binding. Mutations in *gyrB* are a less frequent cause of fluoroquinolone resistance in clinical isolates than *gyrA* mutations and elicit a lower level of resistance [6]. In Gram-positive bacteria such as *S. aureus*, topoisomerase IV is frequently the primary target of fluoroquinolones, and resistance mutations are usually found in the two genes (*parC* and *parE*) that encode the subunits of this protein.

Antibiotic resistance due to target alteration by mutation usually involves only a small number of discrete genetic changes; however, more extensive genetic remodeling of antibiotic targets can occur as a result of recombination. This route to antibiotic resistance is particularly evident in bacterial species that are naturally transformable and are therefore capable of taking up naked DNA from the environment and incorporating it into their chromosomes by recombination. Two such species, *Streptococcus pneumoniae* and *Neisseria gonorrhoeae*, have been shown to assimilate DNA fragments of the genes encoding the targets of the β-lactam antibiotics (the penicillin binding proteins [PBPs]) from other bacterial species into their own [7, 8]. The resulting "mosaic" PBP genes can include DNA fragments from β-lactam insensitive PBPs, thereby rendering the PBPs encoded by the recombinant genes (and their bacterial hosts) resistant to β-lactam antibiotics.

The processes of mutation and recombination may also act together to evolve an antibiotic-susceptible target to a more resistant form. As indicated earlier, mutation-based target alteration is not a common route to resistance against antibiotics targeting the ribosome, since the multi-copy nature of the alleles encoding the ribosomal RNA (rRNA) in most bacterial pathogens acts to mask the resistance phenotype. However, resistance to linezolid – a synthetic antibacterial drug of the oxazolidinone class that inhibits bacterial protein synthesis though interaction with 23S rRNA – has emerged *via* genetic target alteration in bacterial genera that harbor four to six copies of the 23S rRNA gene [9]. Such linezolid-resistant isolates carry resistance mutations in domain V of the 23S rRNA, and these mutations are usually found in two to five of the alleles encoding this rRNA species. These multiple resistance alleles are not the result of independent mutational events but are instead the consequence of "gene conversion"; following emergence of a resistance mutation in a single 23S rRNA allele, this mutation is then introduced to one or more of the drug-susceptible alleles by homologous recombination within the cell, thereby converting them to the resistant form [10].

2.1.1.2 Overproduction of Antibiotic Target

Antibiotics interact with their targets in a set ratio; usually, one molecule of antibiotic binds to one molecule of the target. By increasing production of the target, bacteria can raise the concentration of antibiotic required to inhibit sufficient target molecules to prompt an antibacterial effect, thereby resulting in resistance.

Bacterial mutants resistant to antibiotics as a result of target overproduction may be readily selected in the laboratory. By contrast, this mechanism of resistance is relatively uncommon in clinical isolates, suggesting that outside the protected environment of the Petri dish the hyperproduction of drug targets usually has negative consequences for a bacterium that place it at a competitive disadvantage. Nonetheless, there are a few well-documented examples of antibiotic resistance arising in clinical isolates through target overproduction. One of these concerns resistance to the synthetic antibacterial trimethoprim in the bacterium *E. coli*. Trimethoprim competitively inhibits dihydrofolate reductase (DHFR), an enzyme with a crucial role in the synthesis of nucleotide precursors, inhibition of which ultimately leads to cessation of DNA replication. Point mutations arising within the promoter region of the gene encoding DHFR can drive increased transcription, thereby elevating expression of the enzyme. The excess DHFR effectively acts to titrate trimethoprim and a higher concentration of the drug is therefore required to reduce cellular DHFR activity to a level that will inhibit bacterial growth [11].

A more complex example of target overproduction has been identified as a cause of low-level ("intermediate") resistance to the antibiotic vancomycin in *S. aureus*. Prior to the introduction of several new antistaphylococcal drugs in the early years of the twenty-first century, the glycopeptide vancomycin was considered the drug of last resort for treating infections caused by multidrug-resistant *S. aureus* strains, including methicillin-resistant *S. aureus* (MRSA). It was therefore of considerable concern when in 1997 an MRSA isolate exhibiting an intermediate level of vancomycin resistance was detected in Japan [12]. Vancomycin-intermediate *S. aureus* (VISA) strains have subsequently been found worldwide, and the unusual mechanism underlying this resistance phenotype has been characterized. In vancomycin-susceptible *S. aureus* (VSSA), the drug acts as an inhibitor of cell wall biosynthesis by binding to the D-alanyl-D-alanine (usually abbreviated to D-ala-D-ala) termini of lipid-linked peptidoglycan precursors and sterically hindering their incorporation into the growing peptidoglycan cell wall [13]. Because staphylococcal cell wall synthesis is focussed at the septa during cell division, vancomycin must diffuse through the wall to the tip of the division septa to reach its primary site of action (Figure 2.3). VISA strains produce a dramatically thickened cell wall and in most isolates the peptidoglycan also shows reduced cross-linking [15]. Both circumstances lead to an increase in the presence of muropeptides with free D-ala-D-ala termini in the cell wall, thereby providing a multitude of additional sites within the peptidoglycan mesh to which vancomycin can bind. These additional binding sites essentially act as "decoys," trapping vancomcyin in the outer layers of the cell wall, well away from the sites of nascent peptidoglycan synthesis (Figure 2.3). Consequently, the concentration of vancomycin reaching the primary site of cell wall synthesis is reduced and the antibacterial effect of the drug is attenuated [14].

2.1.1.3 Enzymatic Modification of Antibiotic Target
Antibiotic resistance can result from enzyme-catalyzed chemical modification of an antibiotic target. Indeed, enzymatic modification of rRNA represents an important mechanism of resistance to antibiotics that inhibit bacterial protein synthesis by targeting the ribosome. Most of these antibiotic classes bind the large (50S) subunit of the ribosome, making contact with the 23S rRNA near the peptidyltransferase centre (PTC)

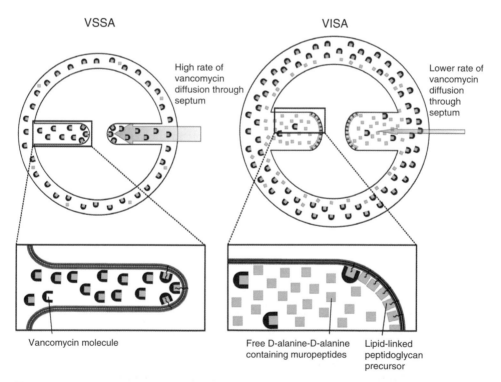

Figure 2.3 Overproduction of the antibiotic target results in reduced susceptibility to vancomycin in vancomycin-intermediate *Staphylococcus aureus* (VISA) strains. VISA cells produce a thicker cell wall with an excess of free D-ala-D-ala termini compared to vancomyin-susceptible *S. aureus* (VSSA) strains. These termini bind vancomycin, thereby reducing the concentration of the drug that reaches the target site at the tip of the division septa. *Source:* Adapted from [14].

or the peptide exit tunnel. Modification of single nucleotides within the 23S rRNA near this region by mono- or dimethylation (addition of one or two methyl groups) can prevent these antibiotics from binding, either because the methylation event prompts local conformational change that alters the antibiotic binding pocket or by direct steric hindrance. For example, methyltransferase enzymes of the erythromycin rRNA methylase (Erm) family mediate resistance to macrolide, lincosamide, and streptogramin B antibiotics (MLS_B phenotype) *via* methylation of a single adenine of the 23S rRNA (nucleotide position 2058 [*E. coli* numbering]). The Erm family represent the most widespread mechanism of resistance to macrolides and lincosamides in pathogenic bacteria, and to date over 40 members of this family have been reported [16]. Another methyltransferase, termed Cfr, methylates nucleotide A2053 (*E. coli* numbering) in the 23S rRNA (Figure 2.4). Although originally identified as the first example of horizontally-acquired resistance to linezolid in the staphylococci, Cfr also mediates resistance to several other

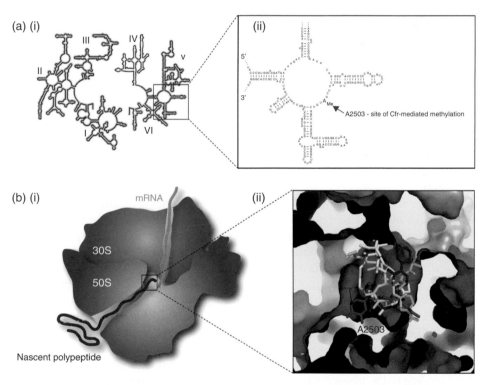

Figure 2.4 Cfr-mediated methlyation of 23S rRNA provides resistance to several different classes of antibiotics that bind the ribosome at the peptidyltransferase centre (PTC). (a) The site of Cfr-mediated methylation in domain V (pink) of 23S rRNA (b) The PTC resides in the large ribosomal subunit, adjacent to the peptide exit tunnel, and catalyzes peptide bond formation and peptide release; several structurally unrelated classes of antibiotic bind into the PTC, causing cessation of translation. Representative members of the oxazolidinones, phenicols, lincosamides, pleuromutilins, and group A streptogramins are shown bound at this site (linezolid [red]; chloramphenicol [green]; clindamycin [pink]; tiamulin [aqua]; the dalfopristin [yellow]). Methylation of A2503 (red surface) by Cfr alters the binding pocket of these five antibiotic classes, reducing their ability to bind and inhibit the ribosome, and thereby mediating concurrent resistance to all of these classes. *Source:* Part (a) adapted from [17].

clinically-important and structurally-unrelated antibiotic classes that share overlapping binding sites with linezolid (phenicols, lincosamides, pleuromutilins, and group A streptogramins) [18] (Figure 2.4).

Target modification is also an important mechanism of resistance to antibiotics of the polymyxin class, such as colistin (polymyxin E). While toxic side effects associated with this antibiotic class led to it falling out of favor for routine clinical use several decades ago, colistin has re-emerged as a drug of last resort for treating infections caused by extensively drug-resistant Gram-negative bacteria. In 2015, the first horizontally-transmissible colistin resistance determinant was found among the commensal bacteria of livestock in China [19]. This determinant, termed *mcr-1*, encodes a phosphoethanolamine transferase that mediates modification of the lipopolysaccharide (LPS) component of the outer membrane (OM) of Gram-negative bacteria. The LPS is the initial target for colistin on the bacterial cell surface, with antibiotic binding proceeding *via* electrostatic interaction between the negatively-charged LPS and the positively-charged colistin molecule. By catalyzing the transfer of phosphoethanolamine onto lipid A, the lipid component of LPS, MCR-1 alters the charge of LPS from negative to positive [20]. As a result, colistin can no longer interact effectively with the target to exert its antibacterial effect.

2.1.1.4 Protection of Antibiotic Target

In common with antibiotic resistance resulting from enzymatic modification of an antibiotic target, the phenomenon of target protection requires the action of a resistance protein upon the target. However, the latter mechanism can be distinguished from the former in that it does not involve permanent chemical modification of the target. Instead, target protection requires that there is continuous or frequently repeated direct physical interaction between the resistance protein and the antibiotic target to bring about resistance.

The best characterized example of target protection provides resistance to the tetracyclines, a class of antibiotic that binds to the 16S rRNA in the 30S subunit of the ribosome and prevents entry of incoming aminoacyl-tRNA species into the ribosomal A (acceptor) site during protein synthesis. A family of determinants termed the ribosomal protection proteins (RPPs) mediate resistance to tetracyclines in both Gram-positive and Gram-negative bacteria [21]. To date, more than 10 RPPs have been identified (http://faculty.washington.edu/marilynr/tetweb1.pdf), and detailed study of the two most prevalent examples (Tet[M] and Tet[O]) has provided much of our understanding of this family of resistance proteins [22, 23]. RPPs exhibit homology to bacterial housekeeping G-proteins that participate in protein synthesis (particularly elongation factor-G [EF-G] that associates transiently with the ribosomal A-site to bring about translocation), and their site and mode of binding to the ribosome closely mimic that of EF-G. However, while domain IV of the EF-G protein acts to reach into the decoding site of the ribosome where it prompts translocation by driving the movement of A-site tRNA to the P-site, the equivalent domain of the RPPs directly contacts the binding site of tetracycline, thereby acting to displace the drug from the ribosome [22, 23].

RPP-mediated tetracycline resistance was for many years the sole documented example of antibiotic resistance resulting through target protection. However, several other instances of target protection have recently been identified, and it is now clear that this type of mechanism plays an important role in mediating clinically-significant antibiotic resistance. In Gram-positive bacteria, the ARE (antibiotic resistance) ABC-F proteins employ target protection to elicit resistance to essentially all clinically-deployed antibiotic classes that target the 50S subunit of the ribosome [24]. Like the tetracycline RPPs, the ARE ABC-F proteins share homology with a translation factor; in this case, the energy dependent translational throttle protein (EttA) from *E. coli* that acts to restrict protein translation in response to low cellular ATP levels. Both ARE ABC-F proteins and EttA consist of two ABC domains separated by an inter-domain linker, the latter of which is often longer in ARE ABC-F proteins. EttA binds to the E (exit)-site of the ribosome and inserts its linker region toward the PTC, thereby preventing peptide bond formation and stalling translation [25, 26]. It has been proposed that ARE ABC-F proteins bind the ribosome in a similar manner to EttA, but that the extended linker region of the former affords deeper penetration toward the PTC, where it can prompt dissociation of bound antibiotic [24, 27].

Target protection is not always the result of antibiotic displacement. The antibiotic fusidic acid (FA) binds to EF-G when the latter is resident in the A-site of the ribosome, once translocation has occurred. Binding of FA acts to stabilize the ordinarily transient association between this translation factor and the ribosome, thereby occluding the A-site and consequently inhibiting protein synthesis. The FusB-type proteins represent the major source of FA resistance in the staphylococci and bind to EF-G to protect it from the inhibitory action of the drug [28, 29] (Figure 2.5). These proteins do not appear to interfere with the association between drug and target, and indeed, their binding site on EF-G involves domains distinct from those participating in binding to FA (Figure 2.5). Instead, FusB-type proteins have been shown to actively drive the release of EF-G from the ribosome in both the absence and the presence of the antibiotic (Figure 2.5) [28]. Thus, the action of the FusB-type proteins on EF-G (driving dissociation from the ribosome) directly counteracts that of FA (favoring retention on the ribosome), thereby providing resistance to this antibiotic [28, 30].

2.1.2 Resistance as a Consequence of Target Bypass

As described above, alteration of an antibiotic target by a variety of means can lead to resistance. A variation on this theme is target bypass, a phenomenon wherein a bacterium acquires through HGT the genetic instructions necessary to synthesize an alternate, antibiotic-insusceptible version of the drug target (Figure 2.1).

A classic example of target bypass is methicillin resistance in *S. aureus*. Methicillin and other members of the β-lactam class bind and inactivate the transpeptidase function of the PBPs, thereby inhibiting cross-linking of peptidoglycan in the bacterial cell wall. MRSA strains possess a gene of the *mec* family (typically *mecA*), the encoded product of which is a transpeptidase designated PBP2a that is refractory to inhibition by most β-lactam antibiotics [31]. In the presence of β-lactams, the native PBPs of *S. aureus* become rapidly inactivated by the drug, leading to cessation of cell wall cross-linking and ultimately bringing about cell death. However, in MRSA, the PBP2a enzyme does not become inactivated by β-lactams and is able to take over the essential cross-linking function, allowing continued survival and growth even in the presence of the antibiotic. The inhibition of β-lactam-susceptible PBPs proceeds in two stages: first, the formation of an initial non-covalent complex between the drug and the PBP, and subsequently, acylation of the crucial active site serine, an event that inactivates the enzyme. It has been demonstrated that the lack of susceptibility of the PBP2a enzyme to β-lactams is

(a)

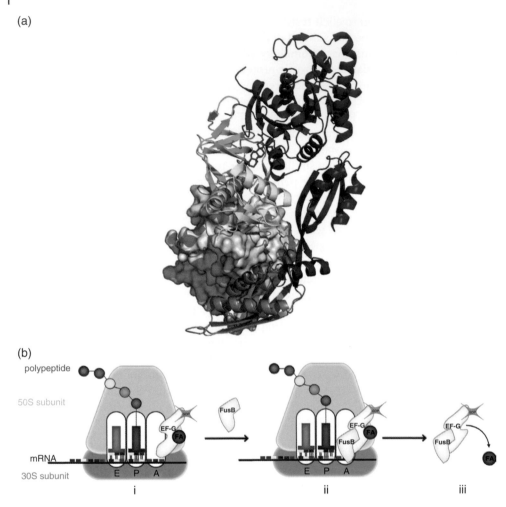

(b)

Figure 2.5 **The FusB-type resistance proteins protect EF-G from the antibiotic fusidic acid.** (a) Model of a FusB-type protein (shown in gray) bound to EF-G (shown in rainbow colors, with individual domains of the protein colored distinctly). Fusidic acid (FA) (shown in pink) is shown bound to EF-G, well away from the site of interaction of EF-G with FusB-type proteins. (b) Mechanism of FA resistance mediated by FusB-type proteins. (i) FA exerts its antibacterial effect by immobilizing EF-G (in GDP-bound form) on the ribosome, thereby stalling protein synthesis, (ii) FusB-type proteins bind to EF-G and drive its dissociation from the ribosome, allowing protein synthesis to continue. Since binding of FA to EF-G only occurs on the ribosome, FA is thought to spontaneously dissociate from EF-G once the latter has been dislodged from the ribosome (iii), and the EF-G molecule is once again free to participate in subsequent rounds of protein synthesis [28, 30].

due to inefficient acylation of the active site serine, rather than a failure of β-lactams to bind and form the initial non-covalent drug–target complex [32].

A more sophisticated example of target bypass is provided by the *vanA* operon (Figure 2.6), a five-gene cluster conferring high-level resistance to vancomycin in Gram-positive bacteria. This type of vancomycin resistance was first reported in enterococci in 1988 [33], and since 2002 there have been isolated reports of horizontal transfer of the

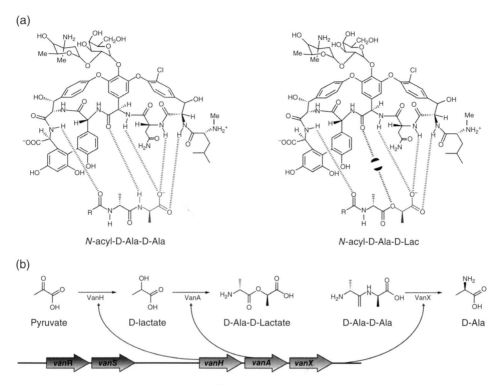

Figure 2.6 Structural and genetic basis of horizontally-acquired vancomycin resistance in Gram-positive bacteria. (a) Vancomycin binds to its target, the D-ala-D-ala termini of peptidoglycan precursors, *via* a network of hydrogen bonds. Vancomycin resistance results from substitution of D-ala-D-ala with D-ala-D-lac; the latter exhibits a *c.* 1000-fold decrease in affinity for vancomycin owing to loss of a hydrogen bond and gain of electronic repulsion mediated by the ester oxygen. **(b)** The *vanA* operon consists of five genes, two of which encode regulatory proteins (shown in red) that upregulate expression of the operon in the presence of vancomycin. The other three genes (shown in green) encode a pyruvate reductase (VanH), a D-Ala-D-lac ligase (VanA) and a D-ala-D-ala dipeptidase (VanX). These enzymes collectively reprogram peptidoglycan synthesis to incorporate D-ala-D-lac termini into peptidoglycan precursors in place of D-ala-D-ala.

determinants responsible from enterococci into *S. aureus.* The *vanA* operon encodes five proteins, three of which work in concert to substitute the terminal D-alanine residue of peptidoglycan precursor molecules in the cell with the chemically- and structurally-related D-lactate (Figure 2.6). VanH catalyzes the generation of D-lactate from the cellular pyruvate pool, and this compound is subsequently ligated to D-alanine by the D-alanyl-D-lactate ligase (VanA), thereby generating the dipeptide for incorporation into the muropeptide precursors (Figure 2.6). VanX is a D-ala-D-ala dipeptidase that acts to hydrolyze the D-ala-D-ala dipeptide generated by the native cell wall biosynthetic machinery, thereby ensuring that muropeptide precursors incorporated into peptidoglycan essentially all contain a terminal D-lactate residue instead of D-alanine [34]. The other two proteins encoded by the operon (Figure 2.6) constitute a two-component regulatory system consisting of VanS, a membrane-associated kinase that is activated upon sensing vancomycin, and VanR, which when phosphorylated by VanS activates transcription of the three structural genes that together act to reprogram peptidoglycan biosynthesis. Replacement of the amide bond in the D-ala-D-ala dipeptide with the ester linkage of D-ala-D-lac prevents formation of a crucial hydrogen bond between the drug and its target, and introduces electronic repulsion of the drug owing to the lone pair electrons of the ester oxygen of D-ala-D-lac (Figure 2.6) [35]. The net effect of these changes is an approximately 1000-fold decrease in the affinity of vancomycin for the muropeptide, and cell wall synthesis and bacterial growth are consequently able to continue in the presence of concentrations of the drug that are therapeutically achievable in the patient.

The *vanA* operon is found in environmental bacteria that synthesize glycopeptides, where it functions as a mechanism of self-protection to prevent the organism from inhibiting its own cell wall biosynthesis. Based on the highly similar genetic architecture of vancomycin-resistance operons in gycopeptide-producer organisms and those found in vancomycin-resistance clinical isolates, and the sequence identity of the encoded proteins, these environmental organisms appear to represent the original source of the vancomycin-resistance operons present in bacterial pathogens [36].

2.1.3   Resistance as a Consequence of Antibiotic Inactivation

Enzymatic inactivation of an antibiotic, rendering it less able or unable to bind to its target, may occur in two ways (Figure 2.1). Antibiotic degradation involves rearrangement or loss of a key chemical feature that is essential for antibacterial activity. Antibiotic modification is similar, but in this instance the antibacterial action is abrogated as a result of enzymatic addition of one or more chemical moieties to the antibiotic that interfere with its ability to interact with the target.

2.1.3.1   Antibiotic Degradation

The most extensively studied proteins mediating resistance through antibiotic degradation are the β-lactamases, a family of enzymes that inactivate members of the β-lactam class of antibiotics. Since the β-lactams remain the most widely used class of antibiotic worldwide, and the β-lactamases represent the predominant mechanism of resistance to these drugs in bacterial pathogens, it follows that these enzymes are among the most clinically-important mediators of antibiotic resistance.

The common structural feature of all β-lactams, and the portion of the molecule responsible for the antibacterial action, is a highly reactive four-membered β-lactam ring (Figure 2.7). This ring exhibits structural analogy to the D-ala-D-ala portion of the muropeptide precursors of peptidoglycan recognized by PBP enzymes (Figure 2.7); this feature of β-lactams allows them to bind PBPs in place of the native substrate, acylate the crucial active site serine, and thereby render the enzyme non-catalytic. β-lactamases act to hydrolytically open the β-lactam ring, an event that removes the structural analogy with D-ala-D-ala and thereby renders the antibiotic inactive.

According to the Ambler classification scheme, β-lactamases are categorized by their primary protein structure into four groups, designated A, B, C, and D (Figure 2.7) [38]. Ambler classes A, C, and D are known as the serine β-lactamases, as their catalytic mechanism employs a serine residue in the active site of the enzyme for hydrolytic attack on the β-lactam ring. The serine β-lactamases show architectural and mechanistic similarities to the PBPs (Figure 2.7), implying an evolutionary relationship between the antibiotic target and the corresponding resistance protein [39]. The class B

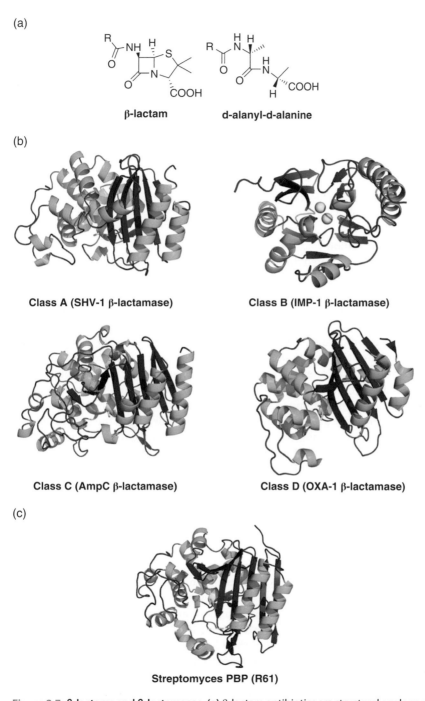

(a)

β-lactam   **d-alanyl-d-alanine**

(b)

Class A (SHV-1 β-lactamase)   **Class B (IMP-1 β-lactamase)**

Class C (AmpC β-lactamase)   **Class D (OXA-1 β-lactamase)**

(c)

Streptomyces PBP (R61)

Figure 2.7 **β-lactams and β-lactamases. (a)** β-lactam antibiotics are structural analogs of d-alanyl-d-alanine, a property that enables them to bind the transpeptidase domain of the PBP enzymes in place of the native substrate. **(b)** Structural comparison of representative members of the four classes of β-lactamases. Catalytic serine residues are shown in yellow for classes A, C and D; catalytic Zn^{2+} ions are shown in yellow in the class B enzyme. **(c)** The target of the β-lactams, the PBPs, are evolutionarily-related to the serine β-lactamases; a PBP from Streptomyces R61 is shown that exhibits clear structural similarities to class A, C, and D β-lactamases. *Source:* Adapted from [37].

metallo-β-lactamases are mechanistically distinct from the serine enzymes (Figure 2.7), utilizing instead at least one divalent zinc ion to hydrolyze the amide bond of the β-lactam ring [37]. Members of this latter class exhibit a very broad substrate profile and are capable of hydrolyzing types of β-lactam drug (*e.g.*, the carbapenems) that the majority of serine β-lactamases cannot. It is this ability to inactivate most types of β-lactam drugs that makes the metallo-β-lactamases of particular concern. The most high-profile member of these is the New Delhi metallo-β-lactamase-1 (NDM-1) [40], an enzyme first identified in 2008 in clinical isolates of *Enterobacteriaceae* originating from India, and subsequently detected in countries worldwide. Although numerous other metallo-β-lactamases have been identified in clinical isolates of bacterial pathogens, NDM-1 is particularly worrying because the gene encoding this enzyme is associated with genetic elements that enable it to become rapidly disseminated to other strains and bacterial species *via* HGT [41].

2.1.3.2 Antibiotic Modification

Antibiotics can become inactivated through enzymatic addition of a chemical moiety that negatively impacts interaction with the target. Common chemical modifications to antibiotics by bacterial pathogens include phosphorylation, acetylation, and adenylation; less frequently, other types of modification may also occur (*e.g.*, ribosylation, glycosylation).

Resistance to the aminoglycoside antibiotics in both Gram-positive and Gram-negative pathogens is predominantly mediated by modification of the drug through *N*-acetylation, *O*-adenylation, or *O*-phosphorylation (Figure 2.8). Addition of these bulky chemical groups to any one of several OH or NH_2 groups on the aminoglycoside molecule disrupts the interaction of the compound with its target, the ribosomal RNA. Over 100 aminoglycoside modifying enzymes (AMEs) catalyzing such reactions have been identified, and these are grouped into three classes according to the type of chemical modification catalyzed: aminoglycoside acetyltransferases (AACs), aminoglycoside nucleotidyltranferases (ANTs), and aminoglycoside phosphotransferases (APHs). In addition to indicating what type of modification they catalyze, AMEs are named to reflect the position on the antibiotic where they catalyze the modification event; for example, an AAC enzyme that modifies the drug at the 6′ site is abbreviated to AAC(6′). Roman numerals are used to specify the substrate profile of the enzyme, while a lower case letter distinguishes enzymes that are functionally equivalent but evolutionarily distant, *e.g.*, AAC(6′)-Ia and AAC(6′)-Ib are proteins that differ in their amino acid sequence but exhibit identical substrate profiles.

Some AMEs are bifunctional enzymes, possessing the ability either to perform more than one type of chemical modification or to modify more than one type of substrate. The enzyme AAC (6′)-APH(2″) is one such bifunctional AME, and is one of the most common mediators of aminoglycoside resistance in MRSA. The N-terminal portion of this enzyme possesses AAC activity, while the C-terminal domain has APH activity. The dual catalytic activity of this enzyme appears to have arisen as a result of fusion of two ancestral genes to yield a single, bifunctional enzyme possessing a broad substrate profile that encompasses nearly all members of the aminoglycoside class (with the exception of streptomycin and spectinomycin) [42]. Another bifunctional enzyme known as AAC(6′)-Ib-cr has, in addition to aminoglycoside modifying activity, evolved the ability to inactivate through modification members of the structurally unrelated antibacterial drug class, the fluoroquinolones (Figure 2.8). In contrast to AAC

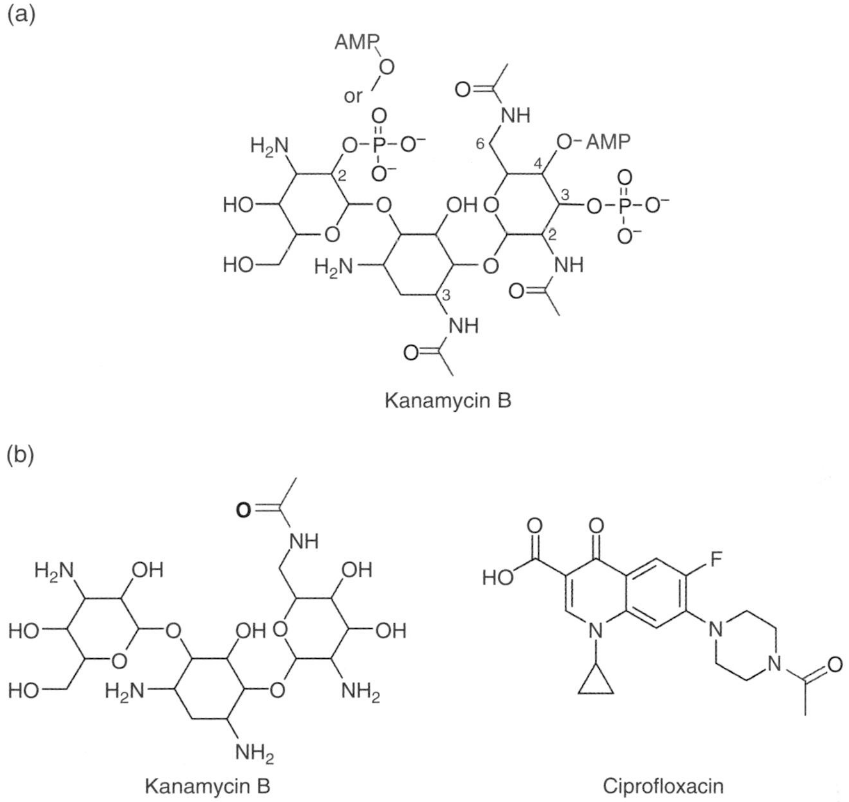

(a)

Kanamycin B

(b)

Kanamycin B   Ciprofloxacin

Figure 2.8 Examples of antibiotic inactivation mediated by the aminoglycoside modifying enzymes. (a) Sites and types of chemical modification on the representative aminoglycoside, kanamycin B: sites subject to enzymatic acetylation (red), adenylation (green) and phosphorylation (blue) are indicated. **(b)** AAC(6′)-Ib-cr is a bifunctional enzyme capable of acetylating (in red) two chemically-distinct classes of antibacterial drugs; the aminoglycosides (exemplified by kanamycin B, left) and the fluoroquinolones (exemplified by ciprofloxacin, right).

(6′)-APH(2″), this enzyme is not a composite of two proteins with distinct enzymatic activities, but the bifunctionality is instead due to point mutations that have arisen in the ancestral AME gene (aac[6′]-Ib). These mutations encode two amino acid substitutions ($W_{102}R$ and $D_{179}Y$) in AAC(6′)-Ib that are together responsible for expanding the substrate profile of the enzyme to encompass the fluoroquinolones. AAC(6′)-Ib-cr is the first reported instance of an enzyme capable of inactivating members of two structurally unrelated antibacterial drug classes and, furthermore, represents the first example of enzymatic inactivation of a fully synthetic antibacterial agent (the fluoroquinolones) in bacterial pathogens [43].

2.1.4 Resistance as a Consequence of Restricted Antibiotic Accumulation

Antibiotics whose targets reside within the bacterial cell must traverse one or more membranes to reach their site of action. By restricting the accumulation of an antibiotic inside the cell, a bacterium may experience only low (subinhibitory) antibiotic concentrations at the target, even under conditions where extracellular concentrations of antibiotic are high. This effect may be achieved in two ways (Figure 2.1); (i) the bacterial envelope may limit ingress of an antibiotic into the cell or (ii) antibiotic molecules that have gained access to the cell may be exported back out to the extracellular milieu by efflux transporter proteins. These two mechanisms, which are often found working in concert to limit drug accumulation in bacteria, are important mediators of both intrinsic and acquired antibiotic resistance. Indeed, the lack of antibacterial activity against Gram-negative bacteria of a considerable proportion of antibiotics that are active against Gram-positives can be attributed to these two processes working individually or together to restrict intracellular antibiotic accumulation (Figure 2.9).

2.1.4.1 Restricted Antibiotic Entry into Bacteria Owing to Permeability Barriers

The membranes that form part of the bacterial cell envelope present a barrier to the unrestricted ingress of antibiotics into the cell, and therefore represent a source of intrinsic antibiotic resistance; the OM of Gram-negative bacteria, a second membrane that surrounds the peptidoglycan cell-wall and the cytoplasmic membrane, is particularly

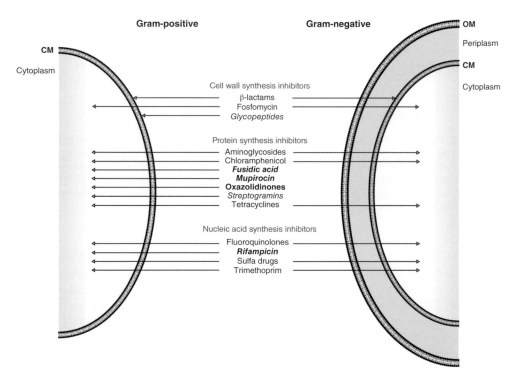

Figure 2.9 Decreased accumulation of antibiotics, mediated by outer membrane (OM) impermeability and efflux, largely accounts for the intrinsic resistance of Gram-negative bacteria to antibacterial drugs that are active against Gram-positive bacteria. The intrinsic resistance of Gram-negative bacteria like *Escherichia coli* to drugs shown in **bold** is primarily mediated by efflux, whereas *italicized* drugs fail to efficiently traverse the outer membrane. Intrinsic resistance to compounds shown in ***bold and italics*** results from a combination of both mechanisms. CM – cytoplasmic membrane.

important in this regard (Figure 2.9). In contrast to the cytoplasmic membrane, which in both Gram-positive and Gram-negative bacteria consists predominantly of a bilayer of phospholipids, the OM of Gram-negative bacteria typically comprises an asymmetric bilayer composed of an inner leaflet of phospholipids and an outer leaflet of LPS. The lipid component of LPS (lipid A) differs from typical phospholipids as it possesses six saturated fatty acid chains per molecule, rather than two saturated or unsaturated chains. The presence of numerous fatty acid chains in lipid A increases the propensity for hydrogen bonding between neighboring LPS molecules, thereby promoting strong lateral interactions in the membrane. In addition, the carboxyl groups of these fatty acid chains act in tandem with phosphate groups of the oligosaccharide portion of adjacent LPS molecules to complex divalent cations, which results in further cross-linking between LPS molecules. These structural features result in a low-fluidity, gel-like membrane, and one across which hydrophobic compounds will diffuse only slowly [44]. Consequently, the Gram-negative OM restricts ingress of hydrophobic antibiotics into the cell, and most are therefore unable to accumulate at sufficiently high concentrations in the vicinity of their intracellular targets to achieve an antibacterial effect.

In common with other biological membranes, the hydrophobic nature of the Gram-negative OM does not permit passage of hydrophilic compounds through the lipid portion of the membrane. Consequently, hydrophilic antibiotics that have an intracellular target in Gram-negative bacteria traverse the OM *via* water-filled beta-barrel protein channels (porins) that span the membrane. *E. coli* produces three major porins (OmpF, OmpC, and PhoE), each of which exhibits a general preference for the size and charge of the solutes it will accept. Crucially, however, there is an upper size limit for diffusion through the porins; antibiotics with a molecular mass over 600 Da are unable to pass through, a phenomenon that helps to explain the intrinsic resistance of *E. coli* to large antibiotic molecules such as vancomycin.

Porins can also participate in acquired antibiotic resistance. Clinical isolates of a variety of Gram-negative species have been identified in which antibiotic resistance has been acquired through porin changes. Bacteria may acquire genetic changes that lead to (i) down-regulation or complete loss of expression of a porin, (ii) a change in porin profile, involving a switch to expression of porins more restrictive to the ingress of a particular antibiotic, or (iii) amino acid substitution(s) in a porin that render it more restrictive to passage of an antibiotic. All three strategies ultimately result in decreased antibiotic permeation through the OM. For example, down-regulation of porins analogous to the OmpC and OmpF porins of *E. coli* contributes to acquired resistance to β-lactam antibiotics in *Enterobacter aerogenes* [45, 46]. In *Pseudomonas aeruginosa,* resistance to the β-lactam imipenem can arise due to loss of OprD, the porin channel through which this drug ordinarily traverses the OM [47]. A switch from the wild-type porin profile to favor the production of less permeable porins contributes to multidrug resistance in *Klebsiella pneumoniae*; in such strains, the OmpF analogue OmpK35 is replaced with the OmpC analogue, OmpK36 [48]. Amino acid substitutions in porins that lead to antibiotic resistance most frequently occur within an extended loop that forms a constriction within the barrel of the porin; substitution of residues in this so-called "eyelet" region results in changes to channel size and specificity, and has been associated with high-level β-lactam resistance in bacterial pathogens such as *N. gonorrhoeae* [49].

2.1.4.2 Restricted Antibiotic Accumulation in Bacteria Owing to Efflux Transporters

Bacteria have evolved an array of membrane-located proteins capable of actively trans-porting intracellular molecules across the membrane(s) and out of the cell. A subset of these efflux transporters include within their substrate profile one or more antibiotic classes and can therefore act to reduce the intracellular accumulation of these compounds. Thus, efflux transporters play an important role in both intrinsic and acquired antibiotic resistance.

The transporter proteins involved in antibiotic resistance are distributed across five protein families; the major facilitator superfamily (MFS), small multidrug resistance (SMR) family, resistance-nodulation-cell division (RND) family, multidrug and toxin extrusion (MATE) family, and ABC family [50]. To achieve active transport of substrates such as antibiotics against a concentration gradient, these transporters require a source of energy. Efflux proteins belonging to the SMR, MFS, MATE, and RND families all uti-lize the proton-motive force generated by cellular metabolism to drive transport, while transporters of the ABC family derive the necessary energy from the hydrolysis of ATP. Considerable diversity exists in the architecture and properties of efflux transporters. Efflux systems can comprise a single protein residing in the cytoplasmic membrane, as is often the case for transporters in Gram-positive bacteria, or, may form larger assemblies that straddle both the cytoplasmic/outer membranes of Gram-negative bacteria. Some efflux transporters are specific for one antibiotic class, while others are capable of recog-nizing and exporting a wide range of structurally diverse molecules.

A major contribution to both intrinsic and acquired antibiotic resistance in Gram-negative bacteria is provided by members of the RND transporter family, the best char-acterized of which is the AcrAB–TolC efflux pump found in *E. coli* (Figure 2.10). This tripartite assembly comprises a transporter protein (AcrB) that resides in the cytoplas-mic membrane, and two accessory proteins: a periplasmic adaptor protein (AcrA), and an OM pore (TolC). Collectively, these proteins form a complex that spans the entirety of the cell envelope and can translocate antibiotics from inside the bacterium out to the extracellular milieu. As alluded to earlier, this system is responsible for the intrinsic resistance that *E. coli* exhibits to various antibiotic classes; laboratory strains of *E. coli* in which the genes encoding this transporter have been deleted show a dramatic increase in susceptibility to antibiotics that are ordinarily only active against Gram-positive species (*e.g.*, oxazolidinones, fusidic acid, mupirocin) (Figure 2.9) [51]. However, such transporters also contribute to acquired resistance; mutations in regulatory genes leading to hyperproduction of the transporter complex can result in increased export of antibiotics. For example, mutations in the *acrR* repressor have been associated with elevated expression of the AcrAB efflux pump, which in conjunction with expression of a low-affinity PBP, is responsible for high-level β-lactam resistance in clinical isolates of *Haemophilus influenzae* [52]. Similarly, deletions within the *acrR* gene have been asso-ciated with fluoroquinolone resistance in *K. pneumoniae* [53]. Overproduction of the comparable MexAB-OprM RND efflux system of *P. aeruginosa* can occur through mutation in any of three repressor genes (namely *mexR* [54], *nalC* [52], and *nalD* [55]), and has been linked to multidrug resistance in clinical isolates.

Acquired antibiotic resistance mediated by efflux may also arise following horizontal acquisition of a gene or genes encoding a transporter capable of exporting anti-biotics from the cell. An important example of horizontally-acquired antibiotic

(a)

(b)

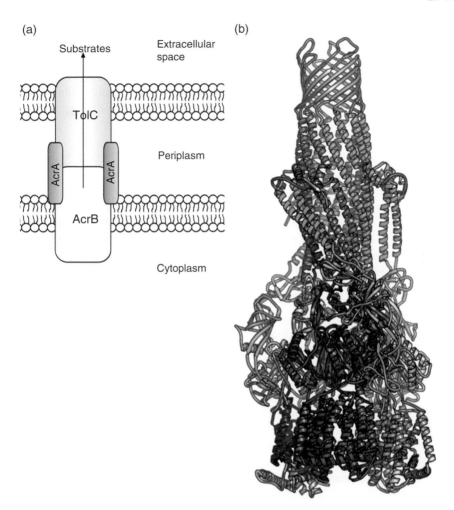

Figure 2.10 **Structure of the archetypal RND efflux transporter, AcrAB–TolC.** (a) The AcrAB–TolC complex spans the entirety of the Gram-negative cell envelope, and acts to expel antibiotics and other substrates from the cell. The complex consists of the inner membrane RND transporter (AcrB), coupled to an outer membrane channel (TolC) *via* a periplasmic membrane fusion protein (AcrA). (b) Structural model of the assembled AcrAB–TolC tripartite complex. Each of the three proteins are present as trimers. AcrB subunits are shown in orange/yellow, TolC subunits in shades of blue, and AcrA in green. Regions of AcrB and TolC that are buried in the cytoplasmic and outer membranes respectively are shown in gray. *Source:* Model from M. Symmons (Cambridge University), personal communication.

transporters are the tetracycline-specific (tet) efflux proteins. Tetracycline was the first antibiotic for which efflux was identified as a mechanism of resistance [56] and over 30 different tetracycline efflux proteins of the MFS type have now been described (http://faculty.washington.edu/marilynr/tetweb1.pdf). These proteins are found in both Gram-positive and Gram-negative bacteria, and represent the predominant mechanism of tetracycline resistance in the latter. The widespread emergence of resistance to tetracyclines through efflux and ribosomal protection was instrumental in driving a systematic search for new tetracycline analogues that could evade these resistance mechanisms. The result was the glycylcyclines, of which the best known is tigecycline; these antibiotics are not susceptible to current tetracycline efflux or ribosomal protection mechanisms. However, it has been established that mutations in pre-existing tet efflux pumps can elicit resistance to the glycylcyclines, thereby revealing a route by which widespread resistance to these drugs could develop with sustained clinical use [57, 58].

2.2 Summary and Concluding Remarks

Research efforts to understand antibiotic resistance mechanisms have intensified in recent years, reflecting an increased appreciation of the importance of resistance as a major threat to our ability to successfully treat bacterial disease, and underpinned by an ongoing revolution in our ability to study biological systems in molecular detail. The result has been a dramatic increase in our knowledge regarding the genetic, biochemical, and structural aspects of antibiotic resistance.

Viewed in the most simplistic terms, all antibiotic resistance mechanisms ultimately work in the same way; they act to prevent, or mitigate the effects of, an antibiotic binding to its cellular target. However, as outlined in this chapter, this outcome can be achieved through diverse means. Antibiotic resistance can result from changes to the nature of the target that reduce its susceptibility to an antibiotic (alteration, over-production, modification, protection and bypass), from enzymatic changes to the antibiotic molecule that reduce its antibacterial activity (degradation or modification), or from cellular structures/processes that prevent an antibiotic from reaching its target in sufficient concentration to cause inhibition of growth (restricted permeability and/or efflux). These different mechanistic types of antibiotic resistance are not mutually exclusive; in some instances, two or more such mechanisms may work in concert to mediate resistance to an antibiotic.

Aside from fundamental understanding, the insights that have been gained into antibiotic resistance mechanisms have important practical implications. In particular, they can assist the development of strategies to circumvent or overcome antibiotic resistance, and thereby help to rejuvenate the clinical activity of existing antibacterial drug classes. A classic approach to "rescuing" an antibacterial drug whose activity has become compromised by resistance is to generate chemical analogues of the agent that are not impacted (or are less impacted) by the resistance mechanisms that already exist in bacterial pathogens. Such an approach, exemplified by the development of the glycylcyclines (*see above*), undoubtedly benefits from a detailed understanding of the mechanism(s) responsible for resistance. This is also true of approaches that seek to develop compounds capable of directly inhibiting antibiotic resistance mechanisms, leaving the antibiotic free to inhibit its target without interference from the latter. This strategy, for which proof-of-principle has been provided by the development and successful clinical deployment of β-lactamase inhibitors such as clavulanic acid and tazobactam, can also realize considerable benefit from elucidation of the molecular and structural details of antibiotic resistance mechanisms. Thus, the study of antibiotic resistance mechanisms represents an important aspect of developing the next generation of antibacterial therapies to effectively treat and cure bacterial disease.

References

1 O'Neill, A.J. and Chopra, I. (2004). Preclinical evaluation of novel antibacterial agents by microbiological and molecular techniques. *Expert Opin. Investig. Drugs.* 13: 1045–1063.

2 Silver, L.L. (2011). Challenges of antibacterial discovery. *Clin. Microbiol. Rev.* 24: 71–109.

3 O'Neill, A.J., Huovinen, T., Fishwick, C.W., and Chopra, I. (2006). Molecular genetic and structural modeling studies of *Staphylococcus aureus* RNA polymerase and the fitness of rifampin resistance genotypes in relation to clinical prevalence. *Antimicrob. Agents Ch.* 50: 298–309.

4 Fàbrega, A., Madurga, S., Giralt, E., and Vila, J. (2009). Mechanism of action of and resistance to quinolones. *Microb. Biotechnol.* 2: 40–61.

5 Froelich-Ammon, S.J. and Osheroff, N. (1995). Topoisomerase poisons: harnessing the dark side of enzyme mechanism. *J. Biol. Chem.* 270: 21429–21432.

6 Hooper, D.C. (1999). Mechanisms of fluoroquinolone resistance. *Drug Resistance Updates: Reviews and Commentaries in Antimicrobial and Anticancer Chemotherapy* 2: 38–55.

7 Spratt, B.G. (1988). Hybrid penicillin-binding proteins in penicillin-resistant strains of *Neisseria gonorrhoeae*. *Nature*. 332: 173–176.

8 Dowson, C.G., Hutchison, A., Brannigan, J.A. et al. (1989). Horizontal transfer of penicillin-binding protein genes in penicillin-resistant clinical isolates of *Streptococcus pneumoniae*. *Proc. Natl Acad. Sci.* 86: 8842–8846.

9 Eliopoulos, G.M., Meka, V.G., and Gold, H.S. (2004). Antimicrobial Resistance to Linezolid. *Clin. Infect. Dis.* 39: 1010–1015.

10 Lobritz, M., Hutton-Thomas, R., Marshall, S., and Rice, L.B. (2003). Recombination proficiency influences frequency and locus of mutational resistance to linezolid in *Enterococcus faecalis*. *Antimicrob. Agents Ch.* 47: 3318–3320.

11 Flensburg, J. and Skold, O. (1987). Massive overproduction of dihydrofolate reductase in bacteria as a response to the use of trimethoprim. *Eur. J. Biochem./FEBS.* 162: 473–476.

12 Hiramatsu, K., Hanaki, H., Ino, T. et al. (1997). Methicillin-resistant *Staphylococcus aureus* clinical strain with reduced vancomycin susceptibility. *J. Antimicrob. Chemother.* 40: 135–136.

13 Perkins, H.R. (1969). Specificity of combination between mucopeptide precursors and vancomycin or ristocetin. *Biochem. J.* 111: 195–205.

14 Pereira, P.M., Filipe, S.R., Tomasz, A., and Pinho, M.G. (2007). Fluorescence ratio imaging microscopy shows decreased access of vancomycin to cell wall synthetic sites in vancomycin-resistant *Staphylococcus aureus*. *Antimicrob. Agents Ch.* 51: 3627–3633.

15 Howden, B.P., Davies, J.K., Johnson, P.D. et al. (2010). Reduced vancomycin susceptibility in *Staphylococcus aureus*, including vancomycin-intermediate and heterogeneous vancomycin-intermediate strains: resistance mechanisms, laboratory detection, and clinical implications. *Clin. Microbiol. Rev.* 23: 99–139.

16 Roberts, M.C. (2008). Update on macrolide–lincosamide–streptogramin, ketolide, and oxazolidinone resistance genes. *FEMS Microbiol. Lett.* 282: 147–159.

17 Petrov, A.S., Bernier, C.R., Hershkovits, E. et al. (2013). Secondary structure and domain architecture of the 23S and 5S rRNAs. *Nucleic Acids Res.* 41 (15): 7522–7535.

18 Long, K.S., Poehlsgaard, J., Kehrenberg, C. et al. (2006). The Cfr rRNA Methyltransferase confers resistance to Phenicols, Lincosamides, Oxazolidinones, Pleuromutilins, and Streptogramin a antibiotics. *Antimicrob. Agents Ch.* 50: 2500–2505.

19 Liu, Y.Y., Wang, Y., Walsh, T.R. et al. (2016). Emergence of plasmid-mediated colistin resistance mechanism MCR-1 in animals and human beings in China: a microbiological and molecular biological study. *Lancet Infect. Dis.* 16: 161–168.

20 Poirel, L., Jayol, A., and Nordmann, P. (2017). Polymyxins: antibacterial activity, susceptibility testing, and resistance mechanisms encoded by plasmids or chromosomes. *Clin Microbiol. Rev.* 30: 557–596.

21 Connell, S.R., Tracz, D.M., Nierhaus, K.H., and Taylor, D.E. (2003). Ribosomal protection proteins and their mechanism of tetracycline resistance. *Antimicrob. Agents Chemother.* 47: 3675–3681.

22 Arenz, S., Nguyen, F., Beckmann, R., and Wilson, D.N. (2015). Cryo-EM structure of the tetracycline resistance protein TetM in complex with a translating ribosome at 3.9-Å resolution. *Proc. Natl Acad. Sci. USA* 112 (17): 5401–5406.

23 Spahn, C.M., Blaha, G., Agrawal, R.K. et al. (2001). Localization of the ribosomal protection protein Tet(O) on the ribosome and the mechanism of tetracycline resistance. *Mol. Cell.* 7: 1037–1045.

24 Sharkey, L.K.R., Edwards, T.A., and O'Neill, A.J. (2016). ABC-F proteins mediate antibiotic resistance through ribosomal protection. *MBio.* 7: e01975–e01915.

25 Boel, G., Smith, P.C., Ning, W. et al. (2014). The ABC-F protein EttA gates ribosome entry into the translation elongation cycle. *Nat. Struct. Mol. Biol.* 21: 143–151.

26 Chen, B., Boel, G., Hashem, Y. et al. (2014). EttA regulates translation by binding the ribosomal E site and restricting ribosome-tRNA dynamics. *Nat. Struct. Mol. Biol.* 21: 152–159.

27 Lenart, J., Vimberg, V., Vesela, L. et al. (2015). Detailed mutational analysis of Vga(A) interdomain linker: implication for antibiotic resistance specificity and mechanism. *Antimicrob. Agents Chemother.* 59: 1360–1364.

28 Cox, G., Thompson, G.S., Jenkins, H.T. et al. (2012). Ribosome clearance by FusB-type proteins mediates resistance to the antibiotic fusidic acid. *Proc. Natl Acad. Sci.* 109: 2102–2107.

29 O'Neill, A.J. and Chopra, I. (2006). Molecular basis of fusB-mediated resistance to fusidic acid in *Staphylococcus aureus. Mol. Microbiol.* 59: 664–676.

30 Tomlinson, J.H., Thompson, G.S., Kalverda, A.P. et al. (2016). A target-protection mechanism of antibiotic resistance at atomic resolution: insights into FusB-type fusidic acid resistance. *Sci. Rep.* 6: 19524.

31 Hartman, B.J. and Tomasz, A. (1984). Low-affinity penicillin-binding protein associated with beta-lactam resistance in *Staphylococcus aureus. J. Bacteriol.* 158: 513–516.

32 Lu, W.-P., Sun, Y., Bauer, M.D. et al. (1999). Penicillin-binding protein 2a from methicillin-resistant *Staphylococcus aureus*: kinetic characterization of its interactions with β-lactams using electrospray mass spectrometry. *Biochemistry-Us.* 38: 6537–6546.

33 Uttley, A.C., Collins, C.H., Naidoo, J., and George, R.C. (1988). Vancomycin-resistant enterococci. *Lancet.* 331: 57–58.

34 Walsh, C. (2003). *Antibiotics: Actions, Origins, Resistance*. Washington, DC: ASM.

35 Bugg, T.D., Wright, G.D., Dutka-Malen, S. et al. (1991). Molecular basis for vancomycin resistance in *Enterococcus faecium* BM4147: biosynthesis of a depsipeptide peptidoglycan precursor by vancomycin resistance proteins VanH and VanA. *Biochemistry-Us.* 30: 10408–10415.

36 Marshall, C.G., Lessard, I.A., Park, I., and Wright, G.D. (1998). Glycopeptide antibiotic resistance genes in glycopeptide-producing organisms. *Antimicrob. Agents Ch.* 42: 2215–2220.

37 Drawz, S.M. and Bonomo, R.A. (2010). Three decades of beta-lactamase inhibitors. *Clin. Microbiol. Rev.* 23: 160–201.

38 Hall, B.G. and Barlow, M. (2005). Revised ambler classification of {beta}-lactamases. *J Antimicrob. Chemother.* 55: 1050–1051.

39 Knox, J.R., Moews, P.C., and Frere, J.M. (1996). Molecular evolution of bacterial beta-lactam resistance. *Chem. Biol.* 3: 937–947.

40 Yong, D., Toleman, M.A., Giske, C.G. et al. (2009). Characterization of a new metallo-beta-lactamase gene, bla(NDM-1), and a novel erythromycin esterase gene carried on a unique genetic structure in *Klebsiella pneumoniae* sequence type 14 from India. *Antimicrob. Agents Ch.* 53: 5046–5054.

41 Kumarasamy, K.K., Toleman, M.A., Walsh, T.R. et al. (2010). Emergence of a new antibiotic resistance mechanism in India, Pakistan, and the UK: a molecular, biological, and epidemiological study. *Lancet Infect. Dis* 10: 597–602.

42 Zhang, W., Fisher, J.F., and Mobashery, S. (2009). The bifunctional enzymes of antibiotic resistance. *Curr. Opin. Microbiol.* 12: 505–511.

43 Robicsek, A., Strahilevitz, J., Jacoby, G.A. et al. (2006). Fluoroquinolone-modifying enzyme: a new adaptation of a common aminoglycoside acetyltransferase. *Nat. Med.* 12: 83–88.

44 Nikaido, H. (2003). Molecular basis of bacterial outer membrane permeability revisited. *Microbiol. Mol. Biol. Rev.* 67: 593–656.

45 Yigit, H., Anderson, G.J., Biddle, J.W. et al. (2002). Carbapenem resistance in a clinical isolate of *Enterobacter aerogenes* is associated with decreased expression of OmpF and OmpC Porin analogs. *Antimicrob. Agents Ch.* 46: 3817–3822.

46 Bornet, C., Davin-Regli, A., Bosi, C. et al. (2000). Imipenem resistance of *Enterobacter aerogenes* mediated by outer membrane permeability. *J. Clin. Microbiol.* 38: 1048–1052.

47 Quinn, J.P., Dudek, E.J., DiVincenzo, C.A. et al. (1986). Emergence of resistance to imipenem during therapy for *Pseudomonas aeruginosa* infections. *J. Infect. Dis.* 154: 289–294.

48 Hasdemir, U.O., Chevalier, J., Nordmann, P., and Pages, J.M. (2004). Detection and prevalence of active drug efflux mechanism in various multidrug-resistant *Klebsiella pneumoniae* strains from Turkey. *J. Clin. Microbiol.* 42: 2701–2706.

49 Gill, M.J., Simjee, S., Al-Hattawi, K. et al. (1998). Gonococcal resistance to β-lactams and tetracycline involves mutation in loop 3 of the Porin encoded at thepenB locus. *Antimicrob. Agents Ch.* 42: 2799–2803.

50 Lomovskaya, O., Zgurskaya, H.I., Totrov, M., and Watkins, W.J. (2007). Waltzing transporters and 'the dance macabre' between humans and bacteria. *Nat. Rev. Drug. Discov.* 6: 56–65.

51 Randall, C.P., Mariner, K.R., Chopra, I., and O'Neill, A.J. (2013). The target of daptomycin is absent from *Escherichia coli* and other gram-negative pathogens. *Antimicrob. Agents Ch.* 57: 637–639.

52 Cao, L., Srikumar, R., and Poole, K. (2004). MexAB-OprM hyperexpression in NalC-type multidrug-resistant *Pseudomonas aeruginosa*: identification and characterization of the nalC gene encoding a repressor of PA3720-PA3719. *Mol. Microbiol.* 53: 1423–1436.

53 Schneiders, T., Amyes, S.G.B., and Levy, S.B. (2003). Role of AcrR and RamA in Fluoroquinolone resistance in clinical *Klebsiella pneumoniae* isolates from Singapore. *Antimicrob. Agents Ch.* 47: 2831–2837.

54 Ziha-Zarifi, I., Llanes, C., Kohler, T. et al. (1999). In vivo emergence of multidrug-resistant mutants of *Pseudomonas aeruginosa* overexpressing the active efflux system MexA-MexB-OprM. *Antimicrob. Agents Ch.* 43: 287–291.

55 Sobel, M.L., Hocquet, D., Cao, L. et al. (2005). Mutations in PA3574 (nalD) Lead to increased MexAB-OprM expression and multidrug resistance in laboratory and clinical isolates of *Pseudomonas aeruginosa*. *Antimicrob. Agents Ch.* 49: 1782–1786.

56 McMurry, L., Petrucci, R.E. Jr., and Levy, S.B. (1980). Active efflux of tetracycline encoded by four genetically different tetracycline resistance determinants in *Escherichia coli*. *Proceedings of the National Academy of Sciences* 77: 3974–3977.

57 Guay, G.G., Tuckman, M., and Rothstein, D.M. (1994). Mutations in the tetA(B) gene that cause a change in substrate specificity of the tetracycline efflux pump. *Antimicrob. Agents Ch.* 38: 857–860.

58 Du, X., He, F., Shi, Q. et al. (2018). The Rapid Emergence of Tigecycline Resistance in blaKPC–2 Harboring Klebsiella pneumoniae, as Mediated in Vivo by Mutation in tetA During Tigecycline Treatment. *Front. Microbiol.* 9: 648.

3

Resistance to Glycopeptide Antibiotics

François Lebreton[1] and Vincent Cattoir [2,3,4]

[1] Departments of Ophthalmology, Microbiology and Immunobiology, Harvard Medical School, Massachusetts Eye and Ear Infirmary, Boston, MA, USA
[2] Inserm Unit U1230, University of Rennes 1, Rennes, France
[3] Department of Clinical Microbiology, Rennes University Hospital, Rennes, France
[4] National Reference Center for Antimicrobial Resistance (Lab Enterococci), Rennes, France

3.1   Introduction

Antibiotics of the glycopeptide family share closely related chemical structures and are naturally obtained from various species of actinomycetes, typically recovered from soil samples [1]. All glycopeptides exhibit a narrow spectrum of activity, being active only against aerobic and anaerobic Gram-positive bacteria. Of note, only vancomycin and teicoplanin are clinically used in the treatment of severe infections caused by resistant Gram-positive pathogens, especially methicillin-resistant *Staphylococcus aureus* (MRSA) and enterococci, or for patients with an allergy to β-lactam antibiotics. In addition, vancomycin is recommended for the treatment of severe and complicated infections caused by *Clostridium difficile* [2]. As opposed to vancomycin that was approved for use by the FDA in 1958 and then used worldwide, teicoplanin is not available in the United States but is used in Europe [3]. Teicoplanin presents some advantages over vancomycin because it presents a lower toxicity, a better tissue penetration, and a prolonged serum half-life [4].

Structurally related to natural glycopeptides, three semisynthetic lipoglycopeptides (*i.e.*, dalbavancin, oritavancin, and telavancin) have been recently developed for the treatment of infections caused by resistant Gram-positive pathogens. Of them, only telavancin is currently approved for the treatment of patients with complicated skin and soft-tissue infections (USA and Canada) or hospital-acquired pneumonia, including ventilator-associated pneumonia, caused by MRSA [5, 6]. Like teicoplanin, telavancin seems to be less toxic than vancomycin with a prolonged half-life allowing for a once-daily dosing [7].

Bacterial Resistance to Antibiotics – From Molecules to Man, First Edition.
Edited by Boyan B. Bonev and Nicholas M. Brown.
© 2020 John Wiley & Sons Ltd. Published 2020 by John Wiley & Sons Ltd.

3.2 Glycopeptides – Structure and Mode of Action

3.2.1 Structure

Initially referred as "Compound 05865" and "Mississipi mud" because of its brown appearance due to product impurities, vancomycin was isolated in 1956 from cultures of *Amycolatopsis oritentalis* (formerly designated *Streptomyces orientalis* then *Nocardia orientalis*), an actinomycete recovered from a soil sample in Borneo, Indonesia [8]. Teicoplanin (originally called teichomycin A2) is produced by fermentation of another actinomycetal species, *Actinoplanes teichomyceticus* (formerly *Streptomyces teichomyceticus*) [9]. Both vancomycin and teicoplanin are complex, large molecules (molecular weights of 1500 and 1900 Da, respectively) with a conserved heptapeptide backbone. However, teicoplanin exhibits a tetracyclic structure (due to the presence of an additional hydroxyphenyl linkage), whereas vancomycin presents a tricyclic structure (Figure 3.1a and b) [10]. Moreover, two sugar moieties are attached to the central core of vancomycin (D-glucose and L-vancosamine) whereas three are present in teicoplanin (D-mannose, *N*-acetylglucosamine, and *N*-acylglucosamine) (Figure 3.1a and b) [11]. Since various fatty acids can be attached to its *N*-acylglucosamine residue, teicoplanin actually corresponds to a mixture of five closely-related compounds, named teicoplanin A2–1 to teicoplanin A2–5 (Figure 3.1b) [4, 11]. Finally, telavancin is a semisynthetic derivative of vancomycin that differs by the addition of a hydrophobic decylaminoethyl moiety attached to the vancosamine sugar (hence the name of lipoglycopeptide) and a hydrophilic phosphomethyl aminomethyl residue present at the 4′ position of the ring 7 (Figure 3.1a) [12].

3.2.2 Mode of Action

Unable to penetrate the cytoplasm, glycopeptides act by inhibiting the biosynthesis of the cell-wall peptidoglycan at the outer face of the cytoplasmic membrane [10]. Like β-lactams, glycopeptides interfere with the late stage of the peptidoglycan formation. However, the molecular target of glycopeptides is not an enzyme (*e.g.*, transglycosylase or transpeptidase) but is actually the substrate of these aforementioned enzymes. Indeed, vancomycin specifically interacts with the C-terminal D-Alanyl-D-Alanine (D-Ala-D-Ala) extremity of peptidoglycan pentapeptidic precursors by forming a stable non-covalent complex (Figures 3.2 and 3.3) [10, 14, 15]. By steric hindrance, the transglycosylation reaction (*i.e.*, polymerization of the polysaccharide backbone) is the first to be inhibited, and the transpeptidation reaction (*i.e.*, peptide cross-linking of the polysaccharide backbone) is subsequently blocked (Figure 3.2) [10]. Because of the inhibition of both reactions, the cell wall biosynthesis is completely stopped and cytoplasmic precursors accumulate, which prevents the synthesis of nucleic acids and proteins. At the molecular level, glycopeptides are non-covalently linked to the D-Ala-D-Ala motif through five hydrogen bounds (Figure 3.3) [10]. Note that the affinity of teicoplanin is about four to fivefold higher than that of vancomycin [4].

Telavancin exerts a dual mechanism of action [16]. First, as other glycopeptides, it binds to the C-terminus of the D-Ala-D-Ala extremity of peptidoglycan precursors, leading to the inhibition of cell wall synthesis. Second, telavancin selectively interacts with the bacterial cell membrane through the presence of its lipophilic chain, leading to increased membrane permeability, efflux of extracellular ATP and potassium, and rapid

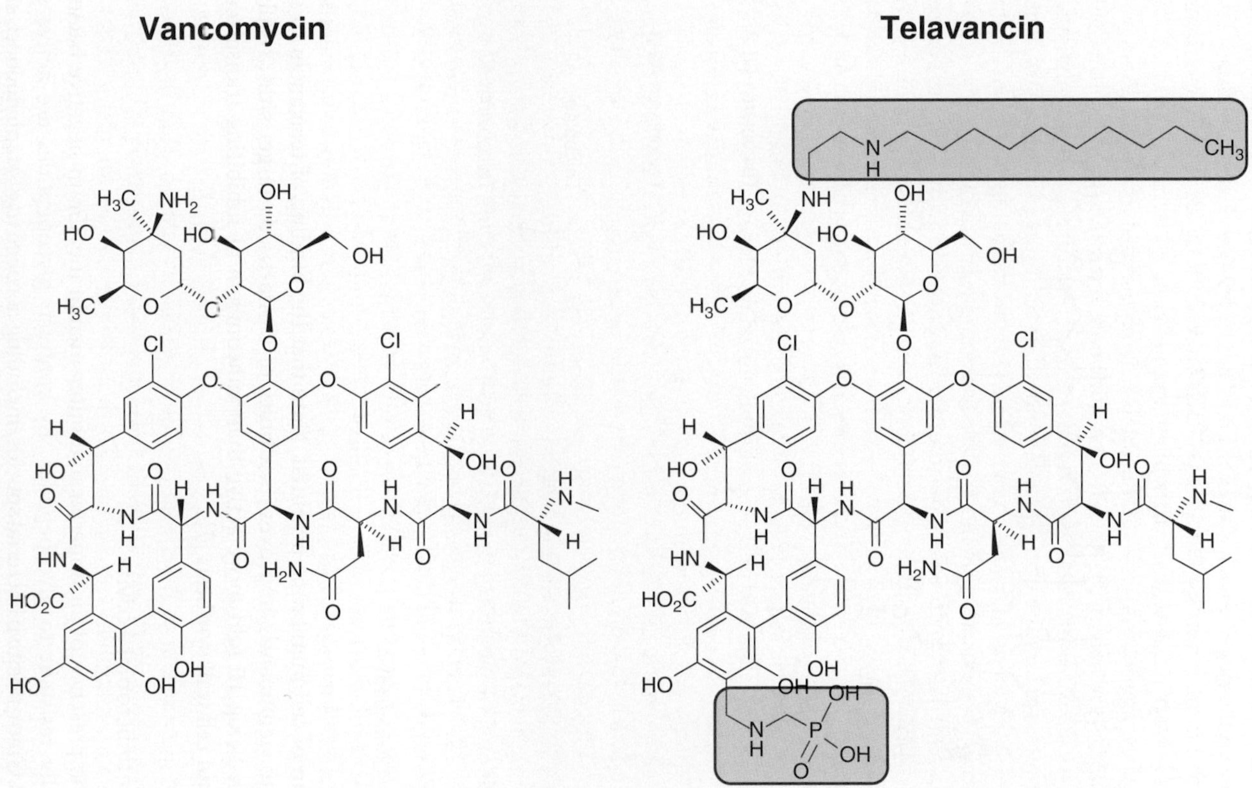

Figure 3.1 **Structure of glycopeptide antibiotics.** (a) Vancomycin and telavancin. The two additional moieties (hydrophobic decylaminoethyl side chain and phosphonomethyl aminomethyl hydrophilic moiety) of telavancin compared to vancomycin are boxed in blue. (b) Teicoplanin. The five different compounds composing teicoplanin (*i.e.*, teicoplanin A2-1 to teicoplanin A2-5) are represented.

(b)

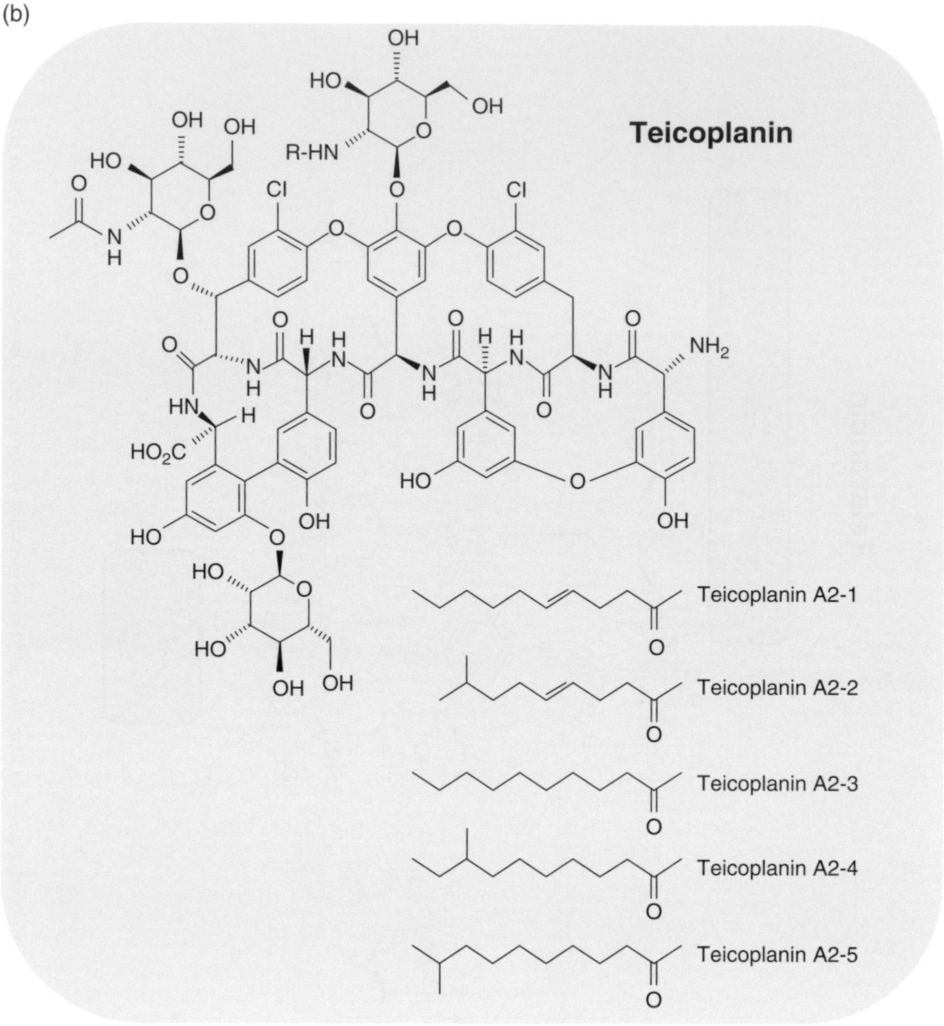

Figure 3.1 (Continued)

dissipation of cell membrane potential. Note that the binding of telavancin to the cell membrane significantly improves its affinity to D-Ala-D-Ala target sites, resulting in telavancin being 10-fold more active than vancomycin at inhibiting transglycosylase activity and cell wall synthesis [7].

3.2.3 Antimicrobial Activity

Due to the presence of an impermeable outer membrane, Gram-negative bacteria are intrinsically resistant to glycopeptides. By contrast, glycopeptides are active against almost all Gram-positive bacterial species, including aerobes (*e.g.*, staphylococci, strepto-cocci, enterococci, *Listeria monocytogenes*, corynebacteria, and *Bacillus* spp.) and anaer-obes (*e.g.*, clostridia, *Propionibacterium* spp., *Actinomyces* spp., and Gram-positive cocci)

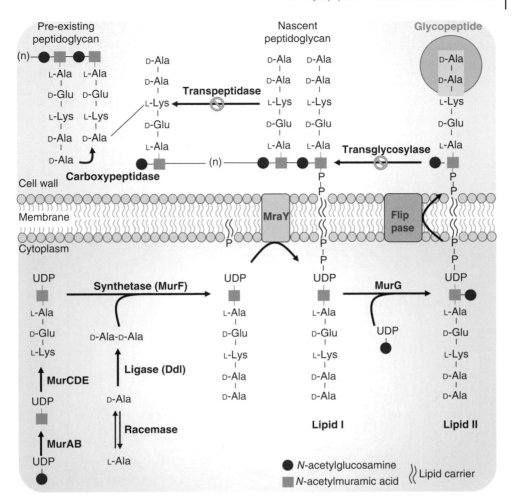

Figure 3.2 **Schematic representation of peptidoglycan biosynthesis and action mechanism of glycopeptides** Peptidoglycan biosynthesis occurs in three steps: (i) a first step in the cytoplasm with the formation of UDP-N-acetyl-muramoyl-pentapeptides from UDP-N-acetyl-muramoyl-tripeptides and D-Ala-D-Ala dipeptides; (ii) a translocation step through the cytoplasmic membrane of pentapeptide precursors as lipid-linked intermediates (Lipid I and Lipid II); and (iii) a final step consisting of the incorporation of precursors to the nascent peptidoglycan by transglycosylation and the formation of cross-links by D,D-transpeptidation into the cell wall. Glycopeptides bind to the D-Ala-D-Ala extremity of pentapeptide precursors and thus inhibit the final steps of peptidoglycan biosynthesis (*i.e.*, transglycosylation and D,D-transpeptidation) [10, 14] Ala, alanine; D, dextrorotatory; Ddl, D-Ala-D-Ala ligase; L, levorotatory; Lys, lysine; Glu, glutamic acid. *Source:* Modified from [13].

(Table 3.1). Whereas *in vitro* activities of natural glycopeptides are similar against most of aerobic Gram-positives, MICs of teicoplanin against coagulase-negative staphylococci and enterococci are usually higher and lower than those of vancomycin, respectively (Table 3.1) [11]. Noteworthy, some Gram-positive bacteria are intrinsically resistant to the family of glycopeptide antibiotics, such as *Enterococcus gallinarum* and *Enterococcus casseliflavus* (see later), hetero-fermentative lactobacilli, pediococci, *Leuconostoc* spp., *Erysipelothrix rhusiopathiae*, *Clostridium ramosum*, and *Clostridium innocuum* [19].

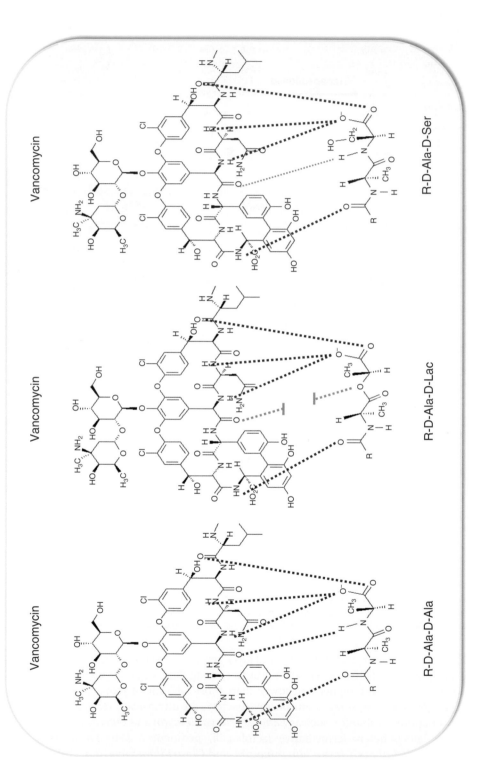

Figure 3.3 **Binding of vancomycin to the different types of the C-terminal end of peptidoglycan precursors.** Interaction between vancomycin and extremities ending in D-Ala-D-Ala (a), D-Ala-D-Lac (b), and D-Ala-D-Ser (c) are represented, with hydrogen bonds shown by dashed lines. In (a), vancomycin strongly binds by five hydrogen bonds to the D-Ala-D-Ala extremity. In (b), the substitution of the -NH group by an oxygen in the D-Ala-D-Lac extremity prevents the formation of a major hydrogen bond (in blue) resulting in a 1000-fold decrease in antibiotic affinity. In (c), the substitution of a methyl group by an -CH₂OH group in the D-Ala-D-Ser extremity induces a steric hindrance to binding leading to a sevenfold reduction in antibiotic affinity.

Table 3.1 *In vitro* activity of vancomycin, teicoplanin, and telavancin.

Pathogen	MIC$_{90}$ (mg/l)		
	Vancomycin	Teicoplanin	Telavancin
Staphylococcus aureus			
Methicillin-susceptible	1	1	0.5
Methicillin-resistant	1	1	0.25
Coagulase-negative staphylococci			
Methicillin-susceptible	2	4	0.5
Methicillin-resistant	2	4	0.5
β-hemolytic streptococci			
Streptococcus pyogenes	0.25	-	0.06
Streptococcus agalactiae	0.5	-	0.06
Streptococcus pneumoniae	0.25	–	0.03
Enterococcus faecalis			
Vancomycin-susceptible	2	0.25	1
Vancomycin-resistant	>256	256	8
Enterococcus faecium			
Vancomycin-susceptible	1	1	0.25
Vancomycin-resistant	>256	128	8
Clostridium spp.	1	–	0.25

Source: From Andersen [17], Saravolatz et al. [18], Zhanel et al. [7].

Glycopeptides exert a time-dependent bactericidal activity. The best parameter predicting clinical efficacy of vancomycin is the ratio "Area under curve/Minimal inhibitory concentration" (AUC/MIC), and it has been demonstrated that there is a highly significant association between clinical cure and an AUC/MIC >400 and between bacteriological eradication and an AUC/MIC >850 [20]. For instance, to achieve the target AUC/MIC >400 for organisms with a vancomycin MIC ≤1 mg/l, a vancomycin trough level of 15–20 mg/l has been proposed [21]. Note that the *in vitro* bactericidal activity of vancomycin is reduced for high-inoculum infections or in the presence of biofilm. As compared to natural glycopeptides, telavancin produces a more rapid bactericidal effect (8 hours) than vancomycin (24 hours) with a prolonged post-antibiotic effect (4–6 hours), which is fourfold longer than that of vancomycin [7]. The AUC/MIC ratio is also the best parameter for predicting the clinical efficacy of telavancin.

3.3 Molecular Bases of Resistance

3.3.1 Glycopeptide Resistance in Enterococci

The first clinical isolates of vancomycin-resistant enterococci (VRE) were reported in Europe in 1988 [22, 23]. Similar strains were later detected in the USA. [24]. Glycopeptide resistance in enterococcal species is due to the presence of operons that contain specific enzymes for the synthesis of low-affinity peptidoglycan precursors (in which the C-terminal D-Ala-D-Ala is replaced by D-Alanyl-D-Lactate [D-Ala-D-Lac] or D-Alanyl-D-Serine [D-Ala-D-Ser]) and the elimination of natural high-affinity precursors produced by the host bacterium [25]. Since 1988, nine operons that confer resistance to glycopeptides have been distinguished. They are named based on the sequence of genes with ligase activity, either a D-Ala-D-Lac ligase (*vanA, vanB, vanD,* and *vanM*) or a D-Ala-D-Ser ligase (*vanC, vanE, vanG, vanL,* and *vanN*) [26–29].

3.3.1.1 Types of Resistance

The classification of glycopeptide resistance is based on the primary sequence of the structural genes for the resistance-mediating ligases (Figure 3.4); however the nine operons have been identified according to genotypic and phenotypic criteria (Table 3.2 and Figure 3.5) [27]. Although all nine types of resistance involve genes encoding related enzymatic functions, they can be distinguished by the location of the genes within the operon and by the various modes of gene expression regulation. For the D-Ala-D-Lac phenotype (*i.e.,* VanA, VanB, VanD, and VanM), the resistance requires the presence of a ligase (VanA-B-D-M), a dehydrogenase (VanH-H_B-H_D-H_M), and a D,D-dipeptidase (VanX-X_B-X_D-X_M). Beside these, the accessory gene *vanY*-Y_B-Y_D-Y_M is shared by all the D-Ala-D-Lac operons and code for a D,D-carboxypeptidase whereas *vanZ* and *vanW* genes are of unknown function in *vanA* and *vanB* operons, respectively (Figure 3.5a). Finally, the two genes located upstream the resistance genes (*vanS*-S_B-S_D-S_M and *vanR*-R_B-R_D-R_M) code for a two-component system that regulates the expression of the resistance operon (Figure 3.5a). On the contrary, in D-Ala-D-Ser resistance operons (*i.e.,* vanC, vanE, vanL and vanN), the regulation part (*vanS*$_C$-S_E-S_L-S_N and *vanR*$_C$-R_E-R_L-R_N) is located downstream of the resistance genes with the exception of *vanG* (Figure 3.5b). In addition, in the VanC-, VanE-, VanG-, VanL- and VanN- type phenotype, a D-Ala-D-Ser ligase (VanC-E-G-L-N), a D,D-dipeptidase/ D,D-carboxypeptidase (VanXY$_C$-XY$_E$-XY$_G$-XY$_L$-XY$_N$), and a serine racemase (VanT$_C$-T_E-T_G-Tm$_L$-Tr$_L$-T_N) are the proteins required for resistance (Figure 3.5b) [27].

The D-Ala-D-Lac VanA-type resistance, characterized by high-level (MIC >64 mg/l) inducible resistance to both vancomycin and teicoplanin (Table 3.2), was the first type

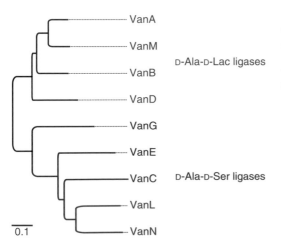

Figure 3.4 **Phylogenetic analysis of vancomycin resistance ligases.** The four (VanA, VanM, VanB, and VanD) D-Ala-D-Lac ligases (in red) clustered separately from the five (VanG, VanE, VanC, VanL, and VanN) D-Ala-D-Ser ligases (in blue).

Table 3.2 Types of resistance to glycopeptides in enterococci.

| | Acquired resistance | | | | | | | | Intrinsic resistance |
| | High level | | Variable | Moderate | Low level | | | | Low level |
	VanA	VanM	VanB	VanD	VanE	VanG	VanL	VanN	VanC1/C2/C3
Susceptibility									
Vancomycin	R	R	r-R	R	r	r	r	r	r
Teicoplanin	R	R	S	r-R	S	S	S	S	S
Transferability	+	+	+	–	–	+	–	+	–
Main species	Efm/Efs[a]	Efm	Efm/Efs	Efm/Efs	Efs	Efs	Efs	Efm	Ega/Eca
Expression	I	?	I	C	I/C	I	I	C	I/C
Genetic location	Plasmid (Chr)	Plasmid (Chr)	Plasmid (Chr)	Chr (plasmid)	Chr	Chr	Chr	Plasmid	Chr
Precursors end	D-Ala-D-Lac	D-Ala-D-Lac	D-Ala-D-Lac	D-Ala-D-Ser	D-Ala-D-Ser	D-Ala-D-Ser	D-Ala-D-Ser	D-Ala-D-Ser	D-Ala-D-Ser

R, high level of resistance (MIC >16 mg/l); r, low level of resistance (MIC = 8–16 mg/l); S, susceptible; Efm, *E. faecium*; Efs, *E. faecalis*; Ega, *E. gallinarum*; Eca, *E. casseliflavus*; I, inducible; C, constitutive; Chr, chromosome.

[a] Also other enterococcus species.

Source: Modified from Cattoir and Leclercq [13].

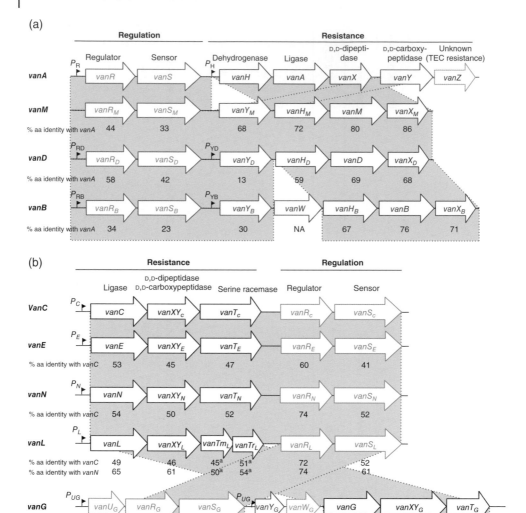

Figure 3.5 Comparison of the *van* gene clusters (modified from [27]). D-Ala-D-Lac (a) and D-Ala-D-Ser (b) resistance operons are represented. Open arrows represent coding sequences (green arrows, regulatory genes; red arrows, genes required for resistance in the D-Ala-D-Lac resistance type; blue arrows genes required for resistance in the D-Ala-D-Ser resistance type; purple arrows, accessory genes; gray arrows, genes of unknown function) and indicate the direction of transcription. The percentages of amino acid (aa) identity between the deduced proteins of reference strains BM4147 (VanA), V583 (VanB), BM4339 (VanD), BM4174 (VanC), BM4405 (VanE), BM4518 (VanG), Efm-HS0661 (VanM), N06–0364 (VanL), and UCN72 (VanN) are indicated under the arrows [26–29]. [a]When compared to the corresponding domains of the VanT$_C$ racemase; [b]*vanYG* contains a frameshift mutation leading to a predicted truncated protein; NA, not applicable; P_X, promoter for gene *X*; TEC, teicoplanin.

of resistance to be described and the most widespread [13, 22]. In 2010, the VanM-type resistance was characterized from *E. faecium* strains isolated in Shanghai. If it shares the phenotype of the VanA-type resistance and is genetically closely related, the prevalence of VanM-positive strains remains very low in the hospital setting [29]. VanB-type strains have variable levels of inducible resistance to vancomycin only (MICs generally from 16 to 32 mg/l) (Table 3.2). Indeed, these strains remain susceptible to teicoplanin, as it is not an inducer for this operon [30]. VanD-type strains are characterized by constitutive resistance to moderate levels of the two glycopeptides [31]. The five D-Ala-D-Ser VanC-, VanE-, VanG-, VanL-, and VanN-type strains are resistant to low levels of vancomycin but remain entirely susceptible to teicoplanin (Table 3.2) [26–28]. Finally, telavancin is poorly active against VanA-type VRE while it has a modest activity against VanB-type VRE [7].

Except for the *vanC* operon that is intrinsic to *E. gallinarum* and *E. casseliflavus*, the other eight operons confer acquired glycopeptide resistance (Table 3.2) [32]. Indeed, these operons have been associated with mobile genetic elements such as plasmids and/ or transposons. The *vanA*, *vanB*, *vanM*, and *VanN* operons are located on plasmids or in the chromosome [28, 32, 33], whereas the *vanC*, *vanD*, *vanE*, *vanG*, and *vanL* operons have so far been found exclusively in the chromosome [26, 27]. The *vanA* and *vanB* operons are usually carried by a Tn3-type transposon (the so-called Tn*1546*) and by the Tn*1547* transposon, respectively [25]. Transposition of Tn*1546* and Tn*1547* into self-transferable plasmids and subsequent transfer by conjugation seems to be responsible for the high-rate dissemination of *vanA* and *vanB* resistance operons. Recently, the *vanE* operon has been described as part of a conjugative element named Tn*6202*, but transfer has not been detected [34]. In addition, VanG and VanN are the only transferable D-Ala-D-Ser resistance types characterized so far (Table 3.2) [27, 28].

The VanA and VanB are the types the most frequently detected in enterococci and have occasionally been detected in coryneform bacteria and streptococci [35]. The *vanA* operon has also been rarely detected in *S. aureus* (see below) [36]. VanE, VanG, and VanL are exceptional and corresponding genes have so far been detected only in the chromosome of *E. faecalis*, while VanN has only been described in *E. faecium*, in a single clinical isolate from France and in five clonal strains isolated from chickens in Japan (Table 3.2) [27, 28, 33].

3.3.1.2 Modification of the Target for Glycopeptides

As detailed earlier, the molecular target of glycopeptide antibiotics is the D-Ala—D-Ala terminus of intermediates in peptidoglycan synthesis (Figure 3.2). The synthesis of peptidoglycan, the major component of the bacterial cell wall, requires several steps [10, 14, 37]. In the cytoplasm, a racemase converts L-alanine to D-alanine (D-Ala), and then two molecules of D-Ala are joined by a specific ligase (named Ddl) creating the dipeptide D-Ala-D-Ala, which eventually forms a UDP-*N*-acetylmuramyl-pentapeptide through several enzymatic reactions. This precursor is then bound to the undecaprenol lipid carrier, which allows its translocation to the outer surface of the cytoplasmic membrane (Figure 3.2). Finally, the *N*-acetylmuramyl-pentapeptide residue is incorporated into the nascent peptidoglycan by transglycosylation with the formation of crossbridges between the peptide side chains by transpeptidation (Figure 3.2).

To achieve resistance, the D-Ala-D-Ala target modification results from the cooperation of several enzymes, required for a complete reprogramming of the peptidoglycan

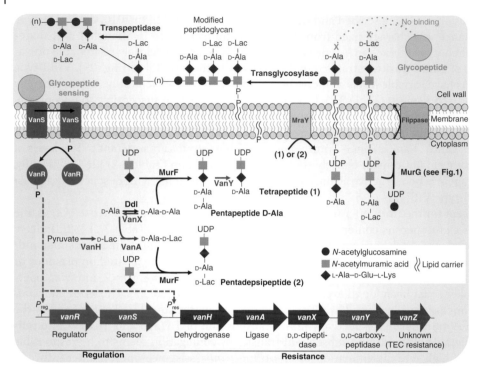

Figure 3.6 VanA-type glycopeptide resistance through synthesis of D-Ala-D-Lac modified peptidoglycan precursors. The acquired *vanA* operon is represented and contains regulatory genes (in green) and genes involved in the resistance mechanism (in red). Glycopeptide resistance occurs through three steps: (i) the VanS sensor auto-phosphorylates in the presence of glycopeptide thus activating the VanR regulator that activates the transcription of the resistance genes in the *vanA* operon, (ii) the modification of the target is achieved by replacing the pentapeptide (D-Ala) by a pentadepsipeptide (produced through the activity of the VanA ligase and the VanH dehydrogenase) that results in reduced affinity for vancomycin due to steric hindrance and, (iii) the elimination of the target by VanX that hydrolyses the dipeptide D-Ala-D-Ala and by VanY that cleaves the D-Ala terminal residue of the pentapeptide (D-Ala) [27]. Ala, alanine; D, dextrorotatory; Ddl, D-Ala: D-Ala ligase; L, levorotatory; Lac, lactate; Lys, lysine; Glu, glutamic acid; P_{reg}, promoter for regulatory genes; P_{res}, promoter for resistance genes.

synthesis, that are encoded by genes organized in the various resistance operons (Figure 3.5) [25]. The VanA-type resistance will be described as a model for the various D-Ala-D-Lac resistance types (*i.e.*, VanA, VanB, VanM, and VanD). The *vanA* gene codes for a ligase (VanA) that catalyzes the formation of an ester bond between D-Ala and D-Lac (Figure 3.6). Since the D-Lac substrate is rare in nature and needs to be synthesized, the *vanA* operon also codes for a dehydrogenase (VanH) that reduces pyruvate to D-Lac. The resulting D-Ala-D-Lac depsipeptide replaces the wild-type D-Ala-D-Ala dipeptide to produce a pentadepsipeptide that is further processed during peptidoglycan synthesis [30]. The substitution of the pentapeptide by the pentadepsipeptide significantly decreases the affinity of the molecule for glycopeptides, leading to clinical resistance (Figure 3.6).

Similarly, the VanC-type will be used as model to illustrate the D-Ala-D-Ser (*i.e.*, VanC, VanE, VanG, VanL, and VanN) resistance (Figure 3.7). Two genes are required for the

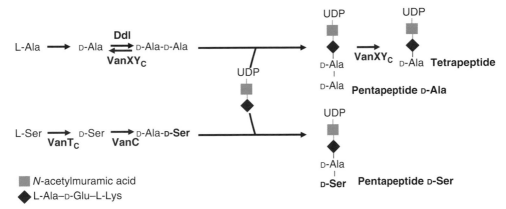

Figure 3.7 **VanC-type glycopeptide resistance through synthesis of D-Ala-D-Ser modified peptidoglycan precursors.** Glycopeptide resistance occurs through two steps: (i) the modification of the target is achieved by replacing the pentapeptide (D-Ala) by a pentapeptide (D-Ser) (produced through the activity of the VanC ligase and the VanT racemase) that results in reduced affinity for vancomycin due to steric hindrance and, (ii) the elimination of the target by VanXY that hydrolyses the dipeptide D-Ala-D-Ala and cleaves the D-Ala terminal residue of the pentapeptide (D-Ala) [27]. Ala, alanine; D, dextrorotatory; Ddl, D-Ala:D-Ala ligase; L, levorotatory; Lac, lactate; Lys, lysine; Glu, glutamic acid.

VanC-type modification of the target: *vanT* encodes the VanT membrane-bound serine racemase that produces D-Ser from L-Ser and the *vanC* gene that code for the ligase VanC. This later enzyme synthesizes a D-Ala-D-Ser dipeptide, which replaces the D-Ala-D-Ala in peptidoglycan precursors to produce a D-Ala-D-Ser depsipeptide (Figure 3.7) [38]. Substitution of the ultimate D-Ala by a D-Ser results in steric hindrance that reduces its affinity for vancomycin [39].

3.3.1.3 Elimination of the Target for Glycopeptides

Despite the production of modified peptidoglycan precursors, resistance is still not expressed at this stage since the synthesis of wild-type (*i.e.*, D-Ala-D-Ala) precursors persists and allows the binding of vancomycin and the subsequent disruption of the cell wall (Figure 3.6). The interaction of a glycopeptide with its target is prevented by eliminating precursors with a terminal D-Ala residue [40]. Elimination of wild-type peptidoglycan precursors is the task of two enzymes in the *vanA* operon: (i) the *vanX* gene codes for a D,D-dipeptidase (VanX) that hydrolyses the dipeptide D-Ala-D-Ala formed by the natural ligase (Ddl), and (ii) the *vanY* gene encoding a D,D-carboxypeptidase (VanY) that eliminates the terminal D-Ala of pentapeptide precursors in case of incomplete elimination of the dipeptide by VanX (Figure 3.6) [41, 42]. In contrast to VanA-type resistance where VanX and VanY activities are encoded by two different genes, VanXY$_C$ (expressed from the *vanC* operon) possesses both D,D-peptidase and D,D-carboxypeptidase activities (Figure 3.7) [43].

3.3.1.4 Regulation of Glycopeptide Resistance

VanS/VanR-type two-component systems control the expression of genes that mediate glycopeptide resistance. More specifically, expression of VanA-, VanB-, VanD-, VanM-, VanC-, VanE-, VanG-, VanL-, and VanN-type resistance is regulated by a VanS/

VanR-type signal transduction system composed of a membrane-bound histidine kinase (VanS, VanS$_B$, VanS$_D$, VanS$_M$, VanS$_C$, VanS$_E$, VanS$_G$, VanS$_L$, or VanS$_N$, respectively) and a cytoplasmic response regulator (VanR, VanR$_B$, VanR$_D$, VanR$_M$, VanR$_C$, VanR$_E$, VanR$_G$, VanR$_L$, or VanR$_N$, respectively) that acts as a transcriptional activator [44]. After a signal associated with the presence of a glycopeptide in the culture medium, the cytoplasmic domain of VanS catalyzes ATP-dependent autophosphorylation on a specific histidine residue and transfers the phosphate group to an aspartate residue of VanR present in the effector domain (Figure 3.6). VanS also stimulates dephosphorylation of VanR in the absence of glycopeptide. The VanS sensor therefore modulates the phosphorylation level of the VanR regulator, by acting either as a phosphatase under non-inducing conditions or as a kinase in the presence of glycopeptides leading to phosphorylation of the response regulator and activation of the resistance genes [45]. Mutations, especially in the gene coding for the VanS sensor, have been associated with the constitutive expression of the *vanA* resistance operon [27]. Similarly, mutations in the *vanS$_B$* sensor gene have been obtained *in vitro* and *in vivo* in animal models following selection by teicoplanin, which have resulted in acquisition of teicoplanin resistance in VanB-type strains (that are usually susceptible to this glycopeptide) due to alterations of VanS$_B$ function [27].

3.3.1.5 Glycopeptide-dependent Strains

A remarkable phenomenon that has developed in some VanA- and VanB-type entero-cocci is the "vancomycin dependence" [46, 47]. These strains are not only resistant to vancomycin or to both vancomycin and teicoplanin, but also require their presence for growth. Variants of glycopeptide-resistant *E. faecalis* and *E. faecium* that grow only in the presence of glycopeptides have been isolated *in vitro*, in animal models, and from patients treated with vancomycin for long periods. In the presence of vancomycin, the *vanA*- or *vanB*-encoded D-Ala-D-Lac ligase is induced, which overcomes the defect in synthesis of peptidoglycan precursors ending in D-Ala-D-Ala because of the lack of a functional Ddl following various mutations in the *ddl* gene and, thus, permits growth of the bacteria. Reversion to vancomycin independence has been observed and occurs as a result of a mutation in either the VanS or VanS$_B$ sensor, which leads to constitutive production of D-Ala-D-Lac and, thus, to teicoplanin resistance, or in the *ddl* gene, which restores synthesis of D-Ala-D-Ala and leads to a VanB phenotype inducible by vancomycin [47].

3.3.2 Glycopeptide Resistance and Reduced Susceptibility in Staphylococci

For many years, there was no indication that vancomycin resistance in *S. aureus* was likely to be a problem. Therefore, initial reports of reduced vancomycin susceptibility in clinical isolates of *S. aureus* from Japan in 1996 generated significant concern in the medical community [48]. Similarly, reports of clinical *S. aureus* isolates that demonstrated reduced teicoplanin susceptibility and the *in vivo* emergence of resistance during teicoplanin therapy came from Europe in the early 1990s, although these strains remained susceptible to vancomycin [49]. Although both the terms glycopeptide-intermediate *S. aureus* (GISA) and vancomycin-intermediate *S. aureus* (VISA) have been used in the literature, the latter term has been more frequently employed and then will be used in this chapter. If this is clear that reduced teicoplanin susceptibility can be present in *S. aureus* without a reduction in vancomycin susceptibility, however VISA strains generally demonstrate reduced teicoplanin susceptibility [50]. According to Clinical and Laboratory Standards Institute (CLSI) guidelines [51], three entities of non-susceptible *S. aureus* can be distinguished from vancomycin-susceptible *S. aureus* (VSSA) (see below): vancomycin-resistant strains (VRSA), strains with an intermediate resistance to vancomycin (VISA), and those displaying a vancomycin heterogeneous intermediate resistance (hVISA).

3.3.2.1 VRSA Strains

After the emergence of VRE in the late 1980s, significant concern existed about the potential for the emergence of VRSA by acquisition of *van* operons from enterococci. First, the transfer of *van* resistance genes from *Enterococcus* species to *S. aureus*, resulting in high-level resistance to vancomycin, was obtained *in vitro* and in an animal model [52]. Most importantly, this transfer also occurs *in vivo*, with the first description of VRSA clinical isolates due to the acquisition of the *vanA* gene from VRE in 2002 [53].

The inducible expression of the *vanA* gene cluster and efficient heterologous expression of the *vanA* operon was characterized in all the VRSA strains thus showing that the mechanism of resistance is like that described in enterococci (Figure 3.6). Indeed, in the absence of glycopeptides in the culture medium, the pentapeptide (D-Ala) was the main late peptidoglycan precursor synthesized by all VRSA strains whereas, after growth in the presence of an inducer, the pentadepsipeptide was the predominant precursor. Analysis of expression of resistance genes revealed that the activities of the VanX and VanY D,D-peptidases, as well as the relative amounts of late peptidoglycan precursors in VRSA isolates, were similar to those in VanA-type enterococci [54].

Among the nine VRSA strains isolated in the USA, the *vanA* gene cluster has been assigned to a plasmid in all cases. Except for VRSA-8 (Table 3.3), a VRE strain was also co-isolated from the same patient along with the VRSA strain, strongly suggesting that MRSA strains have acquired Tn*1546* (carrying *vanA*) from enterococci [55]. Acquisition of vancomycin resistance can result from one or two genetic events: (i) plasmid transfer by conjugation from the *Enterococcus* donor to the *S. aureus* recipient, and (ii) Tn*1546* could transpose from the incoming plasmid into a resident plasmid or the chromosome of the recipient. Indeed, certain enterococcal plasmids can replicate efficiently in staphylococci and be stably maintained in the new host. However, others are less efficient in replication and thus behave as suicide gene delivery vectors, and the incoming DNA (*i.e.*, the *vanA* operon) is rescued by illegitimate recombination. Among the nine VRSA strains isolated in the USA, three (VRSA-3, VRSA-5, and VRSA-6) maintained the enterococcal *vanA* plasmid, whereas Tn*1546* moved from the enterococcal plasmid to a staphylococcal plasmid in others (VRSA-1, VRSA-7, VRSA-8, VRSA-9, and VRSA-10) (Table 3.3) [56].

Finally, a recent study compared the genome of these nine VRSA strains and found that all belong to the clonal cluster 5 (CC5), the predominant lineage responsible for MRSA hospital-acquired infections in the USA [57]. Chromosome and plasmid content comparisons unambiguously showed that each strain independently acquired Tn*1546* carrying a *vanA* operon as discussed earlier. Interestingly, these VRSA (and other CC5-type) strains were found to possess a constellation of traits (*i.e.*, a cluster of unique superantigens and lipoproteins to confound host immunity) that seems to be

Table 3.3 VRSA strains isolated in the United States.

NARSA designation	Patient	Date of isolation (month/year)	State (in the USA)	Vancomycin	Teicoplanin	Genetic location
VRSA-1	1	Jun-02	Michigan	R	R	chromosome
VRSA-2	2	Sep-02	Pennsylvania	R/r	r	NA
VRSA-3	3	Mar-04	New York	R/r	r	plasmid
VRSA-5	4	Feb-05	Michigan	R	R	plasmid
VRSA-6	5	Oct-05	Michigan	R	R	plasmid
VRSA-7	6	Dec-05	Michigan	R	R	chromosome
VRSA-8	7	Oct-06	Michigan	R	R	chromosome
VRSA-9	8	Nov-07	Michigan	R	R	chromosome
VRSA-10	9	Dec-07	Michigan	R	R	chromosome

NA, not available; R, high-level resistance; R/r, moderate level resistance; r, low-level resistance.
Source: From Perichon and Courvalin [54].

optimized for proliferation in precisely the types of polymicrobial infection where transfer could occur [57].

3.3.2.2 The Cell Wall of *S. aureus*

To understand the mechanisms of vancomycin resistance in *S. aureus*, a clear understanding of its cell wall organization is required. Like the enterococci, the cell wall of *S. aureus* is composed of highly cross-linked peptidoglycan, teichoic acids (*e.g.*, wall teichoic acids [WTAs] and lipoteichoic acids [LTAs]), and cell wall-associated proteins (Figure 3.8). By electron microscopy, the cell wall of *S. aureus* appears as thick as 20- to 40-nm-thick [58]. LTAs and WTAs are covalently linked to the peptidoglycan and decorated with D-alanine and *N*-acetylglucosamine residues. The teichoic acids play a role to: (i) help protect the cell envelope as a mechanical barrier to host defense molecules and antibiotics, and (ii) repel positively charged molecules, such as defensins and other cationic antimicrobial peptides (CAMPs), through the positive charge of D-alanine residues (Figure 3.8). In this context, the *dltABCD* operon is controlled by the regulator GraRS (also called ApsRS) that senses and responds to CAMP and regulates the alanylation of teichoic acids in response to the presence of antimicrobial compounds [50].

Many genes appear to be involved in production of *S. aureus* cell wall precursors (Figure 3.8). Important genes include *femA*, *femB*, *femC*, and *femX* genes that code for the penicillin-binding proteins (PBPs) (PbpA, PbpB, PbpC, and PbpD) that are involved in the stepwise synthesis of the pentaglycine bridge essential for bacterial survival; and regulatory genes involved in cell wall biosynthesis, such as *vraSR*. Additional cell wall-associated genes (*i.e.*, the *S. aureus* "cell wall stimulon") that are mainly regulated by *vraSR* have been extensively reviewed [50]. Similarly to the enterococcal cell wall, high-molecular-weight PBPs (PBP1, PBP2, and PBP3) also have an important role in cell wall synthesis because they have a transglycosidase function (to link *N*-acetylglucosamine to *N*-acetylmuramic acid) and a transpeptidase function (to link the penultimate D-Ala to a glycine acceptor in the nascent cell wall). PBPs are of particular interest due to their relevance to antimicrobial therapy (PBPs are the target site for β-lactams) and to antimicrobial resistance (*i.e.*, PBP2a, encoded by *mecA*, is responsible for methicillin resistance in staphylococci) [59].

3.3.2.3 VISA and hVISA Strains

The hVISA strains are possibly the precursors of VISA strains and seem to be induced to homogenous resistance after exposure to cell wall-active antibiotics [60]. Common biochemical and morphological changes can be found in such isolates. The most prominent include: (i) increased cell wall thickness with activated cell wall synthesis, (ii) reduced autolysis, (iii) decreased activity of the staphylococcal global regulator Agr, (iv) reduced lysostaphin susceptibility, and (v) changes in the content of cell wall teichoic acids [50]. Other phenotypic changes include an increased level of production of abnormal muropeptides, an over-expression of PBP2 and PBP2a, reduced PBP4 expression levels, and reduced levels of peptidoglycan cross-linking in most isolates studied [50]. The thickened cell wall appears to be the most consistent feature, even though the exact mechanisms leading to thickening have not been determined. The thicker cell wall of VISA strains may "trap" the glycopeptide molecules, preventing them from reaching their target close to the cell membrane [48]. Reduced autolytic activity is a common feature of hVISA and VISA strains. Some data suggest a possible role for wall teichoic

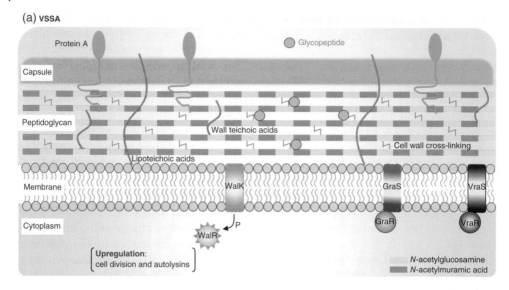

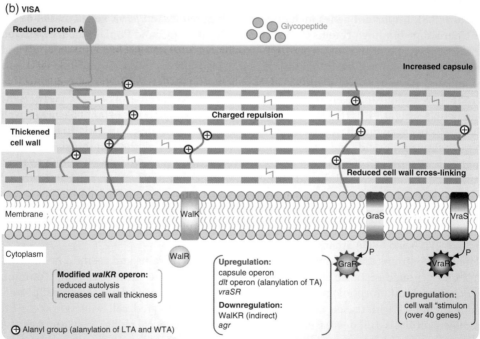

Figure 3.8 Cell wall characteristics of VSSA (a) and molecular determinant for resistance in VISA strains (b) (modified from [50]). VSSA strains in the absence of vancomycin show a normal peptidoglycan with the production of protein A and normal capsular polysaccharide expression (a). VISA strain with mutations in the *graRS*, *vraSR*, or *walKR*two-component systems that might lead to modulations of their respective regulons (b). The consequence of this modification includes cell wall thickening, decreased autolysis, reduced protein A production, increased capsule expression, increased D-alanylation of teichoic acids, and reduced *agr* activity. LTA, lipoteichoic acid; TA, teichoic acid and WTA, wall teichoic acid.

acids of VISA strains suppressing peptidoglycan degradation by autolytic enzymes, while other studies suggested that a reduction in the autolytic activity and altered peptidoglycan hydrolase activity of VISA autolysin extracts are responsible for the reduced autolytic activity [50].

These phenotypic changes are associated with multiple genetic changes that mainly involve several pathways of cell wall metabolism in the presence of glycopeptides (Figure 3.8). Numerous studies aimed at identifying the "cell wall stimulon" of *S. aureus* to investigate the VISA and hVISA phenotypes. Since the involvement of some genes is controversial and still debated [50], only main effectors are discussed here. This stimulon contains genes encoding the two-component system VraSR, which when active positively regulates several genes involved in cell wall synthesis [61]. Even if this stimulon is permanently upregulated for some VISA strains, the VISA phenotype can also be achieved in some strains without the induction of the cell wall stimulon. Among the genes that were observed to be upregulated in VISA strains are those encoding the GraRS two-component regulatory system, so named for its glycopeptide resistance association [62]. The over-expression of GraR or GraS results in a slight increase in the MIC of vancomycin, and a knockout mutation results in hypersusceptibility. The GraRS two-component regulatory system has been shown to control the expression of many genes, including many genes involved in cell wall synthesis [63]. This system also positively regulates a complex network of genes, some of which are associated with virulence in *S. aureus* (Figure 3.8). Another example of this complex connection between the VISA phenotype and virulence regulons involves the production of protein A, product of the *spa* gene. The down-regulation of *spa* is consistently observed in VISA strains. This gene is virtually under the control of numerous transcriptional regulators in which a mutation could potentially simultaneously increase vancomycin resistance and decrease protein A production (Figure 3.8) [50].

Besides transcriptomic and proteomic analyses, recent comparative genomic studies using isogenic VSSA/VISA pairs have identified some of the key *S. aureus* genes that are involved in vancomycin resistance [64, 65]. Among the genes that have been correlated with changes in vancomycin susceptibility patterns, independent studies report nonsynonymous mutations that caused changes in genes of: (i) the *agr* quorum-sensing system, (ii) the *walKR* cell wall regulatory operon, (iii) *graR* that codes for the response regulator of the above-mentioned GraRS two-component regulatory system, and (iv) *vraS*, which was discussed in the previous section of this chapter (Figure 3.8). The literature is abundant and sometimes contradictory; however, it is consistently reported that small changes in the key regulatory genes involved in cell wall metabolism have a profound impact on the vancomycin susceptibility of *S. aureus*.

Mutations, inactivation, and altered expression of *graRS* have been linked to changes in vancomycin susceptibility in *S. aureus* [62, 65]. GraRS is known to control the *dltABCD* operon, which controls the alanylation of wall teichoic acids in response to antimicrobial challenge, indicating that the structure of teichoic acids can change in response to challenges [63]. The *dltABCD* pathway is linked to CAMP resistance in *S. aureus*, and the positive charge of D-alanine residues repels positively charged molecules such as defensins [62, 66]. Importantly, there is also evidence of a link between the D-alanylation state of teichoic acids and vancomycin susceptibility in *S. aureus*, where a *dlt* mutant strain lacking in D-alanine in teichoic acids was shown to have increased vancomycin susceptibility compared to that of the wild-type strain [67]. It is therefore likely that for isolates of hVISA or VISA where the development of resistance is associated with an increased level of expression of *graRS* or mutations in the locus, alterations in susceptibility to antimicrobial peptides are likely to occur, favoring resistance to these agents (Fig. 3.8).

Finally, the *agr* locus is a quorum-sensing system that consists partly of four genes (*agrBDCA*) and numerous studies have linked alterations in *agr* activation or function with vancomycin tolerance, an increased tendency to develop vancomycin resistance, and the presence of the hVISA or VISA phenotype [50].

3.4 Clinical Impact of Resistance

From a clinical point of view, resistance to glycopeptides constitutes a major concern for the treatment of infections caused by clinically relevant Gram-positive organisms, which are *S. aureus* and enterococci. Indeed, glycopeptides represent the treatment of choice for infections caused by MRSA and ampicillin-resistant *E. faecium* strains. In this context, the emergence of clinical isolates with diminished susceptibility or fully resistant to glycopeptides is associated with an increase of mortality and clinical failures. Furthermore, it represents a worrying reduction in therapeutic options because these strains are usually resistant to multiple antibiotics. Also, strict measures of hygiene should be applied and maintained over time to avoid or limit the dissemination of such resistant organisms in hospital settings.

3.4.1 Clinical Impact in Enterococci

Vancomycin-resistant enterococci have become important nosocomial pathogens for which there are limited treatment options, while vancomycin resistance was shown to be an independent factor of death in patients with enterococcal bloodstream infections [68]. For instance, a significant association (odds ratio, 2.52) was reported between vancomycin resistance and mortality in a recent meta-analysis [69]. According to CLSI guidelines, enterococci are categorized as susceptible if vancomycin and teicoplanin MICs are ≤4 mg/l and ≤ 8 mg/l, respectively [51]. In contrast to staphylococci (see later), CLSI recommends both agar diffusion and MIC determination for the detection of glycopeptide resistance in enterococci.

3.4.1.1 Epidemiology

Acquired glycopeptide resistance is mainly found in *E. faecium* followed by *E. faecalis,* while it is much less frequent among other enterococcal species [70]. In addition, VanA and VanB account for the most resistance phenotypes, while VanD is occasionally detected. For instance, of 902 VRE clinical isolates collected from all over France between 2006 and 2008, 94.8% were *E. faecium* and 4.5% were *E. faecalis*, while 65.8%, 33.9%, and 0.2% contained the *vanA, vanB,* and *vanD* genes, respectively [71]. Although they were first described in 1988 in Europe [22, 23], VRE rapidly became epidemic in US hospitals in the 1990s, whereas VRE outbreaks in Europe were uncommon until the early 2000s [13]. This fact can be explained by the difference in epidemiological backgrounds. In Europe, the reservoir of VRE has been the intestinal flora of healthy people in the community, previously colonized through the food chain in which the glycopeptide avoparcin (banned in 1997) was used as a growth promoter for animal feeding [72, 73]. In the USA, where avoparcin was never used, the large and rapid dissemination of VRE in hospital settings is likely related to the extensive use of vancomycin among inpatients. Since then, VRE have disseminated in numerous countries of the five continents, becoming a major threat to public health worldwide. This phenomenon is mainly linked to the emergence and spread of a subpopulation of highly virulent, hospital-adapted *E. faecium* clinical isolates. These strains belong to a clonal complex named CC17, delineated by the multilocus sequence typing (MLST) method [74]. These CC17 strains show high-level resistance to ampicillin and fluoroquinolones (but they are not necessarily resistant to vancomycin), as well as usually possessing a pathogenicity island harboring *esp* and hyl_{Efm} putative virulence genes coding for an enterococcal surface protein and a hyaluronidase, respectively [75, 76]. Interestingly, epidemic hospital-adapted strains emerged around 75 years ago, concomitant with the introduction of antibiotics, from a population that included mainly animal strains, and not from human commensal lines, these two lineages having diverged approximately 3000 years ago [77].

3.4.1.2 Colonization and Transmission

Enterococci are residents of the gastrointestinal tract of humans and other mammals, and make up a small proportion of the gut microbiota. Therefore, VRE can be easily selected for under antibiotic pressure (including anti-anaerobes, cephalosporins, fluoroquinolones, aminoglycosides, and glycopeptides) among hospitalized patients, resulting in a massive and prolonged intestinal colonization [78]. Indeed, VRE become the predominant aerobic species in the gastrointestinal (GI) tract, a phenomenon that precedes infection and serves as a reservoir for transmission to other patients [70]. Tracking VRE colonization in high-risk units appears to be a relevant approach for preventing further VRE transmission [79]. Also, the reservoir of VRE is wide and unclear, with unsuspected fecal carrier patients being one major reservoir. For instance, for every 1 patient infected with VRE there are 2–10 contact patients presenting fecal carriage [13]. It is worth noting that anaerobes are also considered as sources of *van* operons since the transfer of *vanB* gene from *Clostridium symbiosum* to *E. faecalis* and *E. faecium* in the GI tract was demonstrated in a murine model [80]. Taken together, it appears that effective control of VRE dissemination is very challenging, requiring strict prevention strategies.

3.4.2 Clinical Impact in Staphylococci

Although CLSI recommends the broth microdilution (BMD) method for the determination of vancomycin MICs allowing the distinction of VSSA, VISA (including hVISA), and VRSA, this technique is labor intensive and time consuming [51]. In addition, the disk diffusion method is not recommended. Alternative methods have been developed (such as agar dilution, E-test, and automated instruments), even though some differences in MIC results exist. For instance, results obtained by E-test (which is the most commonly used method in routine testing) are consistently onefold dilution higher than those provided by the reference BMD method [50].

3.4.2.1 VRSA

Vancomycin-resistant *S. aureus* strains exhibit MICs of vancomycin ≥16 mg/l, due to the acquisition of the *vanA* operon from VRE (see earlier). VRSA strains remain very uncommon and only a very few strains have been identified so far, mostly in the USA (n = 9), while two additional isolates were reported (but not confirmed) in Iran and India [81]. This indicates that although this mechanism of resistance is a major concern clinically, it is not evolving or spreading rapidly (Table 3.3) [54]. Interestingly, most patients infected by VRSA strains presented chronic skin ulcers, had received vancomycin, and had a history of prior MRSA and VRE infection or colonization [36].

The first MRSA clinical isolate (VRSA-1), exhibiting high-level resistance to glycopeptides (vancomycin MIC >256 mg/l; teicoplanin MIC = 128 mg/l) due to acquisition of the *vanA* operon, was detected in Michigan, USA. Two strains (VRSA-2 and VRSA-3) differ from the others in their levels of resistance to glycopeptides, exhibit moderate resistance to vancomycin (MICs, 32 and 64 mg/l, respectively), and low resistance to teicoplanin (MICs, 4 and 16 mg/l, respectively). In contrast, the remaining VRSA strains exhibited high-level resistance to both glycopeptides (vancomycin MIC >256 mg/l; teicoplanin MIC >32 mg/l) (Table 3.3) [54].

3.4.2.2 VISA and hVISA

According to CLSI guidelines, *S. aureus* strains are categorized as VISA when they present vancomycin MICs of 4–8 mg/l [50]. Notably, these strains remain very uncommon, with a prevalence estimated at 0.1%. hVISA strains correspond to strains phenotypically susceptible to vancomycin by routine laboratory methods but that contain VISA subpopulations, typically present at frequencies of 10^{-6} to 10^{-5} [50]. Since there is a lack of precise definition and no convenient method that is routinely 100% reliable, the exact prevalence of hVISA is difficult to establish and thus remains unclear [50]. Indeed, since a low inoculum is used for MIC determination, the low-frequency subpopulations of hVISA are usually missed. While prevalence rates as low as <1% have been reported in some countries, the hVISA phenotype has been detected up to 50% of clinical MRSA isolates with a vancomycin MIC of 2 mg/l in others [50, 82].

The detection of hVISA relies on the labor-intensive and time-consuming population analysis profile (PAP) method [83]. Although considered to be the gold standard, this method is difficult to apply in routine clinical resistance screenings due to delayed results (three to five days). It is worth noting that a modified protocol was proposed by Wootton et al. in which the area under the curve of a PAP test (PAP/AUC) is determined by comparison to the hVISA reference strain Mu3 [83]. Using this method, strains are categorized as VSSA, hVISA, and VISA for PAP/AUC values of <0.9, 0.9–1.3, and >1.3, respectively. Alternative methods for hVISA detection have been developed and require a higher inoculum, a prolonged incubation, and/or a more nutritious medium. The macromethod E-test (MET) uses a high inoculum (2 McFarland standard) on brain heart infusion (BHI) agar with a prolonged incubation (48 hours). A positive test is reported if teicoplanin MIC is ≥12 mg/l or if MICs of teicoplanin and vancomcyin are both ≥8 mg/l. Sensitivity and specificity of this technique are 69–98% and 89–94%, respectively [50]. Another method is the glycopeptide resistance detection (GRD) E-test using a double-ended E-test strip combining vancomycin and teicoplanin, and a standard inoculum (0.5 McFarland standard) on Mueller–Hinton (MH) blood agar. A positive test is reported if vancomycin or teicoplanin MIC is ≥8 mg/l. Sensitivity and specificity of this technique are 93–94 and 82–95%, respectively [50]. Note that the values determined by these tests are not actual MICs and should not be reported as such to clinicians. Two other agar-screening plates are also used. The first uses a BHI medium with vancomycin at 6 mg/l, inoculated with 10 μl of a 0.5 McFarland standard suspension. The growth of ≥2 colonies at 48 hours is considered as a positive result (sensitivity, 4–12%; specificity, 68–100%). The second test recommends the use of an MH medium supplemented with teicoplanin at 5 mg/l, inoculated with 10 μl of a 2 McFarland standard suspension. The growth of ≥4 colonies at 48 hours is considered a positive result (sensitivity, 65–79%; specificity, 35–95%). Finally, there is no specific molecular-based assay for the detection of hVISA strains.

3.4.2.3 Elevated Vancomycin MIC

In 2006, CLSI revised clinical breakpoints for vancomycin, lowering the susceptibility threshold from 4 to 2 mg/l. However, strains of S. aureus with high MICs in the susceptibility range (i.e., 1 < MIC ≤2 mg/l) have been associated with adverse clinical outcomes in two recent meta-analyses [84, 85]. In the first analysis, 22 studies were included and it was shown that vancomycin MIC was significantly associated with mortality for MRSA infection and treatment failure (odds ratio at 1.64 and 2.69, respectively) [85]. In the second analysis, 33 studies were eligible and high vancomycin MIC (i.e., 1.5 and 2 mg/l) was associated with higher mortality and a higher risk of treatment failure (relative risk at 1.21 and 1.67, respectively) [84]. VSSA, hVISA, and VISA may represent a continuum of small incremental changes ("creep") in vancomycin MIC, while the spectrum of infections causing these three entities is similar [81]. Interestingly, risk factors associated with the emergence of strains exhibiting an elevated vancomycin MIC are like those associated with the development of hVISA, including older patient age, prior vancomycin exposure, prior MRSA bacteremia, and high-inoculum infections. Taken together, nonvancomycin anti-MRSA therapies (see later) may be considered for patients with high-burden MRSA infections that have vancomycin MIC >1 mg/l.

3.4.2.4 Teicoplanin

Only a few studies have been published on the clinical usage of teicoplanin because it is only available in some European countries. In a recent meta-analysis including 24 studies, efficacy and safety of vancomycin versus teicoplanin were compared [86]. Whereas there were no significant differences for all-cause mortality, clinical failure, microbiological failure, and other efficacy outcomes (rates of relapse and superinfection), total adverse events were significantly less common with teicoplanin (relative risk, 0.61).

3.5 Conclusions and Perspectives

Glycopeptides make up a major class of antibiotics, which represent the treatment of choice for resistant Gram-positive organisms. More specifically, a combination of glycopeptide–aminoglycoside is the treatment of choice for serious infections caused by MRSA and enterococci (particularly E. faecium). Due to their extensive clinical use, there is an increase of glycopeptide-intermediate or glycopeptide-resistant strains that have spread in hospital settings as well as in the community. Even though VRSA have been exceptionally isolated so far, strains with elevated vancomycin MICs are increasingly frequently recovered from human infections, and are associated with higher mortality and increased risk of treatment failure. In addition, the number of VRE outbreaks is increasing among European countries, while VRE have become endemic in US hospitals since the 1990s.

Therefore, alternative therapies should be used, such as quinupristin-dalfopristin, linezolid, daptomycin, tigecycline, and ceftaroline. Quinupristin-dalfopristin is an injectable streptogramin, which acts by interfering with bacterial protein synthesis and presents a bactericidal effect against most Gram-positive pathogens [87]. In contrast to *E. faecalis*, which is intrinsically resistant, *E. faecium* is naturally susceptible to this combination; however, resistant *E. faecium* strains can be easily selected for *in vitro* and *in vivo* [88]. Linezolid is a bacteriostatic antibiotic that belongs to the family of oxazolidinones and acts by inhibition of protein biosynthesis [89]. It is active against MRSA and VRE isolates, with MIC_{90} values usually ≤ 2 mg/l. Despite the difficulty of *in vitro* selection, linezolid resistance can nonetheless emerge during therapy (especially but not exclusively after a prolonged therapy) but this remains uncommon [87]. A more worrying mechanism of resistance is the plasmid-mediated resistance due to the Cfr methyltransferase, reported mostly in staphylococci but also in some enterococci [90–92]. Daptomycin is a cyclic lipopeptide antibiotic that exerts a potent and rapid bactericidal activity against Gram-positive organisms through an irreversible interaction with the bacterial cell membrane. More specifically, it demonstrates potency against MRSA and VRE, with MIC_{90} values typically of 0.5 and 2–4 mg/l, respectively [93]. Even though the development of resistance to daptomycin remains very uncommon, several therapy failures have been reported with the emergence of high-level resistance (MIC >32 mg/l), especially in *S. aureus* [94]. Tigecycline is the first representative of the novel glycylcycline class, which exhibits a broad spectrum of activity against Gram-positive and Gram-negative bacteria, acting by inhibition of protein biosynthesis [95]. Resistance is very difficult to select *in vitro* and has not been yet reported in clinical isolates of Gram-positives. Ceftaroline is new broad-spectrum cephalosporin with a potent activity against Gram-positive organisms, including MRSA but not VRE (reminder: enterococci are intrinsically resistant to cephalosporins). Like other β-lactams, it inhibits the formation of the peptidoglycan by high-affinity binding to PBPs, including PBP2a. Most MRSA isolates, including VISA and hVISA, exhibit low MICs (MIC_{90}s usually ≤ 1 mg/l), while high-level resistance has not been yet reported [96]. Finally, other new lipoglycopeptides present interesting properties. For instance, dalbavancin demonstrates potent activity against VISA and VanB-type VRE with a once-weekly dosing, while oritavancin retains activity against VISA, VRSA, and VanA-type VRE with a single dose per treatment course [7].

References

1 Barna, J.C. and Williams, D.H. (1984). The structure and mode of action of glycopeptide antibiotics of the vancomycin group. *Annual Review of Microbiology* 38: 339–357.

2 Surawicz, C.M., Brandt, L.J., Binion, D.G. et al. (2013). Guidelines for diagnosis, treatment, and prevention of *Clostridium difficile* infections. *The American Journal of Gastroenterology* 108: 478–498.

3 Levine, D.P. (2006). Vancomycin: a history. *Clinical Infectious Diseases* 42 (Suppl 1): S5–S12.

4 Parenti, F. (1986). Structure and mechanism of action of teicoplanin. *The Journal of Hospital Infection* 7 (Suppl A): 79–83.

5 Rubinstein, E., Corey, G.R., Stryjewski, M.E., and Kanafani, Z.A. (2011). Telavancin for the treatment of serious gram-positive infections, including hospital acquired pneumonia. *Expert Opinion on Pharmacotherapy* 12: 2737–2750.

6 Torres, A., Rubinstein, E., Corey, G.R. et al. (2014). Analysis of phase 3 telavancin nosocomial pneumonia data excluding patients with severe renal impairment and acute renal failure. *The Journal of Antimicrobial Chemotherapy* 9 (4): 1119–1126.

7 Zhanel, G.G., Calic, D., Schweizer, F. et al. (2010). New lipoglycopeptides: a comparative review of dalbavancin, oritavancin and telavancin. *Drugs* 70: 859–886.

8 Griffith, R.S. (1984). Vancomycin use – an historical review. *The Journal of Antimicrobial Chemotherapy* 14 (Suppl D): 1–5.

9 Parenti, F., Beretta, G., Berti, M., and Arioli, V. (1978). Teichomycins, new antibiotics from *Actinoplanes teichomyceticus* Nov. Sp. I. Description of the producer strain, fermentation studies and biological properties. *Journal of Antibiotics (Tokyo)* 31: 276–283.

10 Reynolds, P.E. (1989). Structure, biochemistry and mechanism of action of glycopeptide antibiotics. *European Journal of Clinical Microbiology & Infectious Diseases* 8: 943–950.

11 Greenwood, D. (1988). Microbiological properties of teicoplanin. *The Journal of Antimicrobial Chemotherapy* 21 (Suppl A): 1–13.

12 Leadbetter, M.R., Adams, S.M., Bazzini, B. et al. (2004). Hydrophobic vancomycin derivatives with improved ADME properties: discovery of telavancin (TD-6424). *Journal of Antibiotics (Tokyo)* 57: 326–336.

13 Cattoir, V. and Leclercq, R. (2013). Twenty-five years of shared life with vancomycin-resistant enterococci: is it time to divorce? *The Journal of Antimicrobial Chemotherapy* 68: 731–742.

14 Courvalin, P. (2009). Glycopeptide and enterococci. In: *Antibiogram* (ed. P. Courvalin, R. Leclercq and L. Rice), 285–294. Paris, France: ESKA.

15 Sheldrick, G.M., Jones, P.G., Kennard, O. et al. (1978). Structure of vancomycin and its complex with acetyl-D-alanyl-D-alanine. *Nature* 271: 223–225.

16 Higgins, D.L., Chang, R., Debabov, D.V. et al. (2005). Telavancin, a multifunctional lipoglycopeptide, disrupts both cell wall synthesis and cell membrane integrity in methicillin-resistant *Staphylococcus aureus*. *Antimicrobial Agents and Chemotherapy* 49: 1127–1134.

17 Anderson, D.L. (2008). Oritavancin for skin infections. *Drugs of Today (Barcelona)* 44: 563–575.

18 Saravolatz, L.D., Stein, G.E., and Johnson, L.B. (2009). Telavancin: a novel lipoglycopeptide. *Clinical Infectious Diseases* 49: 1908–1914.

19 Murray, B.E. and Nannini, E.C. (2010). Glycopeptides (vancomycin and teicoplanin), streptogramins (quinupristin-dalfopristin), and lipopeptides (daptomycin). In: *Principles and Practice of Infectious Diseases* (ed. G.L. Mandell, J.E. Bennet and R. Dolin), 449–468. Philadelphia, PA: Churchill Livingstone.

20 Vandecasteele, S.J., De Vriese, A.S., and Tacconelli, E. (2013). The pharmacokinetics and pharmacodynamics of vancomycin in clinical practice: evidence and uncertainties. *Journal of Antimicrobial Chemotherapy* 68: 743–748.

21 Mohr, J.F. and Murray, B.E. (2007). Point: Vancomycin is not obsolete for the treatment of infection caused by methicillin-resistant *Staphylococcus aureus*. *Clinical Infectious Diseases* 44: 1536–1542.

22 Leclercq, R., Derlot, E., Duval, J., and Courvalin, P. (1988). Plasmid-mediated resistance to vancomycin and teicoplanin in *Enterococcus faecium*. *The New England Journal of Medicine* 319: 157–161.

23 Uttley, A.H., Collins, C.H., Naidoo, J., and George, R.C. (1988). Vancomycin-resistant enterococci. *Lancet* 1: 57–58.

24 Frieden, T.R., Munsiff, S.S., Low, D.E. et al. (1993). Emergence of vancomycin-resistant enterococci in New York City. *Lancet* 342: 76–79.

25 Courvalin, P. (2006). Vancomycin resistance in gram-positive cocci. *Clinical Infectious Diseases* 42 (Suppl 1): S25–S34.

26 Boyd, D.A., Willey, B.M., Fawcett, D. et al. (2008). Molecular characterization of *Enterococcus faecalis* N06-0364 with low-level vancomycin resistance harboring a novel D-Ala-D-Ser gene cluster, *vanL*. *Antimicrobial Agents and Chemotherapy* 52: 2667–2672.

27 Depardieu, F., Podglajen, I., Leclercq, R. et al. (2007). Modes and modulations of antibiotic resistance gene expression. *Clinical Microbiology Reviews* 20: 79–114.

28 Lebreton, F., Depardieu, F., Bourdon, N. et al. (2011). D-Ala-D-Ser VanN-type transferable vancomycin resistance in *Enterococcus faecium*. *Antimicrobial Agents and Chemotherapy* 55: 4606–4612.

29 Xu, X., Lin, D., Yan, G. et al. (2010). *vanM*, a new glycopeptide resistance gene cluster found in *Enterococcus faecium*. *Antimicrobial Agents and Chemotherapy* 54: 4643–4647.

30 Arthur, M., Reynolds, P., and Courvalin, P. (1996). Glycopeptide resistance in enterococci. *Trends in Microbiology* 4: 401–407.

31 Depardieu, F., Reynolds, P.E., and Courvalin, P. (2003). VanD-type vancomycin-resistant *Enterococcus faecium* 10/96A. *Antimicrobial Agents and Chemotherapy* 47: 7–18.

32 Dutka-Malen, S., Molinas, C., Arthur, M., and Courvalin, P. (1992). Sequence of the *vanC* gene of *Enterococcus gallinarum* BM4174 encoding a D-alanine:D-alanine ligase-related protein necessary for vancomycin resistance. *Gene* 112: 53–58.

33 Nomura, T., Tanimoto, K., Shibayama, K. et al. (2012). Identification of VanN-type vancomycin resistance in an *Enterococcus faecium* isolate from chicken meat in Japan. *Antimicrobial Agents and Chemotherapy* 56: 6389–6392.

34 Boyd, D.A. and Mulvey, M.R. (2013). The VanE operon in *Enterococcus faecalis* N00-410 is found on a putative integrative and conjugative element, Tn*6202*. *The Journal of Antimicrobial Chemotherapy* 68: 294–299.

35 Krcmery, V. and Sefton, A. (2000). Vancomycin resistance in gram-positive bacteria other than *Enterococcus* spp. *International Journal of Antimicrobial Agents* 14: 99–105.

36 Sievert, D.M., Rudrik, J.T., Patel, J.B. et al. (2008). Vancomycin-resistant Staphylococcus aureus in the United States, 2002–2006. *Clinical Infectious Diseases* 46: 668–674.

37 Mainardi, J.L., Villet, R., Bugg, T.D. et al. (2008). Evolution of peptidoglycan biosynthesis under the selective pressure of antibiotics in Gram-positive bacteria. *FEMS Microbiology Reviews* 32: 386–408.

38 Reynolds, P.E. and Courvalin, P. (2005). Vancomycin resistance in enterococci due to synthesis of precursors terminating in D-alanyl-D-serine. *Antimicrobial Agents and Chemotherapy* 49: 21–25.

39 Billot-Klein, D., Blanot, D., Gutmann, L., and van Heijenoort, J. (1994). Association constants for the binding of vancomycin and teicoplanin to N-acetyl-D-alanyl-D-alanine and N-acetyl-D-alanyl-D-serine. *Biochemical Journal* 304 (Pt 3): 1021–1022.

40 Reynolds, P.E. (1998). Control of peptidoglycan synthesis in vancomycin-resistant enterococci: D,D-peptidases and D,D-carboxypeptidases. *Cellular and Molecular Life Sciences* 54: 325–331.

41 Arthur, M., Depardieu, F., Cabanie, L. et al. (1998). Requirement of the VanY and VanX D,D-peptidases for glycopeptide resistance in enterococci. *Molecular Microbiology* 30: 819–830.

42 Reynolds, P.E., Depardieu, F., Dutka-Malen, S. et al. (1994). Glycopeptide resistance mediated by enterococcal transposon Tn*1546* requires production of VanX for hydrolysis of D-alanyl-D-alanine. *Molecular Microbiology* 13: 1065–1070.

43 Reynolds, P.E., Arias, C.A., and Courvalin, P. (1999). Gene *vanXY$_C$* encodes D,D -dipeptidase (VanX) and D,D-carboxypeptidase (VanY) activities in vancomycin-resistant *Enterococcus gallinarum* BM4174. *Molecular Microbiology* 34: 341–349.

44 Arthur, M., Molinas, C., and Courvalin, P. (1992). The VanS-VanR two-component regulatory system controls synthesis of depsipeptide peptidoglycan precursors in *Enterococcus faecium* BM4147. *Journal of Bacteriology* 174: 2582–2591.

45 Arthur, M., Depardieu, F., Gerbaud, G. et al. (1997). The VanS sensor negatively controls VanR-mediated transcriptional activation of glycopeptide resistance genes of Tn*1546* and related elements in the absence of induction. *Journal of Bacteriology* 179: 97–106.

46 Fraimow, H.S., Jungkind, D.L., Lander, D.W. et al. (1994). Urinary tract infection with an *Enterococcus faecalis* isolate that requires vancomycin for growth. *Annals of Internal Medicine* 121: 22–26.

47 Van Bambeke, F., Chauvel, M., Reynolds, P.E. et al. (1999). Vancomycin-dependent *Enterococcus faecalis* clinical isolates and revertant mutants. *Antimicrobial Agents and Chemotherapy* 43: 41–47.

48 Hiramatsu, K. (2001). Vancomycin-resistant *Staphylococcus aureus*: a new model of antibiotic resistance. *The Lancet Infectious Diseases* 1: 147–155.

49 Klevens, R.M., Morrison, M.A., Nadle, J. et al. (2007). Invasive methicillin-resistant *Staphylococcus aureus* infections in the United States. *Journal of the American Medical Association* 298: 1763–1771.

50 Howden, B.P., Davies, J.K., Johnson, P.D. et al. (2010). Reduced vancomycin susceptibility in *Staphylococcus aureus*, including vancomycin-intermediate and heterogeneous vancomycin-intermediate strains: resistance mechanisms, laboratory detection, and clinical implications. *Clinical Microbiology Reviews* 23: 99–139.

51 Clinical and Laboratory Standards Institute (2013) Performance standards for antimicrobial susceptibility testing; twenty-third informational supplement. CLSI document M100-S23, Wayne, Philadelphia, PA, USA.

52 Noble, W.C., Virani, Z., and Cree, R.G. (1992). Co-transfer of vancomycin and other resistance genes from *Enterococcus faecalis* NCTC 12201 to *Staphylococcus aureus*. *FEMS Microbiology Letters* 72: 195–198.

53 Chang, S., Sievert, D.M., Hageman, J.C. et al. (2003). Infection with vancomycin-resistant *Staphylococcus aureus* containing the vanA resistance gene. *The New England Journal of Medicine* 348: 1342–1347.

54 Perichon, B. and Courvalin, P. (2009). VanA-type vancomycin-resistant *Staphylococcus aureus*. *Antimicrobial Agents and Chemotherapy* 53: 4580–4587.

55 Weigel, L.M., Clewell, D.B., Gill, S.R. et al. (2003). Genetic analysis of a high-level vancomycin-resistant isolate of *Staphylococcus aureus*. *Science* 302: 1569–1571.

56 Zhu, W., Clark, N.C., McDougal, L.K. et al. (2008). Vancomycin-resistant *Staphylococcus aureus* isolates associated with Inc18-like *vanA* plasmids in Michigan. *Antimicrobial Agents and Chemotherapy* 52: 452–457.

57 Kos, V.N., Desjardins, C.A., Griggs, A. et al. (2012). Comparative genomics of vancomycin-resistant *Staphylococcus aureus* strains and their positions within the clade most commonly associated with methicillin-resistant *S. aureus* hospital-acquired infection in the United States. *mBio* 3: https://doi.org/10.1128/mBio.00112-12.

58 Dmitriev, B.A., Toukach, F.V., Holst, O. et al. (2004). Tertiary structure of *Staphylococcus aureus* cell wall murein. *Journal of Bacteriology* 186: 7141–7148.

59 Pinho, M.G. and Errington, J. (2005). Recruitment of penicillin-binding protein PBP2 to the division site of *Staphylococcus aureus* is dependent on its transpeptidation substrates. *Molecular Microbiology* 55: 799–807.

60 Boyle-Vavra, S., Carey, R.B., and Daum, R.S. (2001). Development of vancomycin and lysostaphin resistance in a methicillin-resistant *Staphylococcus aureus* isolate. *The Journal of Antimicrobial Chemotherapy* 48: 617–625.

61 Kuroda, M., Kuroda, H., Oshima, T. et al. (2003). Two-component system VraSR positively modulates the regulation of cell-wall biosynthesis pathway in *Staphylococcus aureus*. *Molecular Microbiology* 49: 807–821.

62 Cui, L., Lian, J.Q., Neoh, H.M. et al. (2005). DNA microarray-based identification of genes associated with glycopeptide resistance in *Staphylococcus aureus*. *Antimicrobial Agents and Chemotherapy* 49: 3404–3413.

63 Herbert, S., Bera, A., Nerz, C. et al. (2007). Molecular basis of resistance to muramidase and cationic antimicrobial peptide activity of lysozyme in staphylococci. *PLoS Pathogens* 3: e102.

64 Mwangi, M.M., Wu, S.W., Zhou, Y. et al. (2007). Tracking the *in vivo* evolution of multidrug resistance in *Staphylococcus aureus* by whole-genome sequencing. *Proceedings of the National Academy of Sciences of the United States of America* 104: 9451–9456.

65 Neoh, H.M., Cui, L., Yuzawa, H. et al. (2008). Mutated response regulator *graR* is responsible for phenotypic conversion of *Staphylococcus aureus* from heterogeneous vancomycin-intermediate resistance to vancomycin-intermediate resistance. *Antimicrobial Agents and Chemotherapy* 52: 45–53.

66 Peschel, A., Otto, M., Jack, R.W. et al. (1999). Inactivation of the *dlt* operon in *Staphylococcus aureus* confers sensitivity to defensins, protegrins, and other antimicrobial peptides. *Journal of Biological Chemistry* 274: 8405–8410.

67 Peschel, A., Vuong, C., Otto, M., and Gotz, F. (2000). The D-alanine residues of *Staphylococcus aureus* teichoic acids alter the susceptibility to vancomycin and the activity of autolytic enzymes. *Antimicrobial Agents and Chemotherapy* 44: 2845–2847.

68 Lode, H.M. (2009). Clinical impact of antibiotic-resistant gram-positive pathogens. *Clinical Microbiology and Infection* 15: 212–217.

69 DiazGranados, C.A., Zimmer, S.M., Klein, M., and Jernigan, J.A. (2005). Comparison of mortality associated with vancomycin-resistant and vancomycin-susceptible

enterococcal bloodstream infections: a meta-analysis. *Clinical Infectious Diseases* 41: 327–333.

70 Arias, C.A. and Murray, B.E. (2012). The rise of the *Enterococcus*: beyond vancomycin resistance. *Nature Reviews Microbiology* 10: 266–278.

71 Bourdon, N., Fines-Guyon, M., Thiolet, J.M. et al. (2011). Changing trends in vancomycin-resistant enterococci in French hospitals, 2001–08. *The Journal of Antimicrobial Chemotherapy* 66: 713–721.

72 Acar, J., Casewell, M., Freeman, J. et al. (2000). Avoparcin and virginiamycin as animal growth promoters: a plea for science in decision-making. *Clinical Microbiology and Infection* 6: 477–482.

73 Bonten, M.J., Willems, R., and Weinstein, R.A. (2001). Vancomycin-resistant enterococci: why are they here, and where do they come from? *Lancet Infectious Diseases* 1: 314–325.

74 Willems, R.J., Top, J., van Santen, M. et al. (2005). Global spread of vancomycin-resistant *Enterococcus faecium* from distinct nosocomial genetic complex. *Emerging Infectious Diseases* 11: 821–828.

75 Leavis, H., Top, J., Shankar, N. et al. (2004). A novel putative enterococcal pathogenicity island linked to the *esp* virulence gene of *Enterococcus faecium* and associated with epidemicity. *Journal of Bacteriology* 186: 672–682.

76 Rice, L.B., Carias, L., Rudin, S. et al. (2003). A potential virulence gene, hyl_{Efm}, predominates in *Enterococcus faecium* of clinical origin. *Journal of Infectious Diseases* 187: 508–512.

77 Lebreton, F., van Schaik, W., McGuire, A. M. *et al.* (2013) Emergence of epidemic multidrug-resistant *Enterococcus faecium* from animal and commensal strains. *mBio*, **4**.

78 Patel, R. (2003). Clinical impact of vancomycin-resistant enterococci. *The Journal of Antimicrobial Chemotherapy* 51 (Suppl 3): iii13–iii21.

79 Siddiqui, A.H., Harris, A.D., Hebden, J. et al. (2002). The effect of active surveillance for vancomycin-resistant enterococci in high-risk units on vancomycin-resistant enterococci incidence hospital-wide. *American Journal of Infection Control* 30: 40–43.

80 Launay, A., Ballard, S.A., Johnson, P.D. et al. (2006). Transfer of vancomycin resistance transposon Tn*1549* from *Clostridium symbiosum* to *Enterococcus* spp. in the gut of gnotobiotic mice. *Antimicrobial Agents and Chemotherapy* 50: 1054–1062.

81 Holmes, N.E., Johnson, P.D., and Howden, B.P. (2012). Relationship between vancomycin-resistant *Staphylococcus aureus*, vancomycin-intermediate *S. aureus*, high vancomycin MIC, and outcome in serious *S. aureus* infections. *Journal of Clinical Microbiology* 50: 2548–2552.

82 Horne, K.C., Howden, B.P., Grabsch, E.A. et al. (2009). Prospective comparison of the clinical impacts of heterogeneous vancomycin-intermediate methicillin-resistant *Staphylococcus aureus* (MRSA) and vancomycin-susceptible MRSA. *Antimicrobial Agents and Chemotherapy* 53: 3447–3452.

83 Wootton, M., Howe, R.A., Hillman, R. et al. (2001). A modified population analysis profile (PAP) method to detect hetero-resistance to vancomycin in *Staphylococcus aureus* in a UK hospital. *The Journal of Antimicrobial Chemotherapy* 47: 399–403.

84 Mavros, M.N., Tansarli, G.S., Vardakas, K.Z. et al. (2012). Impact of vancomycin minimum inhibitory concentration on clinical outcomes of patients with

vancomycin-susceptible *Staphylococcus aureus* infections: a meta-analysis and meta-regression. *International Journal of Antimicrobial Agents* 40: 496–509.

85 van Hal, S.J., Lodise, T.P., and Paterson, D.L. (2012). The clinical significance of vancomycin minimum inhibitory concentration in *Staphylococcus aureus* infections: a systematic review and meta-analysis. *Clinical Infectious Diseases* 54: 755–771.

86 Svetitsky, S., Leibovici, L., and Paul, M. (2009). Comparative efficacy and safety of vancomycin versus teicoplanin: systematic review and meta-analysis. *Antimicrobial Agents and Chemotherapy* 53: 4069–4079.

87 Eliopoulos, G.M. (2003). Quinupristin-dalfopristin and linezolid: evidence and opinion. *Clinical Infectious Diseases* 36: 473–481.

88 Isnard, C., Malbruny, B., Leclercq, R., and Cattoir, V. (2013). Genetic basis for *in vitro* and *in vivo* resistance to lincosamides, streptogramins A, and pleuromutilins (LS$_A$P phenotype) in *Enterococcus faecium*. *Antimicrobial Agents and Chemotherapy* 57: 4463–4469.

89 Livermore, D.M. (2003). Linezolid *in vitro*: mechanism and antibacterial spectrum. *The Journal of Antimicrobial Chemotherapy* 51 (Suppl 2): ii9–ii16.

90 Diaz, L., Kiratisin, P., Mendes, R.E. et al. (2012). Transferable plasmid-mediated resistance to linezolid due to *cfr* in a human clinical isolate of *Enterococcus faecalis*. *Antimicrobial Agents and Chemotherapy* 56: 3917–3922.

91 Gu, B., Kelesidis, T., Tsiodras, S. et al. (2013). The emerging problem of linezolid-resistant *Staphylococcus*. *The Journal of Antimicrobial Chemotherapy* 68: 4–11.

92 Patel, S.N., Memari, N., Shahinas, D. et al. (2013). Linezolid resistance in *Enterococcus faecium* isolated in Ontario, Canada. *Diagnostic Microbiology and Infectious Disease* 77: 350–353.

93 Steenbergen, J.N., Alder, J., Thorne, G.M., and Tally, F.P. (2005). Daptomycin: a lipopeptide antibiotic for the treatment of serious gram-positive infections. *The Journal of Antimicrobial Chemotherapy* 55: 283–288.

94 Boucher, H.W. and Sakoulas, G. (2007). Perspectives on daptomycin resistance, with emphasis on resistance in *Staphylococcus aureus*. *Clinical Infectious Diseases* 45: 601–608.

95 Peterson, L.R. (2008). A review of tigecycline – the first glycylcycline. *International Journal of Antimicrobial Agents* 32 (Suppl 4): S215–S222.

96 Laudano, J.B. (2011). Ceftaroline fosamil: a new broad-spectrum cephalosporin. *The Journal of Antimicrobial Chemotherapy* 66 (Suppl 3): iii11–iii18.

4

Resistance and Tolerance to Aminoglycosides

Wendy W.K. Mok and Mark P. Brynildsen

Department of Chemical and Biological Engineering, Princeton University, Princeton, NJ, USA

4.1   Introduction

The discovery of the first aminoglycoside (AG), streptomycin, in 1943 was an important landmark in modern medicine [1]. Streptomycin was the first effective anti-tuberculosis therapeutic, and it was the first bioactive small molecule to be isolated from bacteria [2]. Over 70 years since its discovery, streptomycin continues to be active against *Mycobacterium tuberculosis* and remains a first-line antibiotic in combination therapy against drug-resistant *M. tuberculosis* [2]. In the decades that followed the successful isolation of streptomycin, additional AGs were discovered in soil-dwelling *Streptomyces* and *Micromonospora* bacteria [3]. In these *Actinomycetes*, AGs are produced as secondary metabolites, which allow them to kill bacteria and fungi that share the same ecological niche, and thereby gain a competitive advantage [3]. From the 1960s to 1980s, advances in medicinal chemistry led to the emergence of semisynthetic AG derivatives (*e.g.*, dibekacin, amikacin, and netilmicin), which had broader activity than first generation AGs and modifications that conferred defense against resistance mechanisms that had emerged [3].

AGs inhibit the ribosome, perturbing different stages of protein synthesis and ribosome recycling [2]. Many AGs available today have a broad spectrum of antibacterial activity, capable of killing a range of clinically relevant pathogens such as Gram-negative bacilli, *Staphylococcus aureus* (including methicillin-resistant *S. aureus*), and some streptococci [2]. Gentamicin, tobramycin, and amikacin, three of the most commonly prescribed AGs, are routinely administered as surgical prophylaxes, chemotherapy for life-threatening infections, and treatment against biological warfare agents [4]. In addition to their clinical applications, AGs have been widely used in livestock farming and aquaculture, where they were used for therapeutic purposes and as growth promoters in animal feed to control their intestinal microbiota and prevent the spread of zoonotic pathogens [5].

AGs have been shown to exhibit favorable pharmacokinetic and pharmacodynamic properties. For instance, they are not metabolized once consumed and are eliminated by glomerular filtration [6]. They can also kill microorganisms in a concentration-dependent manner and exert their inhibitory effects for a period after antibiotic removal [7]. They

Bacterial Resistance to Antibiotics – From Molecules to Man, First Edition.
Edited by Boyan B. Bonev and Nicholas M. Brown.
© 2020 John Wiley & Sons Ltd. Published 2020 by John Wiley & Sons Ltd.

have synergistic effects when administered with cell wall biosynthesis inhibitors, such as β-lactam antibiotics and vancomycin, and these combinatorial therapies are commonly used to treat infections caused by enterococci and *Pseudomonas aeruginosa* [2]. However, toxicity concerns and absorption issues have limited the use of AGs in the past 30 years [3]. AGs are poorly absorbed when administered orally, and they are associated with severe adverse effects, such as nephrotoxicity and permanent ototoxicity [6]. Consequently, interest in the discovery of new AGs has declined, and the clinical use of AGs has decreased in favor of other classes of broad-spectrum antibiotics with fewer side effects, such as fluoroquinolones, carbapenems, and cephalosporins [3].

The rise of multidrug-resistant superbugs and the dwindling of effective antimicrobials have reinvigorated interest in the use of AGs against serious Gram-negative infections and in the discovery of more resilient AG analogs [3]. An understanding of how AGs disrupt bacterial cellular processes and knowledge of tolerance and resistance mechanisms toward these agents will guide the development of more effective AG analogs and treatment regimens. In this chapter, we will discuss key mechanisms of AG action, bacterial defenses against this class of antibiotics, and research in methods to enhance the efficacy and safety of AG treatment.

4.2 Structure of AGs

Natural and synthetic AGs all contain a six-membered aminocyclitol nucleus that is glycosidically linked to at least one amino sugar ring [8] (Figure 4.1). Streptomycin and its derivatives contain streptamine as its aminocyclitol ring (indicated in purple in

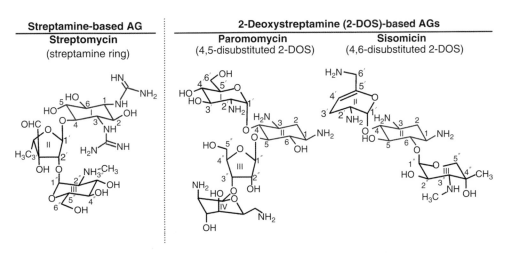

Figure 4.1 Structures of AGs. Streptomycin is an example of an AG with a streptamine core (indicated in purple). Paromomycin and sisomicin both contain 2-deoxystreptamine (2-DOS) cores (indicated in blue). In paromomycin, additional rings are substituted at positions 4 and 5 of the 2-DOS ring, whereas additional rings are substituted at positions 4 and 6 of 2-DOS in sisomicin. The carbons in the streptamine or 2-DOS rings are numbered plainly, whereas those on the additional rings are numbered with prime (′) or double-prime (″) superscripts [9]. (*Source:* Structures generated using ACD/ChemSketch, version 12.00, Advanced Chemistry Development, Inc., Toronto, ON, Canada).

Figure 4.1), whereas many newer generation AGs carry 2-deoxystreptamine (2-DOS) rings (indicated in blue in Figure 4.1). In AGs with 2-DOS rings, the sugars can be attached at positions 4 and 5 of the 2-DOS ring (4,5-disubstituted DOS), as observed in neomycin and its derivatives [8]. Alternatively, the carbohydrate moieties can be attached to the 2-DOS ring at positions 4 and 6 (4,6-disubstituted DOS), as observed in kanamycin, gentamicin, and netilmicin [8]. Differences in ring number, structure, and functional groups can alter interactions between the antibiotic and ribosomal components, producing AGs with distinct ribosomal inhibition profiles and resistance mechanisms [2, 3]. As a result of their structures, AGs are generally water soluble, basic in nature, and positively charged at physiological pH [9].

4.3 Three Stages of AG Uptake

The structures of AGs render them too polar to freely diffuse across the membrane [9], thus the intracellular accumulation of AGs is an energy-dependent processes. Even after over 50 years of investigation, many aspects of AG uptake in bacteria remain elusive [3, 10]. Most AG uptake studies have been conducted with streptomycin and gentamicin, for which radiolabeled or fluorescently labeled analogs are available, and the uptake of these AGs may be mechanistically different from other members of this class of antibiotics. Several studies have shown that AG uptake is dependent on the transmembrane electrical potential ($\Delta\psi$) and on electron flow through membrane-bound respiratory chains [10]. For example, treating *S. aureus* with valinomycin, an ionophore that increases the permeability of potassium ions and results in the collapse of $\Delta\psi$, was found to abolish gentamicin uptake and killing [11]. Furthermore, ubiquinone-deficient *Escherichia coli* mutants were found to exhibit reduced AG accumulation, suggesting that electron transfer through ubiquinone is associated with AG uptake [10].

Kinetic studies examining the uptake of radiolabeled AGs in aerobically cultured bacteria suggested that AG uptake occurs in three phases [10, 12, 13] (Figure 4.2). The first is an ionic binding phase, where the antibiotic binds reversibly and electrostatically to anionic sites on bacterial surfaces, including lipopolysaccharides in Gram-negative organisms, wall teichoic acids in Gram-positive organisms, phospholipid polar head groups, and membrane proteins [10]. Initial adsorption is proposed to be followed by two energy-dependent phases of uptake, as indicated by two kinetic rates [12, 13]. Uptake during the first energy-dependent phase (EDP-I) is suggested to be slow and dependent on a threshold $\Delta\psi$ [10, 12, 13]. It is sensitive to electron transport inhibitors (*e.g.*, cyanide) and proton ionophores (*e.g.*, dinitrophenol [DNP], and carbonyl cyanide *m*-chlorophenylhydrazone [CCCP]) [13]. It has been suggested that EDP-I is dependent on AG concentration and can be bypassed when drug concentrations are high [10]. While some studies have suggested that EDP-I involves a slow rate of uptake [12, 13], others have proposed that it represents enhanced adsorption of AG to cell surfaces rather than internalization of the drug [14]. Like EDP-I, many aspects of the second energy-dependent phase of AG uptake (EDP-II) remain elusive. It is thought that the initial entry of modest amounts of AG can corrupt ribosomes and lead to the release of aberrant proteins, which can target the inner membrane, facilitate the translocation of AGs, and enable the rapid uptake observed in EDP-II [10, 15]. AG uptake during EDP-II is rapid, and this influx of antibiotics can saturate the ribosomes and trigger cell

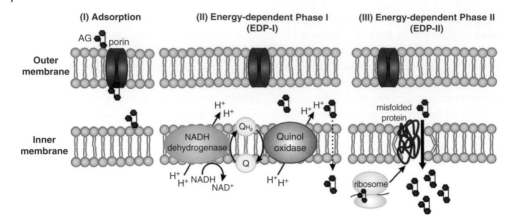

Figure 4.2 **Three stages of AG uptake**. AG uptake is proposed to occur in three stages. In the first stage (adsorption), AGs adsorb onto bacterial cell surfaces in an energy-independent manner. In stage two (EDP-I), increased adsorption or uptake of a small quantity of AG quantity (*via* a mechanism that has yet to be determined) have been shown to occur in a $\Delta\psi$-dependent manner [13, 14]. In the last phase of AG uptake (EDP-II), electron transport and ATP generation are proposed to provide energy needed for the incorporation of misfolded proteins in the membrane [10]. This can subsequently disrupt membrane integrity and promote further AG influx, which can then saturate ribosomes and result in cell death. Enzymatic reactions have been simplified for illustrative purposes. Q: a ubiquinone; QH2: a ubiquinol.

death [10]. It has been suggested that EDP-II relies on energy derived from electron transport and possibly ATP hydrolysis, as respiration inhibitors and oxidative phospho-rylation uncouplers added after 15 minutes of AG accumulation (after the rapid uptake rate was observed) can inhibit further uptake [10, 13]. Protein synthesis is also needed to trigger EDP-II, and translation inhibitors such as chloramphenicol have been shown to abolish this phase of uptake [16]. Further, it has been proposed that electron trans-port may be needed to generate ATP, which is consumed to drive the insertion or trans-location of mistranslated proteins in the cytoplasmic membrane [10].

4.4 Cellular Targets of AGs

AGs bind ribosomal 30S subunits and, in doing so, they can perturb different stages of protein synthesis, including initiation, elongation, termination, and ribosomal recycling [17]. *In vivo*, translation occurs with high fidelity, with an error rate of 10^{-3} to 10^{-4} per codon [18], and the binding of AGs to the ribosome results in a loss of translational fidelity. It has been proposed that the geometry of key residues at the decoding site provides steric discrimination against mismatches. Three universally conserved resi-dues in the 16S rRNA, G530, A1492, and A1493, have been shown to be directly involved in interactions with A-site tRNAs [19]. In the absence of cognate tRNA binding, A1492 and A1493 are stacked in the interior of helix 44 (h44) in the A site [2]. Binding of cog-nate tRNAs induces structural rearrangement of the A site and causes these two bases to flip out, allowing them to form hydrogen bonds with both strands of the minor groove of the codon-anticodon duplex in a manner that is sensitive to Watson–Crick base pairing [19]. Binding also alters the conformation of the base at G530, which is packed in the duplex [19]. Mismatches between mRNA and non-cognate tRNAs at the first two positions would distort the base-pairing geometry, consequently hindering hydrogen bonding between the mRNA–tRNA duplex and residues in 16S rRNA [19].

When AGs with 2-DOS cores interact with the 30S ribosomal subunit, they bind the major groove of h44 [20–23]. Ring I of the neamine core stacks against the base of G1491 and forms two hydrogen bonds with A1408 [22], a residue that is a key selectivity determinant between prokaryotic and eukaryotic ribosomes [24]. Amino and hydroxyl groups on the 2-DOS ring and on additional rings form direct or water-mediated hydrogen bonds with other RNA residues [22]. The interacting groups and number of interactions are variable and are dependent on the structure of the AG [22], but generally, the binding of these AGs stabilizes the extruded conformations of residues A1492 and A1493 that is observed with the docking of cognate tRNAs. This conformational change promotes mistranslation as it reduces the energetic cost for the binding of near-cognate tRNAs [20]. Streptomycin, which carries a streptamine ring instead of a 2-DOS ring, binds close to the binding site of 2-DOS AGs, but it interacts with other nucleotides in 16S rRNA and with amino acids in ribosomal protein S12 to elicit mistranslation [20]. By binding to ribosomes, 2-DOS AGs can also alter the energy barriers to 30S subunit rotations, resulting in substantial translocation defects, ribosome stalling, and impaired ribosome recycling [17, 25].

The release and insertion of misread proteins into the inner membrane of bacteria, which can disrupt membrane integrity, has been proposed as one of the reasons for the bactericidal activity of AGs [15, 26]. The insertion of mistranslated proteins in the membrane may also create channels in the membrane that further promote AG uptake, resulting in irreversible AG accumulation [15]. As AGs accumulate in bacteria, they can further saturate existing ribosomes and inhibit their biogenesis [27], which can contribute to AG-mediated killing. In aerobically cultured *E. coli*, the insertion and translocation of misfolded proteins into and across the inner membrane can stimulate envelope stress (Cpx) and redox (Arc) responsive regulators, which has been suggested to perturb genes involved in metabolism and respiration, resulting in hydroxyl radical formation and cell death [28].

The effect of AGs on cellular physiology is pleiotropic, and AG treatment can perturb cellular processes beyond protein synthesis. When *E. coli* was treated with a subinhibitory concentration of amikacin, impaired cell division was observed, which led to the production of elongated cells [29]. A closer examination revealed that the antibiotic affected the localization of FtsZ and Z-ring assembly, which hindered septation [29]. AGs have also been proposed to affect RNA stability and functions. In *in vitro* studies, certain AGs, especially neomycin B, have been shown to compete with the binding of divalent metal ions and inhibit the activities of ribozymes [30, 31]. AGs have also been demonstrated to inhibit metal ion binding or alter the secondary and tertiary structures of certain tRNAs or transfer-messenger RNAs (tm-RNAs), thereby preventing aminoacylation [32–34].

4.5 Bacterial Resistance against AGs

The soil is a rich source of AG-producing bacteria. For the same reason, it is also a reservoir of intrinsic resistance mechanisms, which protect AG producers from their own antibiotics and from those produced by other microbes colonizing the same environment [35–37]. Widespread clinical and agricultural use of AGs has fueled the selection for and dissemination of these resistance determinants. Within two years of its clinical introduction, streptomycin-resistant bacteria were isolated [35, 38]. The evolution and dissemination of a variety of resistance determinants to other naturally occurring or semisynthetic AGs were detected shortly after their clinical introduction [35].

Resistance involves acquiring heritable genetic changes that result in an increase in the minimal inhibitory concentration (MIC) of the antibiotic, which enables bacteria to grow at antibiotic concentrations that previously inhibited their propagation [39]. These changes enable bacteria to evade AG killing through diverse mechanisms, including: (i) enzymatic modification or inactivation of the drug, (ii) modification of cellular AG targets (such as 16S rRNA methylation), (iii) increase in AG efflux, and (iv) decrease in AG uptake across bacterial membranes.

4.5.1 Modification of AGs

Inactivation of AGs by enzymatic modification represents the most prevalent mechanism of AG resistance observed in Gram-positive and Gram-negative clinical isolates [2, 6]. The ESKAPE (*Enterococcus faecium*, *S. aureus*, *Klebsiella pneumonia*, *Acinetobacter baumannii*, *P. aeruginosa*, and *Enterobacter*) group of pathogens that are associated with nosocomial infections have also been found to carry AG-modifying genes [40]. Following modification of specific amino or hydroxyl groups, AGs bind poorly to ribosomes and fail to trigger EDP-II of uptake, consequently allowing the bacteria to survive the course of treatment [2]. These AG-modifying enzymes (AMEs) can act as AG *O*-phosphotransferases (APHs), AG *N*-acetyltransferases (AACs), or AG *O*-nucleotidyltransferases (ANTs) [2]. These three families of AMEs are further subdivided into classes based on the site of modification and into subclasses based on the resistance profile that the enzyme confers to its host. Each AG can potentially be modified by different subsets of AMEs (Figure 4.3). In this chapter, we will use the nomenclature proposed by Shaw and colleagues [41] when referring to these enzymes, where the three letter name of each enzyme is followed by a number in parentheses that designates its class and position of modification, a roman numeral that designates its subclass, and a lowercase letter that identifies enzymes that are encoded by unique genes. For example, AAC(3)-IIa, IIb, and IIc belong to a class of AG acetyltransferases that modify position 3 of the 2-DOS ring on their targets and confer resistance to gentamicin, tobramycin, sisomicin, netilmicin, and dibekacin, but are encoded by different genes [2].

AMEs catalyze the transfer of specific functional groups to their targets. APHs catalyze the transfer of a phosphate group from ATP or GTP to specific hydroxyl groups on their target AGs [3]. By introducing a negative charge to the molecule, it hinders binding to the ribosomal A site. Of the three families of AMEs, only APHs produce high levels of resistance [2]. With the exception of members in APH(2″) and APH(3′), which can modify multiple AGs, the other five classes of APHs confer resistance to only one AG. AACs catalyze the acetylation of amine groups using acetyl coenzyme A as a substrate [2]. Members of the AAC(6′) class are distributed among Gram-positive and Gram-negative bacteria. They are the most common of all AG-modifying enzymes [3] and can confer resistance to most clinically useful AGs. ANTs, which catalyze the transfer of AMP onto hydroxyl groups using ATP as a substrate, represent the smallest family of AMEs, but they can confer resistance to a number of clinically-important AGs, including

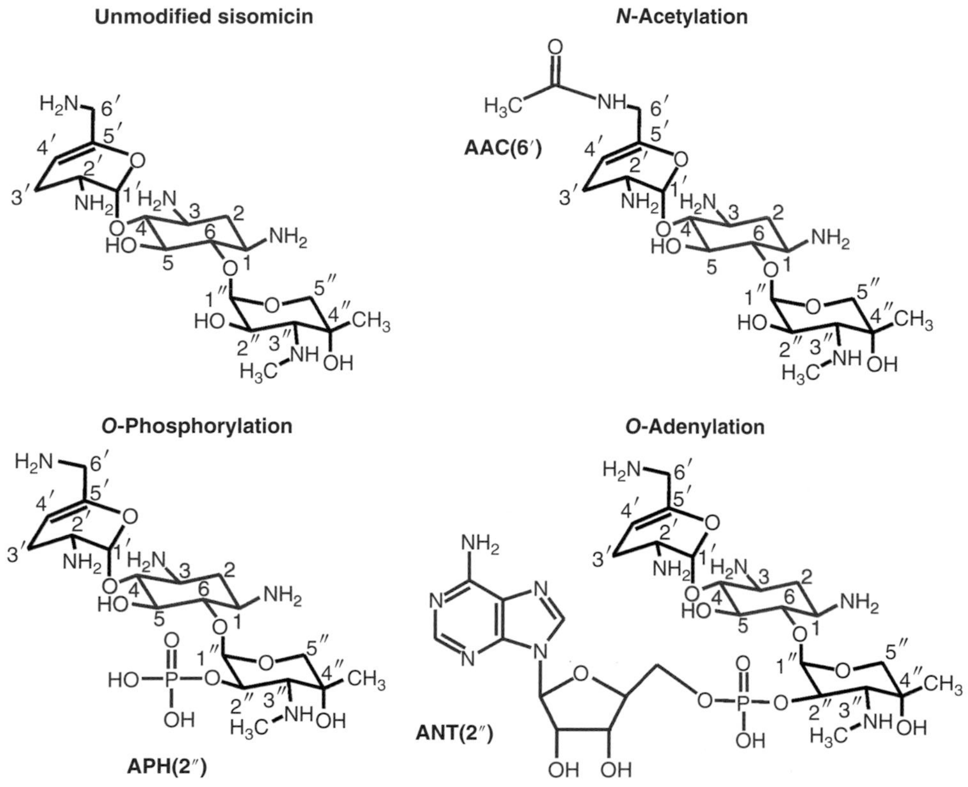

Figure 4.3 **AG modifications.** Individual AGs can be targeted by multiple AMEs. For example, sisomicin is sensitive to *N*-acetylation, *O*-phosphorylation, and *O*-adenylation. (*Source:* Structures generated using ACD/ChemSketch, version 12.00, Advanced Chemistry Development, Inc., Toronto, ON, Canada).

tobramycin, amikacin, gentamicin, and streptomycin [3]. Some AMEs exist as part of a fusion protein with bifunctional activity. For example, AAC(6′)-Ie-APH(2″)-Ia has both acetyltransferase and phosphotransferase activities [42]. This fusion enzyme confers resistance to most AGs except for streptomycin and arbekacin.

It has been proposed that AMEs can be acquired from an AG-producing strain or evolved from cellular metabolic enzymes, such as sugar kinases and acetyltransferases, as a result of selective pressure from AGs [41]. For certain classes of antibiotics, such as β-lactams, a single nucleotide change in the gene encoding the antibiotic-modifying enzyme can alter substrate specificity, readily conferring resistance toward newly introduced drugs [35]. For AGs, however, the most common mode of acquiring a new resistance gene is *via* horizontal gene transfer [35]. Most AMEs, including AAC(6′)-Ie-APH(2″)-Ia, are encoded on or near mobile genetic elements, including transposons, insertion sequences, and plasmids (both conjugative and non-conjugative), which facilitates their dissemination. For AAC(6′)-Ie-APH(2″)-Ia, which is a part of the composite transposon Tn*4001*, the enzyme has been distributed among streptococcal, staphylococcal, and enterococcal isolates [43].

4.5.2 Modification of the 16S rRNA

As an alternative to modifying the antibiotic, resistance to AGs can also be achieved through the accumulation of mutations in the 16S rRNA and ribosomal proteins [44]. To protect their ribosomes from intrinsic AGs, AG-producing *Actinomycetes* harbor genes encoding 16S rRNA methyltransferases (16S-RMTases), which are often encoded in AG biosynthetic gene clusters. 16S-RMTases catalyze the transfer of a methyl group from *S*-adenosylmethionine to AG-binding purines in h44 of the 16S rRNA that are involved in the aminoacyl-tRNA recognition site [44]. In doing so, they prevent the formation of hydrogen bonds between these residues and specific functional groups on AGs, reducing the binding affinities of these compounds to the ribosomal A-site. 16S-RMTases are classified into two subgroups based on the nucleotide and the position they modify. Specifically, N7-G1405 16S-RMTases modify the N7 position of residue G1405 and confer resistance toward AGs with 4,6-disubstituted DOS, but not to 4,5-disubstituted DOS, apramycin, and streptomycin [44]. N1-A1408 RMTases modify the N1 position of residue A1408 and confer resistance toward 4,5- and 4,6-disubstituted DOS as well as apramycin [44] (Figure 4.4). The sensitivities of the different classes of disubstituted DOS to the modifications can be predicted by the hydrogen bonds that

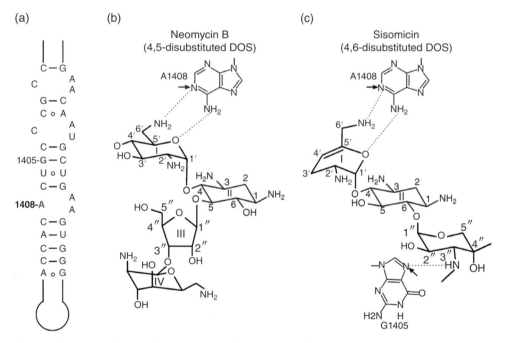

Figure 4.4 Targets of 16S-RMTases. (a) Secondary structure of h44 in the 16S rRNA. The two bases that are modified by 16S-RMTases, G1405 and A1408, are indicated in red. (b) 4,5-Disubstituted DOS like neomycin B interact with the N1 and N6 positions of A1408 (indicated by red lines) and are sensitive to modifications by N1-A1408 RMTases (whose site of modification is indicated by the arrow). (c) 4,6-Disubstituted DOS like sisomicin interact with N1 and N6 of A1408 and with N7 of G1405. Consequently, they are sensitive to both N1-A1408 and N7–1405 RMTases. (*Source:* Structures generated using ACD/ChemSketch, version 12.00, Advanced Chemistry Development, Inc., Toronto, ON, Canada).

form between their sugar rings and the methylated residues. Regardless of structural diversity, ring I of an AG always contacts N1 of A1408 [44]. However, while N7 of G1405 can form hydrogen bonds with ring III of 4,6-disubstituted DOS, the spatial locations of rings III and IV in 4,5-disubstituted DOS are too far for this interaction to occur, and this class of AGs are not affected by N7-G1406 16S RMTases [44].

In addition to AG producers, genes encoding 16S-RMTases have been identified in non-producer strains commonly found in clinical settings, including *P. aeruginosa*, *E. coli*, *K. pneumonia*, *A. baumannii*, and *Serratia marcescens* [44]. In these resistant mutants, 16S-RMTases have been found to confer high levels of resistance with MICs exceeding 512 mg/l toward many clinically important AGs [44] (compared with MICs ranging from 0.5 to 2 mg/l for AGs such as amikacin, kanamycin, tobramycin, and gentamicin in susceptible *P. aeruginosa* strains [45]). Except for NpmA identified in *E. coli* ARS3 isolate from Japan, which acts on N1-A1408, acquired 16S-RMTases belong to the N7-G1405 group [44]. Genes encoding pathogen-derived 16S-RMTases have lower G + C content compared with *Actinomycetes* genes (30–55% compared with 60% G + C), share low sequence similarity with methylases from AG-producers, and are mostly found in mobile genetic elements such as plasmids and transposons [44]. These properties suggest that pathogen-derived 16S-RMTases did not originate from *Actinomyces* AG producers and are readily disseminated *via* horizontal gene transfer [44].

Pathogenic bacteria that produce 16S-RMTases have been isolated worldwide, and a higher prevalence of RMT-positive isolates is observed in Asia [46]. RmtB and ArmA, two widely distributed 16S-RMTases, have been identified in human clinical specimens, household pets, and livestock, suggesting possible transmission of these resistance determinants between human and animals, as well as *via* the food chain [44]. Genes encoding exogenous 16S-RMTases often coexist with genes conferring resistance toward other classes of antibiotics in the same conjugative plasmids and transposable elements [44]. For instance, *rmtB* found on the Tn3 transposon are often associated with the TEM-1 β-lactamase gene and *qepA* encoding a quinolone efflux transporter [44]. Furthermore, several 16S-RMTases have been found to be producers of carbapenemases or extended-spectrum β-lactamases [47]. The coexistence of these genes sparks concern that the dissemination of these plasmids and transposable elements could accelerate the emergence of multidrug-resistant bacteria.

4.5.3   Increased AG Efflux

Another form of defense that bacteria can mount to protect against antibiotic treatment is to limit intracellular drug accumulation. One way this can be achieved is by actively expelling drugs using efflux pumps [48]. Efflux pumps are abundant in bacteria, representing ~6–18% of all transporters encoded in a given bacterial species, and their action can prevent the accumulation of toxic compounds in the microbe [48]. Toxic chemicals and antibiotics like AGs are pumped out of the cell in energy-dependent processes. Efflux pumps involved in AG expulsion use proton motive force (PMF), as opposed to pumps belonging to the ATP-binding cassette (ABC) class of transporters that use energy derived from ATP hydrolysis [48]. Efflux pumps involved in AG elimination are present in several clinically important pathogens, including *P. aeruginosa*, *A. baumannii*, and *Vibrio cholerae*.

Most efflux pumps contributing to AG efflux can expel multiple classes of antibiotics, and in bacteria that carry multidrug efflux pumps, mutations can be acquired to enhance their efficiency [48]. For instance, in *P. aeruginosa*, acquisition of a F1018L mutation in the MexY pump of the MexXY-OprM transporter enhances efflux of AGs, fluoroquinolones, and some β-lactams [49]. Furthermore, bacteria can also accumulate mutations in local or global regulators governing the expression of efflux pump-encoding genes to achieve increased abundance of pumps and greater resistance [48]. In one study, several resistant *P. aeruginosa* isolates from cystic fibrosis patients were found to overproduce MexXY-OprM as a result of mutations in *mexZ*, which encodes a repressor of the *mexXY* operon [50].

Multidrug efflux pumps can also be induced through adaptive resistance. Certain environmental conditions, such as salt concentration, pH, presence of antibiotics, and oxidative stress, have been shown to promote the expression of genes encoding multidrug efflux pumps [48]. The production of the MexXY transporter of the MexXY-OprM efflux pump can be induced by a number of ribosome-targeting agents, including AGs [51]. It is also induced by oxidative stress, which is a clinically relevant observation as *P. aerugionsa* would encounter reactive oxygen species during infection of chronically inflamed lungs of cystic fibrosis patients [52].

4.5.4 Decreased AG Uptake

Another strategy to limit intracellular AG accumulation is to prevent the rapid and irreversible uptake of AG; bacteria have devised ways to achieve this by modifying their membrane permeability. For instance, under conditions of Mg^{2+} limitation in *P. aeruginosa* (as observed in the presence of cation-chelating extracellular DNA that makes up biofilm matrices), the expression of the two-component systems PhoPQ and PmrAB are triggered [53]. They in turn activate a set of genes that modifies the phosphate groups on lipid A of LPS with aminosugars, thereby reducing their net negative charge [53–55]. Such modifications impair the electrostatic interactions between AGs and cell surfaces, which hinder uptake [56].

As AG uptake and its bactericidal activity are dependent on the presence of active and sensitive ribosomes and PMF, this class of antibiotics is inactive toward translationally stalled or energetically compromised bacteria, such as those undergoing fermentative metabolism [2]. Additionally, environmental stressors that slow growth or repress respiratory activity, such as nutrient limitation and exposure to nitric oxide (NO) [57, 58], can impair AG uptake.

4.6 Bacterial Persisters: Small Subpopulations that Can Tolerate AGs

The resistance mechanisms described earlier originate from heritable genetic changes that allow mutants to grow in the presence of a high concentration of an antibiotic and produce substantial increases in MIC toward the drug compared with the MIC observed for susceptible strains [39]. In addition to resistance, bacteria can survive a course of antibiotic treatment through tolerance, which can be achieved phenotypically or through the acquisition of mutations [39]. Unlike resistance, tolerance is transient and does not produce an increase in MIC toward the antibiotic [39]. Whereas resistance and tolerance can occur at the population-level, persistence is a phenomenon where small subpopulations of cells survive extraordinary concentrations of antibiotics that kill the rest of their kin [59, 60] (Figure 4.5a). The existence of persisters is observed when survival data exhibits an initial rapid killing regimen, which is attributed to the death of antibiotic-susceptible cells, followed by a second slower killing phase, which reflects the killing of persisters [59] (Figure 4.5a, panel II).

(a). Persistence

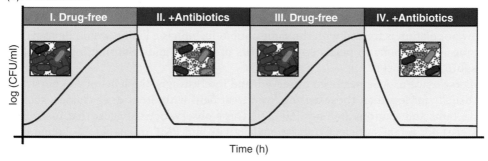

(b). Resistance

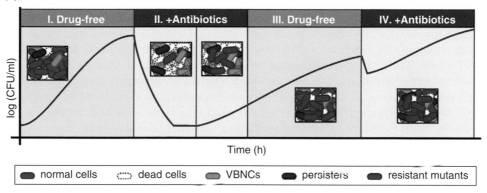

Figure 4.5 **Comparing persistence and resistance.** (a) Persistence. (I) Bacterial persisters (cells in red in the inset) are rare phenotypic variants in a population that also contains cells growing normally (in blue), dead cells (indicated by dashed lines), and viable-but-non-culturable cells (VBNCs; in orange). Persisters can form stochastically or in response to stress. (II) When the culture is treated with a bactericidal antibiotic, cells growing normally are killed, whereas persisters and VBNCs survive the course of treatment. The survival of persisters during treatment produces a second phase of killing (kinetic death-rate), which is depicted as a plateau here, though its slope need not be zero. CFU, colony-forming units. (III) Following antibiotic removal, persisters can resume growth and re-establish a population, which consists of persisters and non-persisters similar to the original population in panel I. (IV) Treatment of this new population would also result in biphasic killing. (b) Resistance. Contrary to persistence, antibiotic resistance arises from heritable genetic changes, which enable mutants to survive and grow during antibiotic treatment. When the original population (I) is treated with a bactericidal antibiotic (II), susceptible cells are killed, leaving behind persisters, VBNCs, and resistant mutants. While resistant mutants continue to grow during treatment, persisters resume growth only once the antibiotic is removed, which gives rise to wild-type cells (III). (IV) As the progeny of resistant mutants genetically retain resistance to the same antibiotic, they are unaffected by the next course of treatment, whereas wild-type cells that arose from persisters are killed.

Persisters have been found to be highly heterogeneous; a single culture can contain distinct persister subpopulations, which can form in response to unique stresses and have different tolerance mechanisms [61–64]. Persisters can arise from cells that are either growing or non-growing prior to antibiotic stress, but they arise at a higher frequency from those that are non-growing (in the same culture, ∼1% of non-growing cells compared with 0.01% of growing cells gave rise to persisters) [65]. During treatments, primary antibiotic targets, which are often essential growth processes, are impaired in persisters, thereby preventing them from antibiotic corruption [66], although exceptions exist [67]. Contrary to resistance (Figure 4.5b), persistence is a transient phenotypic state and cells revert to normal physiology, which allows them to repopulate environments once the antibiotic is removed [39] (Figure 4.5a, panel III). When this new population is treated with the same antibiotic, biphasic killing would be observed again, which points to persisters as culprits of chronic and relapsing infections [68] (Figure 4.5a, panel IV).

Persistence to AGs has been observed, and interestingly, the level of AG persisters is usually far less than those arising from treatment with other drug classes, such as β-lactams and fluoroquinolones [69–72]. These observations suggest that many persister states retain sufficient translational activity and PMF to enable AG killing [73]. However, if that is not the case, such as for persisters in stationary phase cultures, it has been shown that supplementation with specific metabolites can potentiate the activity of AGs [69, 71]. Despite this effectiveness against persisters, it was recently demonstrated that after two to three cycles of AG treatment, under conditions that are similar to the once-daily AG dosing regimen, point mutations that increased persistence by around three orders of magnitude started to emerge in an *E. coli* population [74]. In clones that evolved on amikacin, cross-tolerance to other AGs and to a fluoroquinolone was observed [74]. Similar increases in persistence have also been reported in planktonic cultures and biofilms of ESKAPE pathogens following repeated cycles of AG challenge [75]. These findings emphasize that even though treatment with AGs produce lower persister levels than other classes of antibioitcs, repeated courses of treatment run the risk of dramatically increasing peristence to AG in a short time.

4.7 Overcoming Bacterial Survival Mechanisms

Microbes have devised a variety of resistance and adaptive mechanisms to evade AG killing. To overcome their defenses, strategies to develop AG analogs that are resilient to modification, inhibitors that can act against AMEs, or adjuvants that can stimulate AG uptake are being explored [76]. Among the different methods to inhibit AMEs, modification of existing AGs into non-substrates has offered the most clinical success [76]. The structure of AGs can be redesigned so that it would release the enzyme-appended functional group spontaneously after the drug is modified, which would allow the molecule to self-regenerate [77]. AGs have also been designed so that they would generate intermediates that can irreversibly alter the active site of an AME once modified [2]. For example, by replacing the 2′ amine group with a nitro group in neamine and kanamycin B, the compounds can quickly eliminate phosphate groups once they are modified by APH(3′), resulting in the formation of reactive species that target the active site of APH(3′) [78]. Alternatively, functional groups can be modified to block the access of these enzymes.

Plazomicin (formerly referred to as ACHN-490) is a derivative of sisomicin that is currently in two phase III clinical trials for complicated urinary tract infection and pyelonephritis [46]. Plazomicin carries a hydroxyl-aminobutyric acid at the N1 position, which creates steric hindrance and offers protection against AAC [3], APH(2″), and ANT(2″) enzymes [46] (Figure 4.6). The presence of a hydroxyethyl group at position 6′ shields the amino group from AAC(6′) enzymes. Due to these modifications and the lack of 3′ and 4′ hydroxyl groups in sisomycin, which prevents modifications by ANT(4′, 4″) and APH(3′) enzymes, the only AG modifying enzyme that can elevate the MIC of plazomicin is AAC(2′). AAC(2′) is only found in the chromosome of *Providencia stuartii*, an opportunistic human pathogen, and is not known to have transferred to other species [46]. Importantly, the modifications on plazomicin do not decrease its bactericidal activity relative to sisomicin [46]. Plazomicin is active against multidrug-resistant Enterobacteriaceae, including carbapenem-resistant Enterobacteriaceae.

Another approach to mitigate AG resistance is to directly inhibit the expression or the enzymatic activity of AMEs. In the laboratory setting, antisense oligonucleotides have been employed to silence transcripts encoding AME [76]. In one example, nuclease-resistant locked nucleic acid (LNA)-DNA oligonucleotides, which consist of an antisense sequence that is specific to *aac(6′)-Ib* and a sequence that interacts with ribozyme RNaseP, were generated [79]. These oligonucleotides were able to target the *aac(6′)-Ib* mRNA for RNaseP cleavage *in vitro*. When they were administered to a hyperpermeable *E. coli* strain that carried *aac(6′)-Ib*, the LNA/DNA oligonucleotide inhibited growth in the presence of amikacin, suggesting that the expression of *aac(6′)-Ib* was dampened. However, the internalization of these oligonucleotides remains a hurdle impeding the application of this technique in clinical bacteria.

The crystal structures of several APHs with diverse substrate specificities and regio-specificities have been elucidated and used to guide the identification and design of potential inhibitors. These structures revealed that while the antibiotic-binding sites of APHs are diverse, their NTP binding sites are highly conserved, and they share structural

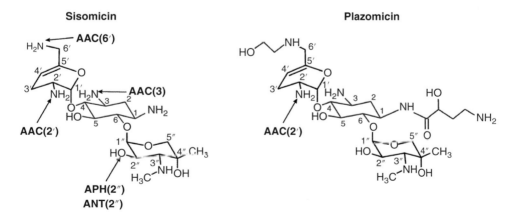

Figure 4.6 Structure and AME susceptibility of plazomicin. Plazomicin is a derivative of sisomicin, which has been modified with a hydroxylaminobutyric group at position N1 and a hydroxyethyl group at the 6′ position. These modifications protect plazomicin from AMEs that can modify sisomicin, with the exception of AAC(2′). (*Source:* Structures generated using ACD/ChemSketch, version 12.00, Advanced Chemistry Development, Inc., Toronto, ON, Canada).

similarities with eukaryotic protein kinases [80]. Structural similarities between the NTP binding site of APHs and eukaryotic protein kinases have prompted the screening of 80 known protein kinase inhibitors against 14 APHs [80]. The inhibitors examined in the screen displayed a range of affinities for APHs, as well as broad and narrow inhibition profiles. Subsequent structural analyses of two APHs in complex with kinase inhibitors with distinct structural scaffolds revealed differences in the binding mode of these inhibitors to APHs compared with their binding to eukaryotic protein kinases [80, 81]. This information can be used to guide the design of selective inhibitors that can be used as adjuvants to AG therapy. In addition to APHs, inhibitors against ANTs [82] and AACs [83] have been identified, as have inhibitors that can act on two mechanistically distinct enzymes [84]. Given the diversity of enzymes, more research will have to be conducted to develop inhibitors that can target multiple classes of AMEs.

The co-administration of adjuvants is a viable approach to prolong the efficacy of existing AGs [85]. The synergism between AGs and β-lactams has long been recognized, where the synergistic effect is attributed to the ability of β-lactam antibiotics to damage the cell membrane and increase AG permeability [86]. AG uptake and killing in bacterial persisters were found to be potentiated by sugars, amino acids, and changes in pH, demonstrating that compounds without antibacterial activity can be explored as adjuvants in AG therapy [69, 71, 87]. As an additional benefit, metabolite-enabled AG killing in persisters can be leveraged to probe persister metabolism [69, 88], which could aid in the discovery of novel methods to eradicate this subpopulation. Another way to increase intracellular AG accumulation is to prevent efflux, which can be achieved with the co-administration of efflux pump inhibitors. For instance, when the efflux pump inhibitor MP 601384, which has specificity toward AG-accommodating RND efflux systems, was used in conjunction with a panel of AGs, the MIC for susceptible and resistant isolates of opportunistic pathogen *Burkholderia vietnamiensis* decreased two to sixfold [89]. The interruption of regulators of efflux pumps offers another path toward enhancing AG sensitivity. Recently, it was found that rifampicin, an RNA polymerase inhibitor, can perturb the function of the AmgRS two-component system and undermine the expression of the MexXY efflux system, consequently potentiating 4,5-disubstituted DOS AG killing in *P. aeruginosa* [90].

4.8 Summary and Conclusion

The administration of AG antibiotics is an attractive solution to address pressing and mounting challenges imposed by multidrug-resistant pathogens and antibiotic failure. Most AGs available today are bactericidal and active against a broad spectrum of microbes [2]. Furthermore, they have been shown to be active against bacterial persisters that are tolerant to other classes of antibiotics, particularly fluoroquinolones and β-lactams, suggesting that AGs can be incorporated as a part of a treatment regime for chronic infections [71, 72].

AGs have been used clinically for over 70 years, and the safety and efficacy of this class of drugs toward different bacterial species have been well studied. Notably, the use of AGs is limited by their associated nephrotoxicity and ototoxicity [2]. To minimize their toxic effects, AGs can be reformulated to enable alternative routes of delivery, which can increase concentration at the site of infection, lower systemic absorption, and reduce serum drug concentration [91]. For instance, aerosolized gentamicin, tobramycin, and amikacin are already used as inhalants in cystic fibrosis patients [91]. AGs have also been encapsulated in nanoparticles [92] or liposomes [93] to enable controlled and site-specific release.

Despite the decline in their use in the past three decades, microbes still harbor diverse mechanisms to resist and tolerate AG therapy. Fortunately, analogs and adjuvants that can evade resistance mechanisms and enhance AG potency have shown promise in killing resistant mutants, tolerant populations, and persisters. Undoubtedly, further knowledge of how resistance and tolerance toward AG is achieved will be beneficial for devising ways to maximize the efficacy of this established class of antibiotics, which is increasingly important as the number of novel antibiotic classes approved for use continues to decline.

Acknowledgments

This work was supported in part by the Army Research Office (W911NF-15-1-0173), the Charles H. Revson Foundation, and Princeton University. The content is solely the responsibility of the authors and does not necessarily represent the official views of the funding agencies.

References

1 Schatz, A., Bugie, E., and Waksman, S. (1944). Streptomycin, a substance exhibiting antibiotic activity against gram-positive and gram-negative bacteria. *Proc. Soc. Exp. Biol. Med.* 55: 6669.

2 Vakulenko, S.B. and Mobashery, S. (2003). Versatility of aminoglycosides and prospects for their future. *Clin. Microbiol. Rev.* 16 (3): 430–450.

3 Becker, B. and Cooper, M.A. (2013). Aminoglycoside antibiotics in the 21st century. *ACS Chem. Biol.* 8 (1): 105–115.

4 Avent, M.L., Rogers, B.A., Cheng, A.C., and Paterson, D.L. (2011). Current use of aminoglycosides: indications, pharmacokinetics and monitoring for toxicity. *Intern. Med. J.* 41 (6): 441–449.

5 Marshall, B.M. and Levy, S.B. (2011). Food animals and antimicrobials: impacts on human health. *Clin. Microbiol. Rev.* 24 (4): 718–733.

6 Ramirez, M.S. and Tolmasky, M.E. (2010). Aminoglycoside modifying enzymes. *Drug Resist. Updat.* 13 (6): 151–171.

7 Craig, W.A. (1998). Pharmacokinetic/pharmacodynamic parameters: rationale for antibacterial dosing of mice and men. *Clin. Infect. Dis.* 26 (1): 1–10.

8 Wright, G.D., Berghuis, A.M., and Mobashery, S. (1998). Aminoglycoside antibiotics. Structures, functions, and resistance. *Adv. Exp. Med. Biol.* 456: 27–69.

9 Wright, G.D. and Thompson, P.R. (1999). Aminoglycoside phosphotransferases: proteins, structure, and mechanism. *Front. Biosci.* 4: D9–D21.

10 Taber, H.W., Mueller, J.P., Miller, P.F., and Arrow, A.S. (1987). Bacterial uptake of aminoglycoside antibiotics. *Microbiol. Rev.* 51 (4): 439–457.

11 Gilman, S. and Saunders, V.A. (1986). Accumulation of gentamicin by *Staphylococcus aureus*: the role of the transmembrane electrical potential. *J. Antimicrob. Chemother.* 17 (1): 37–44.

12 Bryan, L.E. and Van Den Elzen, H.M. (1975). Gentamicin accumulation by sensitive strains of *Escherichia coli* and *Pseudomonas aeruginosa. J. Antibiot. (Tokyo)* 28 (9): 696–703.

13 Bryan, L.E. and Van den Elzen, H.M. (1976). Streptomycin accumulation in susceptible and resistant strains of *Escherichia coli* and *Pseudomonas aeruginosa. Antimicrob. Agents Chemother.* 9 (6): 928–938.

14 Nichols, W.W. and Young, S.N. (1985). Respiration-dependent uptake of dihydrostreptomycin by *Escherichia coli*. Its irreversible nature and lack of evidence for a uniport process. *Biochem. J.* 228 (2): 505–512.

15 Davis, B.D. (1987). Mechanism of bactericidal action of aminoglycosides. *Microbiol. Rev.* 51 (3): 341–350.

16 Hurwitz, C., Braun, C.B., and Rosano, C.L. (1981). Role of ribosome recycling in uptake of dihydrostreptomycin by sensitive and resistant *Escherichia coli. Biochim. Biophys. Acta* 652 (1): 168–176.

17 Borovinskaya, M.A., Pai, R.D., Zhang, W. et al. (2007). Structural basis for aminoglycoside inhibition of bacterial ribosome recycling. *Nat. Struct. Mol. Biol.* 14 (8): 727–732.

18 Kramer, E.B. and Farabaugh, P.J. (2007). The frequency of translational misreading errors in *E. coli* is largely determined by tRNA competition. *RNA* 13 (1): 87–96.

19 Ramakrishnan, V. (2002). Ribosome structure and the mechanism of translation. *Cell* 108 (4): 557–572.

20 Carter, A.P., Clemons, W.M., Brodersen, D.E. et al. (2000). Functional insights from the structure of the 30S ribosomal subunit and its interactions with antibiotics. *Nature* 407 (6802): 340–348.

21 Fourmy, D., Recht, M.I., Blanchard, S.C., and Puglisi, J.D. (1996). Structure of the A site of *Escherichia coli* 16S ribosomal RNA complexed with an aminoglycoside antibiotic. *Science* 274 (5291): 1367–1371.

22 François, B., Russell, R.J., Murray, J.B. et al. (2005). Crystal structures of complexes between aminoglycosides and decoding A site oligonucleotides: role of the number of rings and positive charges in the specific binding leading to miscoding. *Nucleic Acids Res.* 33 (17): 5677–5690.

23 Hobbie, S.N., Pfister, P., Bruell, C. et al. (2006). Binding of neomycin-class aminoglycoside antibiotics to mutant ribosomes with alterations in the A site of 16S rRNA. *Antimicrob. Agents Chemother.* 50 (4): 1489–1496.

24 Recht, M.I., Douthwaite, S., and Puglisi, J.D. (1999). Basis for prokaryotic specificity of action of aminoglycoside antibiotics. *EMBO J.* 18 (11): 3133–3138.

25 Feldman, M.B., Terry, D.S., Altman, R.B., and Blanchard, S.C. (2010). Aminoglycoside activity observed on single pre-translocation ribosome complexes. *Nat. Chem. Biol.* 6 (3): 244.

26 Giudice, E. and Gillet, R. (2013). The task force that rescues stalled ribosomes in bacteria. *Trends Biochem. Sci.* 38 (8): 403–411.

27 Magnet, S. and Blanchard, J.S. (2005). Molecular insights into aminoglycoside action and resistance. *Chem. Rev.* 105 (2): 477–498.

28 Kohanski, M.A., Dwyer, D.J., Wierzbowski, J. et al. (2008). Mistranslation of membrane proteins and two-component system activation trigger antibiotic-mediated cell death. *Cell* 135 (4): 679–690.

29 Possoz, C., Newmark, J., Sorto, N. et al. (2007). Sublethal concentrations of the aminoglycoside amikacin interfere with cell division without affecting chromosome dynamics. *Antimicrob. Agents Chemother.* 51 (1): 252–256.

30 Mikkelsen, N.E., Brännvall, M., Virtanen, A., and Kirsebom, L.A. (1999). Inhibition of RNaseP RNA cleavage by aminoglycosides. *Proc. Natl. Acad. Sci. U. S. A.* 96 (11): 6155–6160.

31 Stage, T.K., Hertel, K.J., and Uhlenbeck, O.C. (1995). Inhibition of the hammerhead ribozyme by neomycin. *RNA* 1 (1): 95–101.

32 Mikkelsen, N.E., Johansson, K., Virtanen, A., and Kirsebom, L.A. (2001). Aminoglycoside binding displaces a divalent metal ion in a tRNA-neomycin B complex. *Nat. Struct. Biol.* 8 (6): 510–514.

33 Walter, F., Pütz, J., Giegé, R., and Westhof, E. (2002). Binding of tobramycin leads to conformational changes in yeast tRNA(Asp) and inhibition of aminoacylation. *EMBO J.* 21 (4): 760768.

34 Corvaisier, S., Bordeau, V., and Felden, B. (2003). Inhibition of transfer messenger RNA aminoacylation and trans-translation by aminoglycoside antibiotics. *J. Biol. Chem.* 278 (17): 14788–14797.

35 Davies, J. (1994). Inactivation of antibiotics and the dissemination of resistance genes. *Science* 264 (5157): 375–382.

36 Riesenfeld, C.S., Goodman, R.M., and Handelsman, J. (2004). Uncultured soil bacteria are a reservoir of new antibiotic resistance genes. *Environ. Microbiol.* 6 (9): 981–989.

37 D'Costa, V.M., McGrann, K.M., Hughes, D.W., and Wright, G.D. (2006). Sampling the antibiotic resistome. *Science* 311 (5759): 374–377.

38 Waksman, S.A., Reilly, H.C., and Schatz, A. (1945). Strain specificity and production of antibiotic substances: V. Strain resistance of Bacteria to antibiotic substances, especially to streptomycin. *Proc. Natl Acad. Sci. USA* 31 (6): 157–164.

39 Brauner, A., Fridman, O., Gefen, O., and Balaban, N.Q. (2016). Distinguishing between resistance, tolerance and persistence to antibiotic treatment. *Nat. Rev. Microbiol.* 14 (5): 320–330.

40 Ramirez, M.S., Nikolaidis, N., and Tolmasky, M.E. (2013). Rise and dissemination of aminoglycoside resistance: the *aac(6′)-Ib* paradigm. *Front. Microbiol.* 4: 121.

41 Shaw, K.J., Rather, P.N., Hare, R.S., and Miller, G.H. (1993). Molecular genetics of aminoglycoside resistance genes and familial relationships of the aminoglycoside-modifying enzymes. *Microbiol. Rev.* 57 (1): 138–163.

42 Chow, J.W., Kak, V., You, I. et al. (2001). Aminoglycoside resistance genes *aph(2″)-Ib* and *aac(6′)- Im* detected together in strains of both *Escherichia coli* and *Enterococcus faecium*. *Antimicrob. Agents Chemother.* 45 (10): 2691–2694.

43 Vakulenko, S.B., Donabedian, S.M., Voskresenskiy, A.M. et al. (2003). Multiplex PCR for detection of aminoglycoside resistance genes in enterococci. *Antimicrob. Agents Chemother.* 47 (4): 1423–1426.

44 Wachino, J. and Arakawa, Y. (2012). Exogenously acquired 16S rRNA methyltransferases found in aminoglycoside-resistant pathogenic gram-negative bacteria: an update. *Drug Resist. Updat.* 15 (3): 133–148.

45 Andrews, J.M. (2001). Determination of minimum inhibitory concentrations. *J. Antimicrob. Chemother.* 48 (Suppl 1): 5–16.

46 Armstrong, E.S. and Miller, G.H. (2010). Combating evolution with intelligent design: the neoglycoside ACHN-490. *Curr. Opin. Microbiol.* 13 (5): 565–573.

47 O'Hara, J.A., McGann, P., Snesrud, E.C. et al. (2013). Novel 16S rRNA methyltransferase RmtH produced by *Klebsiella pneumoniae* associated with war-related trauma. *Antimicrob. Agents Chemother.* 57 (5): 2413–2416.

48 Fernández, L. and Hancock, R.E. (2012). Adaptive and mutational resistance: role of porins and efflux pumps in drug resistance. *Clin. Microbiol. Rev.* 25 (4): 661–681.

49 Vettoretti, L., Plésiat, P., Muller, C. et al. (2009). Efflux unbalance in *Pseudomonas aeruginosa* isolates from cystic fibrosis patients. *Antimicrob. Agents Chemother.* 53 (5): 1987–1997.

50 Vogne, C., Aires, J.R., Bailly, C. et al. (2004). Role of the multidrug efflux system MexXY in the emergence of moderate resistance to aminoglycosides among *Pseudomonas aeruginosa* isolates from patients with cystic fibrosis. *Antimicrob. Agents Chemother.* 48 (5): 1676–1680.

51 Jeannot, K., Sobel, M.L., El Garch, F. et al. (2005). Induction of the MexXY efflux pump in *Pseudomonas aeruginosa* is dependent on drug-ribosome interaction. *J. Bacteriol.* 187 (15): 5341–5346.

52 Fraud, S. and Poole, K. (2011). Oxidative stress induction of the MexXY multidrug efflux genes and promotion of aminoglycoside resistance development in *Pseudomonas aeruginosa*. *Antimicrob. Agents Chemother.* 55 (3): 1068–1074.

53 Mulcahy, H., Charron-Mazenod, L., and Lewenza, S. (2008). Extracellular DNA chelates cations and induces antibiotic resistance in *Pseudomonas aeruginosa* biofilms. *PLoS Pathog.* 4 (11): e1000213.

54 Macfarlane, E.L., Kwasnicka, A., and Hancock, R.E. (2000). Role of *Pseudomonas aeruginosa* PhoP- phoQ in resistance to antimicrobial cationic peptides and aminoglycosides. *Microbiology* 146 (Pt 10): 2543–2554.

55 Miller, A.K., Brannon, M.K., Stevens, L. et al. (2011). PhoQ mutations promote lipid A modification and polymyxin resistance of *Pseudomonas aeruginosa* found in colistin-treated cystic fibrosis patients. *Antimicrob. Agents Chemother.* 55 (12): 5761–5769.

56 Hancock, R.E., Farmer, S.W., Li, Z.S., and Poole, K. (1991). Interaction of aminoglycosides with the outer membranes and purified lipopolysaccharide and OmpF porin of *Escherichia coli*. *Antimicrob. Agents Chemother.* 35 (7): 1309–1314.

57 Gusarov, I., Shatalin, K., Starodubtseva, M., and Nudler, E. (2009). Endogenous nitric oxide protects bacteria against a wide spectrum of antibiotics. *Science* 325 (5946): 1380–1384.

58 McCollister, B.D., Hoffman, M., Husain, M., and Vázquez-Torres, A. (2011). Nitric oxide protects bacteria from aminoglycosides by blocking the energy-dependent phases of drug uptake. *Antimicrob. Agents Chemother.* 55 (5): 2189–2196.

59 Gefen, O. and Balaban, N.Q. (2009). The importance of being persistent: heterogeneity of bacterial populations under antibiotic stress. *FEMS Microbiol. Rev.* 33 (4): 704–717.

60 Lewis, K. (2010). Persister cells. *Annu. Rev. Microbiol.* 64: 357–372.

61 Allison, K.R., Brynildsen, M.P., and Collins, J.J. (2011). Heterogeneous bacterial persisters and engineering approaches to eliminate them. *Curr. Opin. Microbiol.* 14 (5): 593–598.

62 Amato, S.M. and Brynildsen, M.P. (2015). Persister heterogeneity arising from a single metabolic stress. *Curr. Biol.* 25 (16): 2090–2098.

63 Ma, C., Sim, S., Shi, W. et al. (2010). Energy production genes *sucB* and *ubiF* are involved in persister survival and tolerance to multiple antibiotics and stresses in *Escherichia coli*. *FEMS Microbiol. Lett.* 303 (1): 33–40.

64 Zhang, Y. (2014). Persisters, persistent infections and the Yin-Yang model. *Emerg. Microbes Infect.* 3 (1): e3.

65 Orman, M.A. and Brynildsen, M.P. (2013). Dormancy is not necessary or sufficient for bacterial persistence. *Antimicrob. Agents Chemother.* 57 (7): 3230–3239.

66 Maisonneuve, E. and Gerdes, K. (2014). Molecular mechanisms underlying bacterial persisters. *Cell* 157 (3): 539–548.

67 Völzing, K.G. and Brynildsen, M.P. (2015). Stationary-phase persisters to ofloxacin sustain DNA damage and require repair systems only during recovery. *MBio* 6 (5): e0073100715.

68 Cohen, N.R., Lobritz, M.A., and Collins, J.J. (2013). Microbial persistence and the road to drug resistance. *Cell Host Microbe* 13 (6): 632–642.

69 Orman, M.A. and Brynildsen, M.P. (2013). Establishment of a method to rapidly assay bacterial persister metabolism. *Antimicrob. Agents Chemother.* 57 (9): 4398–4409.

70 Orman, M.A. and Brynildsen, M.P. (2015). Inhibition of stationary phase respiration impairs persister formation in *E. coli. Nat. Commun.* 6: 7983.

71 Allison, K.R., Brynildsen, M.P., and Collins, J.J. (2011). Metabolite-enabled eradication of bacterial persisters by aminoglycosides. *Nature* 473 (7346): 216–220.

72 Luidalepp, H., Jõers, A., Kaldalu, N., and Tenson, T. (2011). Age of inoculum strongly influences persister frequency and can mask effects of mutations implicated in altered persistence. *J. Bacteriol.* 193 (14): 3598–3605.

73 Mok, W.W., Park, J.O., Rabinowitz, J.D., and Brynildsen, M.P. (2015). RNA futile cycling in model persisters derived from MazF accumulation. *MBio* 6 (6).

74 Van den Bergh, B., Michiels, J.E., Wenseleers, T. et al. (2016). Frequency of antibiotic application drives rapid evolutionary adaptation of *Escherichia coli* persistence. *Nat. Microbiol.* 1: 16020.

75 Michiels, J.E., Van den Bergh, B., Verstraeten, N. et al. (2016). In vitro emergence of high persistence upon periodic aminoglycoside challenge in the ESKAPE pathogens. *Antimicrob. Agents Chemother.* 60 (8): 4630–4637.

76 Labby, K.J. and Garneau-Tsodikova, S. (2013). Strategies to overcome the action of aminoglycoside – modifying enzymes for treating resistant bacterial infections. *Future Med. Chem.* 5 (11): 1285–1309.

77 Haddad, J., Vakulenko, S., and Mobashery, S. (1999). An antibiotic cloaked by its own resistance enzyme. *J. Am. Chem. Soc.* 121: 11922–11923.

78 Roestamadji, J., Grapsas, I., and Mobashery, S. (1995). Mechanism-based inactivation of bacterial aminoglycoside 3′-phosphotransferases. *J. Am. Chem. Soc.* 117: 80–84.

79 Soler Bistué, A.J., Mart'n, F.A., Vozza, N. et al. (2009). Inhibition of *aac(6′)-Ib*-mediated amikacin resistance by nuclease-resistant external guide sequences in bacteria. *Proc. Natl Acad. Sci. USA* 106 (32): 13230–13235.

80 Shakya, T., Stogios, P.J., Waglechner, N. et al. (2011). A small molecule discrimination map of the antibiotic resistance kinome. *Chem. Biol.* 18 (12): 1591–1601.

81 Stogios, P.J., Spanogiannopoulos, P., Evdokimova, E. et al. (2013). Structure-guided optimization of protein kinase inhibitors reverses aminoglycoside antibiotic resistance. *Biochem. J.* 454 (2): 191–200.

82 Hirsch, D.R., Cox, G., D'Erasmo, M.P. et al. (2014). Inhibition of the ANT(2″)-Ia resistance enzyme and rescue of aminoglycoside antibiotic activity by synthetic α-hydroxytropolones. *Bioorg. Med. Chem. Lett.* 24 (21): 4943–4947.

83 Chiem, K., Jani, S., Fuentes, B. et al. (2016). Identification of an inhibitor of the aminoglycoside 6′-N-acetyltransferase type Ib [AAC(6′)-Ib] by glide molecular docking. *Medchemcomm* 7 (1): 184–189.

84 Welch, K.T., Virga, K.G., Whittemore, N.A. et al. (2005). Discovery of non-carbohydrate inhibitors of aminoglycoside-modifying enzymes. *Bioorg. Med. Chem.* 13 (22): 6252–6263.

85 Bollenbach, T. (2015). Antimicrobial interactions: mechanisms and implications for drug discovery and resistance evolution. *Curr. Opin. Microbiol.* 27: 1–9.

86 Plotz, P.H. and Davis, B.D. (1962). Synergism between streptomycin and penicillin: a proposed mechanism. *Science* 135 (3508): 1067–1068.

87 Lebeaux, D., Chauhan, A., Létoffé, S. et al. (2014). pH-mediated potentiation of aminoglycosides kills bacterial persisters and eradicates *in vivo* biofilms. *J. Infect. Dis.* 210 (9): 1357–1366.

88 Mok, W.W., Orman, M.A., and Brynildsen, M.P. (2015). Impacts of global transcriptional regulators on persister metabolism. *Antimicrob. Agents Chemother.* 59 (5): 2713–2719.

89 Jassem, A.N., Zlosnik, J.E., Henry, D.A. et al. (2011). In vitro susceptibility of *Burkholderia vietnamiensis* to aminoglycosides. *Antimicrob. Agents Chemother.* 55 (5): 2256–2264.

90 Poole, K., Gilmour, C., Farha, M.A. et al. (2016). Potentiation of aminoglycoside activity in *Pseudomonas aeruginosa* by targeting the AmgRS envelope stress-responsive two-component system. *Antimicrob. Agents Chemother.* 60 (6): 3509–3518.

91 Ratjen, F., Brockhaus, F., and Angyalosi, G. (2009). Aminoglycoside therapy against *Pseudomonas aeruginosa* in cystic fibrosis: a review. *J. Cyst. Fibros.* 8 (6): 361–369.

92 Imbuluzqueta, E., Gamazo, C., Lana, H. et al. (2013). Hydrophobic gentamicin-loaded nanoparticles are effective against *Brucella melitensis* infection in mice. *Antimicrob. Agents Chemother.* 57 (7): 3326–3333.

93 Schiffelers, R., Storm, G., and Bakker-Woudenberg, I. (2001). Liposome-encapsulated aminoglycosides in pre-clinical and clinical studies. *J. Antimicrob. Chemother.* 48 (3): 333–344.

5

Tetracyclines

Mode of Action and their Bacterial Mechanisms of Resistance

Marilyn C. Roberts

Department of Environmental and Occupational Health Sciences, University of Washington, Seattle, WA, USA

5.1 Introduction of Tetracyclines

Tetracyclines are one of the oldest classes of antibiotics used and the first broad spectrum class of antibiotics. Tetracyclines interact with the bacterial ribosomes by reversibly attaching to the ribosome that blocks protein synthesis. The first generation compounds, chlortetracycline discovered in 1945 followed by oxytetracycline in 1950, and tetracycline in 1953. Chlortetracycline was first isolated from the soil microbe *Streptomyces aureofaciens* in 1948 [1]. The core structure includes a four-ring system labeled A–D where rings A–C include saturated carbon centers and ring D is aromatic. Bacterial resistance of the first generation lead to the development of chemically modified derivatives with better pharmacological properties, including doxycycline in 1967 and minocycline in 1972. Meta-substitution of ring D resulted in the third generation of tetracycline, tigecycline, which was approved by Food and Drug Administration (FDA) in 2005 and Europe in 2006 [2].

Minocycline has better pharmacokinetic profiles and a longer half-life than tetracycline. It has excellent tissue penetration and high levels of bioavailability and is now more often used than tetracycline in human medicine. Other derivatives such as oxytetracycline are often used in animals. Tetracyclines have been prescribed at low dosage for long term use (>10 years) for acne vulgaris and is generally well tolerated. Tetracyclines are able to pass through the blood–brain barrier and thus can be used to treat central nervous system (CNS) infections. The most common side effects of nausea, vertigo, and mild dizziness may disappear when the antibiotics are no longer used. If used long term (more than six months), monitoring every three months for liver toxicity, pigmentation, and systemic lupus erythematosus is recommended. Both first and second generation tetracyclines have good safety records and, in general, these drugs have limited major adverse side effects, though they should not be taken by people who are exposed to the sun for long periods, growing children, or pregnant women. Tetracyclines may interfere with some medications including birth control pills. Tetracyclines are an

Bacterial Resistance to Antibiotics – From Molecules to Man, First Edition.
Edited by Boyan B. Bonev and Nicholas M. Brown.
© 2020 John Wiley & Sons Ltd. Published 2020 by John Wiley & Sons Ltd.

important group of antibiotics for treatment and prevention, including treating exposure to weaponized bacterial pathogens.

Third generation semi-synthetic glycylcycline tigecycline is an analog of minocycline but inhibits bacterial translation by binding to the 30S ribosomal subunit and blocking the entry of amino-acyl to RNA molecules into the A site of the ribosome. This is a different mechanism of action than either the first or second generation tetracyclines and thus bacteria which carry active efflux or ribosomal protection tetracycline acquired genes are usually much more susceptible than they are with other tetracyclines [3]. However, mutations to existing *tet* genes can increase resistance to tigecycline.

Tetracyclines are a broad spectrum group of antibiotics that have activity against a wide range of Gram-negative and Gram-positive bacteria, including those causing anthrax, intercellular chlamydiae and rickettsiae, cell-wall free mycoplasma, community-acquired pneumonia, Lyme disease, cholera, syphilis, *Yersinia pestis*, and periodontal infections. More recently, filarial parasite species such as *Brugia malayi*, *Dirofilaria* spp., *Loa loa*, *Onchocerca volvulus*, *Wuchereria bancrofti*, and others that harbor the endosymbiotic bacterium of the genus *Wolbachia* spp. can be successfully treated with doxycycline even if the parasite is resistant to anti-parasitic drugs. The antibiotic stops microfilarial production by eliminating the endosymbiont bacteria which leads to sterilization of the adult female worm [4].

Tetracyclines are bacteriostatic antibiotics, which means under laboratory (*in vitro*) conditions bacteria are no longer able to replicate but in general are not killed and if the antibiotic is washed out the bacteria are able to grow. The antibiotic must enter the bacterial cell in concentrations that are able to bind to the bacterial 30S ribosomal subunit of the mRNA translation complex. The tetracyclines allosterically inhibit binding of the amino-acyl tRNA to the growing peptide and which at high concentrations within the cell inhibits protein synthesis. The use of tetracycline therapy has declined with the increase in tetracycline resistant bacteria. Resistance is primarily due to the acquisition of new genes, most of which code for active efflux of the tetracyclines or a protein that protects the ribosome from the action of tetracycline. Bacterial resistance to tetracycline will be discussed in length later in this chapter.

Forty years after the discovery of the first tetracycline it was reported that tetracyclines were able to modulate the host response including inhibition of pathologic matrix metalloproteinase activity. Tetracyclines also have non-antibiotic properties including anti-inflammatory, and antiapoptotic activities, inhibitory effects on proteolysis, angiogeneis and tumor metastasis, and have been used for long-term treatment of rosacea and other skin conditions not thought to be due to bacteria. There are a list of other noninfectious diseases and conditions that tetracycline can be used for and a review on the non-antibiotics properties of tetracycline are detailed in [5]. A randomized double-blind study found that minocycline given to children and adolescents with fragile X syndrome had greater global improvement than placebo treatment over a three month treatment period [6]. What the antibiotic is doing to improve these children is not clear but is of great interest. Early studies suggested that tetracyclines may inhibit cancer cell growth. Exploration of tetracyclines to expand activity to cancer targets have been carried out [7] and there has been one randomized phase II trial [8].

Over the past 70 years the number of tetracycline resistant pathogenic, opportunistic, and commensal bacteria has increased as has the identification of tetracycline resistant environmental bacteria. In environments associated with human impact such as farms and sewage treatment plants, a number of different tetracycline resistant new species and genera have been identified [9, 10] that may persist and act as potential reservoirs for transmission of specific tetracycline resistant bacteria and/or *tet* genes to humans, animals, and other environments.

A major non-human use of tetracycline in the USA and some other countries is as growth promoters. The antibiotic is given to animals at subtherapeutic levels in their feed. In the 1940s, it was discovered that *Streptomyces auerofaciens* biomass remaining after fermentation had a growth promotion effect on chickens. It was thought that the growth promotion effect was due to low levels of chlortetracycline left in the mixture and this lead to the development and wide use of antibiotics, especially tetracyclines, as antibiotic feed additives. The levels used were at subthera-peutic levels, often for the life span of the animal. These levels are now known to efficiently select for increased drug resistant normal flora within the animal as well as drug resistance in the surrounding environmental microbiota. This practice has becoming increasingly controversial because using low levels of antibiotics are not necessary for growth promotion and that the practice contributes to the emergence of antibiotic resistance in both animal and human bacteria, which in turn can reduce the therapeutic options for disease treatment. Because of this concern, the European Union has banned the use of antibiotics for growth promotion, which has resulted in significant reductions of antibiotic use in agriculture in these countries, without a significant loss in meat production [11].

In June 2015, the US FDA published a final rule, known as the Veterinary Feed Directive (VFD), that extends the use of VFDs to an increased number of medically important antimicrobials used in food animal production (FDA FACT SHEET: VFD Final Rule and Next Steps. [cited 2017 May 30] https://www.fda.gov/AnimalVeterinary/ DevelopmentApprovalProcess/ucm449019.htm). The rule became effective on October 1, 2015, and may have impacted use and prescribing of medically important antibiotics in food animals in the years prior to implementation and will going forward. In the EU evidence supporting this idea is derived from the experience of the EU. On July 1, 1989, an EU-wide ban on the use of four growth-promoting antibiotics, spiramycin, tylosin, bacitracin zinc, and virginiamycin, came into effect. The result of this ban was a dramatic drop in the sales of antimicrobial growth-promoting agents.

5.2 Bacterial Resistance

5.2.1 Mechanism of Bacterial Resistance

Antibiotic resistance genes have co-evolved with antibiotic producing organisms in the soil and in other environments, with many genes originating in antibiotic producing microbes. Similar *tet* resistance genes can be found in these tetracycline producers. However, there is also good evidence to indicate that antibiotic resistant bacteria and genes pre-dates the use due to human activities, though it is clear that human activities have greatly increased the overall burden of drug resistance in the bacterial populations. This is illustrated by the study of Hughes and Datta [12] which found that from a collection of *Enterobacteriaceae*, isolated between 1917 and 1954, 24% carried conjugative plasmids but only 2% were tetracycline resistant and all isolates were from the genera *Proteus*. None of the *Salmonella, Shigella, Escherichia,* or *Klebsiella* isolates were positive for tetracycline resistance (Tc^r). However by the mid-1950s, Tc^r and multi-drug resistant *Escherichia coli* and *Shigella* were described which were later determined be due to the presence of plasmid-mediated antibiotic resistance [13]. A lack of tetracycline resistance genes was also found in early enterococci [14] and *Neisseria gonorrhoeae* [15]. All these studies indicate that increased antibiotic resistance genes over time corresponds with the increased use of antibiotic in the past 70 years.

Bacteria can become resistant to antibiotics in a variety of ways including: (i) random chromosomal mutations that lead to changes that alter the protein's properties or in some cases prevents expression of a protein; (ii) acquisition of new DNA by natural transformation often creating mosaic genes which are composed of the host's bacterial and foreign DNA; (iii) acquisition of new DNA on mobile elements usually by conjugation but occasionally by transformation. Some species and genera of bacteria are able to participate in natural transformation. These bacteria have receptors that recognize specific DNA fragments, normally from related strains or species, and are usually taken into the cell then integrate the foreign DNA into their own genomes, where it can be expressed. The DNA may be parts of genes, complete genes, or defined mobile elements. Some species of bacteria are able to acquire foreign DNA by using a bacteria phage for transmission which packages host bacterial DNA into the phage protein coat which, when injected into a new host bacterium, allows the foreign DNA to be incorporated into the host's genome [16].

The most common and normally the most important way bacteria become antibiotic resistant is by acquisition of new genes associated with mobile elements (plasmids, transposons, and integrons). Conjugation is the most common way tetracycline resistant genes are acquired or moved through bacterial communities. Mobile elements allow multiple genes, which may code for other antibiotic resistance genes, virulence factors, toxins, metal resistance, and/or use of alterative carbon sources, to be linked and selection of any one gene on the element will select for the entire element. This is thought to be one reason why genes that code for resistance to a particular antibiotic are maintained in the bacterial population even when the antibiotic is no longer being used. Mobile elements are the main driving force in horizontal gene transfer with and between ecosystems whether associated with humans, animals, and/or the environment. If a particular gene is part of a broad host range mobile element it is much more likely to spread through bacterial ecosystems.

Conjugative gene exchange is the main process of horizontal gene transfer between bacteria strains, species, and genera. It is a key element in bacterial evolution and the spread of antibiotic resistance genes within and between different ecosystems and their associated bacteria. Conjugation is primarily responsible for the lateral transfer of most antibiotic resistance genes. Conjugation can lead to a rapid dissemination of antibiotic resistance genes within and between bacterial communities and between different ecosystems. Conjugation also allows DNA to be directly transferred from one living bacterium to another by direct cell-to-cell contact. Conjugation can occur between the same or different strains, species and/or genera of bacteria and it is clear that once a gene is in a bacterial community it can be spread to many unrelated ecosystems. Conjugation can also occur between bacteria and eukaryotic cells [17].

Today, most *tet* genes are associated with conjugative, nonconjugative, and mobilizable plasmids, transposons, and conjugative transposons (CTns). It is currently assumed that mobility and the type of element that a specific *tet* gene is associated with directly influences the *tet* genes host range and ability to spread to new genera and multiple ecosystems. Yet some *tet* genes (such as the *tet*(E) gene, which do not appear to be associated with mobile elements, though found on large non-conjugative plasmids) have been identified in multiple genera from a variety of different ecosystems around the world [18, 19].

Mobile elements include plasmids, CTns, transposons, and integrons. All of these elements have been involved in lateral gene exchange between strains, species, genera, from one ecosystem to another, and between humans and the environment and *vice versa*. These elements are able to carry a variety of different antibiotic resistance genes, as well as genes that confer resistance to disinfectants and heavy metals, genes that produce toxins and virulence factors, and/or genes for using alternative energy sources. All the genes carried on one of these elements may influence the carriage of the entire mobile unit and/or help maintain the genes in the bacterial population [20]. Because the various *tet* genes are found in all these different genetic elements (except for integrons) they have been linked to a variety of other genes (see later). It is likely that new mobile elements will continue to be described allowing different *tet* genes to be linked with other genes.

Plasmids were first described in the early 1950s and most are circular DNA molecules that range in size and can replicate independently of the bacterial chromosome [13] or became integrated into the chromosome [21]. The first Tc^r bacteria were identified in Japan and carried conjugative plasmids with multiple resistance genes [13]. Plasmids may include transposons and/or integrons and carry a variety of different antibiotic and/or heavy metals resistance genes, genes that code for toxins, genes that code for the degradation of various compounds, as alternative carbon sources and/or code for other virulence factors. These genes normally move as a unit from bacterium to bacterium and selection for any one gene may allow all the genes to be maintained in the population [18].

A total of 61 distinct *tet* genes have now been identified and shown to confer tetracycline resistance in their host (Table 5.1). This includes 13 *tet* genes that code for enzymes that break tetracycline, 34 *tet* genes that code for efflux that pumps tetracycline out of the cell, 13 ribosomal protection genes that protect the ribosomes from the action of tetracyclines, and one gene with an unknown resistance mechanism. Many of these *tet* genes are associated with plasmids, transposons, and CTns that may be integrated into the host chromosomes or on plasmids. Currently, 136 different genera have been identified that carry one or more *tet* genes. The most widely distributed *tet* gene is the ribosomal protection gene, *tet*(M), which has been identified in 80 different genera (41 Gram-positive and 39 Gram-negative genera). This wide host range is thought to be due to *tet*(M) association with wide-host range CTns that allow for a much wider host range than other *tet* genes (range 1–46 genera) (See Tables 5.2 and 5.3). Under laboratory conditions the *tet*(M) gene can be transferred to both Gram-positive and Gram-negative bacteria, while in general plasmids have a much more restricted host range. Gram-positive *tet* genes are found in both Gram-positive and Gram-negative bacteria, while the Gram-negative *tet* genes are not found naturally in Gram-positive bacteria (Tables 5.2 and 5.3). This is not unique to the *tet* genes since similar distributions are found with genes conferring antibiotic resistance to other classes of antibiotics such as macrolides ([22], http://faculty.washington.edu/marilynr).

Table 5.1 Mechanism of resistance for characterized *tet* and *otr* genes.

Efflux (34)	Ribosomal Protection (13)	Enzymatic (13)	Unknown (1)
tet(A), *tet*(B), *tet*(C), *tet*(D), *tet*(E), *tet*(59)	*tet*(M), *tet*(O), *tet*(S), *tet*(W), *tet*(32)	*tet*(X)	*tet*(U)
tet(G), *tet*(H), *tet*(J), *tet*(V), *tet*(Y)	*tet*(Q), *tet*(T), *tet*(36), *tet*(61)	*tet*(37)	
tet(Z), *tet*(30), *tet*(31), *tet*(33), *tet*(57)	*otr*(A), *tetB*(P), *tet*	*tet*(34)	
tet(35)	*tet*(44)	*tet*(47), *tet*(48), *tet*(49), *tet*(50)	
tet(39), *tet*(41)		*tet*(51), *tet*(52) *tet*(53), *tet*(54)	
tet(K), *tet*(L), *tet*(38), tet(45), tet(58)		*tet*(55), *tet*(56)	
tetA(P), *tet*(40)			
otr(B), *otr*(C)			
tcr3			
tet(42)	**Mosiac Ribosomal Protection (11)**		
tet(43)	*tet*(O/32/O, *tet*(O/W/32/O), *tet*(O/32/O		
tetAB(46)[a]	*tet*(O/W/32/O/W/O), *tet*(W/32/O), *tet*(O/W)		
tetAB(60)[b]	*tet*(W/32/O/W/O), *tet*(O/W/O), *tet*(O/W/32/O)		
tet(62)	*tet*(S/M), *tet*(W/N/W)		

[a] Representing two different genes.
[b] Representing two different genes both needed for resistance.
Source: http://faculty.washington.edu/marilynr.

The *tet* plasmids occur in a variety of sizes, from the small 4.45 kb *tet*(K) positive *Staphylococcus aureus* pT181 plasmid to large plasmids of ≥300 kb. Many of the large *tet* plasmids are conjugative or capable of mobilization; however, the *tet*(E) gene has been found on large plasmids ~170 kb that do not demonstrate these tendencies under laboratory conditions [23, 24]. Even though mobility has not been demonstrated under laboratory conditions, the *tet*(E) gene has been identified in 11 different Gram-negative genera, many of which are associated with water and/or aquaculture from around the world (Table 5.2). How the *tet*(E) gene has spread between bacteria in nature is a mystery. There distribution indicates that we cannot yet reproduce all the possible mechanisms of gene exchange that occurs in nature under laboratory conditions [23, 24].

Transposons are discrete pieces of DNA that are able to move from one location in the bacterial genome to another location. Transposons are flanked by terminal inverted or direct repeats. Insertion sequences (IS elements) are a family of small elements that range from 768 to 1426 bp and flank composite transposons, while unit transposons do not carry insertion sequences. Transposons vary in the genes they carry including the

Table 5.2 Distribution of *tet* resistance genes among Gram-negative bacteria.

Efflux				Ribosomal Protection and/or Efflux and/or Enzymatic			
One Gene n=11		Two or More Genes n=11		One Gene n=12		Two or More Genes n=48	
Aggregatibacter	*tet*(B)	*Bordetella*	*tet*(A)(C)(31)	*Acidaminococcus*[a]	*tet*(W)	*Acinetobacter*	*tet*(A)(B)(C)(D)(G)(H)(L)(M)(O)(W)(Y)(39)
Agrobacterium	*tet*(30)	*Mannheimia*	*tet*(B)(G)(H)(L)	*Capnocytophaga*	*tet*(Q)	*Actinobacillus*[b]	*tet*(B)(H)(L)(O)
Alteromoas	*tet*(D)	*Moraxella*	*tet*(B)(H)	*Comamonas*	*tet*(X)	*Aeromonas*	*tet*(A)(B)(C)(D)(E)(G)(H)(L) (M)(O)(T)(Y)(31)(34)
Bibersteinia	*tet*(H)	*Ochrobactrum*	*tet*(A)(B)(G)(L)	*Delftia*	*tet*(X)	*Anaerovibrio*[a]	*tet*(O)(Q)
Chlamydia	*tet*(C)	*Plesiomonas*	*tet*(A)(B)(D)	*Eikenella*	*tet*(M)	*Alcaligenes*	*tet*(A)(E)(M)(30)(39)(M)
Erwinia	*tet*(B)	*Roseobacter*	*tet*(B)(C)(E)(G)	*Epilithonimonas*	*tet*(X)	*Bacteroides*[a]	*tet*(M)(Q)(W)(X)(36)
Gillamella	*tet*(H)	*Salmonella*	*tet*(A)(B)(C) (D)(G)(L)	*Hafnia*	*tet*(M)	*Brevundimonas*	*tet*(B)(D)(G)(O)(T)(W)(39)
Francisella	*tet*(C)	*Variovorax*	*tet*(A)(L)	*Kingella*	*tet*(M)	*Burkholderia*	*tet*(D)(O)
Histophilus	*tet*(H)	*Yersinia*	*tet*(B)(D)	*Legionella*	*tet*(56)	*Butyrivibrio*	*tet*(O)(W)
Laribacter	*tet*(A)	*Halomonas*	*tet*(C)(D)	*Sphingobacterium*	*tet*(X)	*Campylobacter*	*tet*(O)(44)
Treponema[a,c]	*tet*(B)	*Burkholderia*	*tet*(D)(O)	*Wautersiella*	*tet*(X)	*Chryseobacterium*	*tet*(A)(D)(T)(W)
				Ralstonia	*tet*(M)	*Citrobacter*	*tet*(A)(B)(C)(D)(L)(M)(O)(S)(W)
						Edwardsiella	*tet*(A)(D)(M)
						Enterobacter	*tet*(A)(B)(C)(D)(G)(L)(M)(39)(X)
						Escherichia	*tet*(A)(B)(C)(D)(E)(G)(I)(L)(M)(W)(Y)(X)

(Continued)

Table 5.2 (Continued)

Efflux			Ribosomal Protection and/or Efflux and/or Enzymatic		
One Gene n=11	Two or More Genes n=11		One Gene n=12	Two or More Genes n=48	
				Flavobacterium	*tet*(A)(E)(L)(M)
				Fusobacterium[a]	*tet*(G)(L)(M)(O)(Q)(W)
				Gallibacterium	*tet*(B)(H)(K)(L)(31)
				Haemophilus	*tet*(B)(K)(M)
				Klebsiella	*tet*(A)(B)(C)(D)(E)(L)(M)(S)(W)(X)
				Kurthia	*tet*(L)(M)
				Lawsonia	*tet*(M)(W)
				Megasphaera	*tet*(O)(W)
				Mitsuokella	*tet*(Q)(W)
				Morganella	*tet*(A)(D)(J)(L)(M)
				Myroides	*tet*(L)(X)
				Neisseria	*tet*(B)(M)(O)(Q)(W)
				Pantoea	*tet*(B)(M)
				Pasteurella	*tet*(B)(D)(H)(G)(L)(M)(O)
				Photobacterium	*tet*(B)(D)(M)(Y)
				Porphyromonas[a]	*tet*(Q)(W)
				Prevotella[a]	*tet*(M)(Q)(W)
				Proteus	*tet*(A)(B)(C)(E)(H)(G)(L)(J)(M)
				Providencia	*tet*(B)(E)(G)(M)(39)(57)
				Pseudoalteromonas	*tet*(B)(M)

Genus	tet genes
Pseudomonas	*tet*(A)(B)(C)(D)(E)(G)(L)(M)(O)(T)(W)(X)(34)(39)(42)
Psychrobacter	*tet*(H) (M)(O)(39)
Rahnella	*tet*(A)(L)(M)
Rhizobium	*tet*(A)(B)(D)(M)(O)(T)(W)
Riemerella	*tet*(A)(B)(M)(O)(Q)(X)
Selenomonas[a]	*tet*(M)(Q)(W)
Serratia	*tet*(A)(B)(C)(E)(M)(S)(X)(34)(41)
Shewanella	*tet*(B)(D)(G)(M)(O)(T)(W)
Shigella	*tet*(A)(B)(C)(D)(M)
Stenotrophomonas	*tet*(B)(H)(M)(O)(T)(35)(39)
Subdolgranulum[a]	*tet*(Q)(W)
Veillonella[a]	*tet*(A)(L)(M)(Q)(S)(W)
Vibrio	*tet*(A)(B)(C)(D)(E)(G)(K)(M)(34)(35)

Carrying Mosaic genes n = 2

Megasphaera tet(O/W), *tet*(O/W/O)

Riemerella tet (O/W/32/O)

Source: http://faculty.washington.edu/marilynr

[a] Anaerobic genus.

[b] *Actinobacillus actinomycetemcomitans* is now *Aggregatibacter actinomycetemcomitans*

[c] *T. denticola* anaerobic but not all species in genus are anaerobes

Table 5.3 Distribution of tetracycline resistance genes among Gram-positive bacteria, *Mycobacterium*, *Mycoplasma*, *Nocardia*, *Streptomyces*, and *Ureaplasma*.

One Determinant n = 27		Two Determinants n = 8		Three or More Determinants n = 20	
Abiotrophia	*tet*(M)	*Aerococcus*	*tet*(M)(O)	*Actinomyces*	*tet*(L)(M)(W)
Afipia	*tet*(M)	*Arthrobacter*	*tet*(33)(M)	*Bacillus*	*tet*(K)(L)(M)(O)(T)(W)(39)(42)(45) *otr*(A)
Amycolatopsis	*tet*(M)	*Gardnerella*	*tet*(M)(Q)	*Bifidobacterium*[a]	*tet*(L)(M)(O)(W)
Anaerococcus	*tet*(M)	*Gemella*	*tet*(M)(O)	*Bhargavaea*	*tet*(L)(M)(45)
Bacterionema	*tet*(M)	*Granulicatella*	*tet*(M)(O)	*Clostridium*[a]	*tet*(K)(L)(M)(O)(P)(Q)(W)(36)(40)(44)
Brachybacterium	*tet*(M)	*Lactococcus*	*tet*(M)(S)	*Corynebacterium*	*tet*(M)(Z)(33)(W)(39)
Catenibacterium[a]	*tet*(M)	*Mobiluncus*[a]	*tet*(O)(Q)	*Enterococcus*	*tet*(K)(L)(M)(O)(S)(T)(U)(58)(61)
Cellulosimicrobium	*tet*(39)	*Savagea*	*tet*(L)(M)	*Eubacterium*[a]	*tet*(K)(M)(O)(Q)(32)
Cottaibacterium	*tet*(M)			*Lactobacillus*	*tet*(K)(L)(M)(O)(Q)(S)(W)(Z)(36)
Erysipelothrix	*tet*(M)			*Listeria*	*tet*(K)(L)(M)(S)
Finegoldia[a]	*tet*(M)			*Microbacterium*	*tet*(M)(O)(42)
Geobacillus	*tet*(L)			*Mycobacterium*	*tet*(K)(L)(M)(O)(V) *otr*(A)(B)
Helcococcus	*tet*(M)			*Nocardia*	*tet*(K)(L)(M)(O)
Leifsonia	*tet*(O)			*Paenibacillus*	*tet*(L)(M)(O)(42)
Lysinibacillus	*tet*(39)			*Peptostreptococcus*[a]	*tet*(K)(L)(M)(O)(Q)
Micrococcus	*tet*(42)			*Sporosarcina*	*tet*(K)(L)(M)
Mycoplasma	*tet*(M)			*Staphylococcus*	*tet*(K)(L)(M)(O)(S)(U)(W)(38)(42)(43)(44)(45)
Oceanobacillus	*tet*(L)			*Streptococcus*	*tet*(K)(L)(M)(O)(Q)(S)(T)(U)(W)(32)(40)AB(46)

Genus	Gene(s)	Genus	Gene(s)
Pediococcus	*tet*(L)		
Rhodococcus	*tet*(O)		
Robinsoniella	*tet*(L)		
Roseburia	*tet*(W)		
Ruminococcus[a]	*tet*(Q)		
Ureaplasma	*tet*(M)		
Vagococcus	*tet*(L)	*Streptomyces*	*tet*(K)(L)(M)(W) *otr*(A)(B)(C), *tcr*3, *tet*
Virgibacillus	*tet*(L)	*Trueperella*	*tet*(K)(L)(M)(W)(33)
Trueperella	*tet*(33)		

Carrying Mosaic genes n = 4

Bifidobacterium *tet*(O/W/32/O/W/O), *tet*(W/32/O), *tet*(O/W)

Clostridium[a] *tet*(O/32/O)

Lactobacillus *tet*(W/32/O/W/O)

Streptococcus *tet*(O/W/32/O), *tet*(O/32/O), *tet*(S/M)

From pig manure *tet*(W/N/W)

---

[a] Anaerobic genus

genes used for insertion and excision into the bacterial genome, and in the antibiotic resistance genes present and other genes such as for heavy metal resistance or disinfection. The typical composite transposon carries a gene that codes for an enzyme, such as a site-specific recombinase or resolvase, which is involved in excision and integration within the host genome, one or more other genes which code for antibiotic resistance, and/or other genes which are flanked by IS elements.

The best characterized of the transposons is Tn*10* which carries the *tet*(B) gene that codes for the most widely distributed Gram-negative tetracycline efflux protein (35 Gram-negative genera) (Table 5.2) [25]. The Tn*10* transposon is most frequently associated with Gram-negative plasmids and occasionally found in the chromosome of Gram-negative bacteria such as *Haemophilus influenzae* [26]. Transposons are found on plasmids and multiple copies may be distributed throughout the bacterial chromosome. The ability to be on the chromosome and/or plasmids allows flexibility, as well as, stability for these antibiotic resistant genes. Insertion of a transposon within the chromosome may lead to mutations and the loss of gene function or modification of gene expression. Only non-lethal mutations allow the host to survive these events.

CTns are self-transmissible integrating elements that may have broad host ranges. The CTns carry all the genes required to move from one bacterial cell to another by cell-to-cell contact. CTns have fewer restrictions in moving between unrelated bacteria than do plasmids because they lack incompatibility exclusion systems. As a result, multiple copies of the same or related CTns can be found within a single bacterium [27]. They may be found on the chromosome as well as in plasmids. The first CTns were identified in the 1980s and were members of the Tn*916*-Tn*1545* transposons family which normally carries the *tet*(M) gene. This family is the most promiscuous of the CTns described and is one of the best characterized. The Tn*916*-Tn*1545* transposons family integrates site-specifically in some species and relatively nonspecifically in other species [28]. The Tn*916* transposon is 18 kb and has relatively few restriction sites. Low-dose exposure to tetracycline has been shown to promote increased conjugal transfer of *tet*(M) transposons and mobilize co-resident plasmids [29].

The Tn*916*-Tn*1545* CTns are adaptable and able to form composite elements by integration of one transposon within another transposon. Both transposons encode for their own transfer and the complete composite element may be transferred to another bacterium, or the embedded transposon can be transferred separately. Composite transposons may have multiple mobile elements, various types of IS sequences, as well as, regions from plasmids and genes from different genera of bacteria. For example, the 65 kb composite Tn*5385* transposon carries resistance genes for penicillin, erythromycin, gentamicin, streptomycin, tetracycline, and mercury resistance and has genetic elements related to those found in three different transposons isolated from enterococci, streptococci, and staphylococci [28].

Conjugative elements carrying *tet* genes continue to evolve [30]. In recent years, these transposons have acquired an increasing number of different antibiotic resistance genes and genes for heavy metal resistance [31, 32]. The *tet*(W) genes are associated with two different transposons and are found in anaerobes such as *Bifidobacterium* and in aerobic *Arcanobacterium pyogenes* ([33, 34]). *Bacteroides* spp. have CTns ranging from 65 to >150 kb. One of the best studied is the 65 kb CTnDOT carrying the *tet*(Q) gene and normally linked to the *erm*(F) gene which codes for a rRNA methylase and confers resistance to macrolides, lincosamides, and streptogramin B (Tables 5.2 and 5.3). In the presence of tetracycline, the frequency of conjugal transfer increases and the CTnDOT is able to mobilize co-resident nonCTns, mobilizable plasmids, and unlinked integrated nonreplicating *Bacteroides* units (NBUs). CTnDOT-like elements have been identified in a variety of different Gram-negative and Gram-positive genera, and now found in >80% of *Bacteroides* spp. [35, 36].

Integrons are two-component systems that include an integrase (*int1*) and an *attI* sequence which is the site for integration of cassettes with containing different antibiotic, heavy metal, and/or disinfectant resistance genes [37]. Integrons are found as part of plasmids, chromosomes and transposons. The *tet* genes are not normally part of integrons but can be found between integrons [38].

5.3 Mutation

Mutations occur during normal replication in all living organisms that use DNA or RNA for their genetic code. Mutations usually result in a low to moderate increase in the level of antibiotic resistance and usually require multiple mutations to confer high-level resistance to tetracyclines. Mutations can also occur due to exposure to ultraviolet light, chemicals, and/or deletion/insertion of mobile elements, but also due to low-level exposure to antibiotics which occurs when antibiotics are used as growth promoters in animal production. Mutations are passed on to daughter cells during cell division and more rarely by transformation and/or transduction. Mutations can occur in specific structural proteins, in RNA molecules, and/or in genes that regulate expression of other proteins. Mutations normally confer a low to moderate level of tetracycline resistance in bacteria.

5.4 Tetracycline Resistance Genes

For this chapter, Gram-positive bacteria also include cell-wall free *Mycoplasma, Ureaplasma*, plus *Mycobacterium, Nocardia,* and *Streptomyces* (Table 5.3). Bacteria may become resistant to tetracyclines, by mutation, while most bacteria become tetracycline resistant because they acquire new genes that: (i) pump tetracycline out the cell (efflux) $n = 34$; (ii) protect the ribosome from the action of tetracyclines $n = 13$; (iii) enzymatically deactivate tetracyclines $n = 13$; (iv) unknown $n = 1$ (Table 5.1). The modern era for tetracycline resistance genes began with the paper by Mendez et al. [39] which demonstrates that there were a few distinctive tetracycline resistant (*tet*) genes found among Gram-negative Tc^r bacteria. Since then 61 different types of *tet* genes have been identified based on the definition that a different *tet* gene must have ≤79% amino acid identity with all previous characterized *tet* genes at the time the gene is characterized and named (http://faculty.washington.edu/marilynr).

Two genes are considered part of the same type/class and given the same gene designation when they shared at ≥80% amino acid sequence identity over the entire length of the protein. No distinctions are made if two genes have >80% amino acid identity. For a *tet* gene to be assigned a new designation, it must be in a bacterial host that is resistant to tetracycline due to the presence of this gene. A potentially new gene must be completely sequenced and its amino acid composition compared with all currently known classes of *tet* genes and shown to confer tetracycline resistance normally in the natural host. It is inadequate to demonstrate that a gene sequence has characteristics in common with previously characterized *tet/otr* genes. Data mining of genomes that look for DNA and amino acid similarities to known *tet/otr* genes, is not adequate for recognition of a new gene or a new species carrying a known gene. The information on the proposed new gene designation is then submitted to the *tet* nomenclature clearing house to Dr. Marilyn C. Roberts (marilynr@uw.edu) in Seattle, WA. She will determine if a new *tet* name is warranted and will provide it. Then the new resistance gene name will be put up on the website (http://faculty.washington.edu/marilynr/), which is updated twice a year. Obtaining the new *tet* gene name should be done prior to submitting the sequence to GenBank or a manuscript for publication. This is critical to reduce confusion and make sure various data bases that use the GenBank system has the correct gene names.

A website (http://faculty.washington.edu/marilynr) with all recognized tetracycline resistance genes has been established and provides tables like the ones in this chapter. Currently, 136 different genera including 54 different Gram-positive and 82 different Gram-negative genera have been identified that carried tetracycline resistance genes (Tables 5.2 and 5.3). The tables are based on the work of Dr. Roberts' laboratory, as well as from published papers and abstracts presented at scientific meetings. Many of the new genera have been associated with the environment and account for a significant fraction of the genera added to the tables over the past 10 years. However, information from GenBank or published manuscripts are not included where the information has not been adequately validated. This is especially true of more recent microbiome work, which may suggest that strictly Gram-negative *tet* genes are associated with Gram-positive bacteria. In addition, as illustrated by the identification of ten enzymatic *tet* genes (*tet*[37], *tet*[47]– *tet*[55]), genes that have been found from microbiome studies of complex samples but not associated with particular bacteria are now becoming more common [40, 41]. As microbiome work continues this will provide a challenge in understanding the role some of these new resistance genes play in the microbial environment.

5.5 Efflux

The first tetracycline resistant efflux proteins were identified in the 1950s in Japan where they were later hypothesized to be located on conjugative plasmids [13]. Today there are 34 genetically distinct efflux genes characterized coding for drug-H^+ energy-dependent trans-membrane sequence (TMS) proteins which span the lipid bilayer of the inner cell membrane 9–14 times. Some researchers have divided these genes into seven different groups based on the number of predicted transmembrane helices and nearest class homologies. The largest group includes most of the genes that have 12 predicted TM helices [2]. There is a conserved sequence motif, GXXXXRXGRR, which is found in tetracycline efflux proteins, glucose uinporters, and sugar/H+ symporters. Most studies on tetracycline efflux proteins have focused on the Tet(B) protein from Tn*10* transposon, though it is assumed that most of the other genes have similar characteristics. However, most efflux proteins normally export tetracycline and doxycycline but not minocycline or tigecycline out of the cell. One exception is the Gram-negative *tet*(B) gene that confers resistance and exports tetracycline, doxycycline, and minocycline. The other exception is *tetAB*(60) gene, which confers resistance to tigecycline as well as other tetracyclines. However more recently other tet genes with mutations may confer resistance to tigecycline.

The efflux proteins are able to reduce intracellular concentrations of tetracyclines. This allows the majority of the ribosomes to continue to function and produce proteins and hence the bacterium is able to grow in the presence of tetracycline. The efflux proteins work by exchanging a proton for the tetracycline–cation complex against a concentration gradient and require intact cells to function. The efflux genes are the most commonly found *tet* genes in aerobic and facultative Gram-negative bacteria. Upstream of the structural efflux gene is a divergently transcribed repressor gene that produces a protein which binds to the palindromic operator in the promoters for both the repressor and the structural *tet* gene, resulting in prevention of the initiation of transcription. At ~1 nM tetracycline, a tetracycline–divalent cation complex interacts with the repressor protein releasing it from the DNA and thus, transcription of both genes occurs. Most efflux genes have this arrangement and a more detailed description of mechanism of tetracycline resistance due to the efflux proteins can be found in Palm et al. [42]. Two exceptions are *tetAB*(46) and *tetAB*(60) genes. Each code for two open reading frames that produce Protein A and Protein B, which produce a heterodimeric efflux transporter [43]. Both genes are required for resistance. The *tetAB*(60) gene was isolated from a human saliva metagenomic library and thus we do not know what the host bacteria is or its location within the host genome. Information on these two genes can be found in their GenBank numbers, KX000272 and KX000273. In contrast, the *tetAB*(46) gene was isolated from *Streptococcus* [43].

The *tet*(C) gene is usually associated with plasmids in most genera listed in Table 5.2. However, the *tet*(C) gene is located in the chromosome in Tc^r *Chlamydia suis,* an obligate intracellular bacteria from the intestinal tract of pigs in Europe and the USA [44, 45]. The Tc^r *C. suis* isolates contained a 13 kb segment of foreign DNA including a truncated repressor gene, *tetR*(C), and a functional *tet*(C) gene with a 13 kb region that has a high degree of identity with a pRAS3.2 plasmid from *Aeromonas salmonicida,* a microbe which does not grow in pigs [46]. This is the first report of a known acquired *tet* gene in an obligate intracellular bacterium and the degree to which the *tet* genes have spread through the bacterial populations and different ecosystems. It is also the only known examples of a mobile characterized antibiotic resistant gene found in an obligate intracellular bacterium.

The distribution of the efflux genes in various Gram-positive and Gram-negative genera are detailed in Tables 5.2 and 5.3 updates can be found at http://faculty.washington.edu/marilynr/.

5.6 Ribosomal Protection

Thirteen ribosomal protection genes have been characterized. The genes have been divided into three base groups related to their amino acid sequences rather than G + C% content as is done with the efflux genes [2]. The ribosomal protection genes code for cytoplasmic proteins of ~ 72.5 kDa that protect the ribosomes from the action of tetracycline *in vitro* and *in vivo*. Unlike the efflux genes, the ribosomal protection genes confer resistance to tetracycline, doxycycline, and minocycline but not tigecycline [47]. The proteins have sequences similar to the ribosomal elongation factors EF-G and

EF-TU and are grouped in the translation factor superfamily GTPases [48]. A model based on Tet(O)-mediated Tc^r biochemical and structural data for both Tet(M) and Tet(O) proteins has been proposed. In this model, the ribosomes without tetracycline function normally and when tetracycline is added to the growth media, it binds to the ribosomes, altering their conformational state, which interrupts the elongation cycle and stops protein synthesis. The ribosomal protection proteins are thought to interact with the base of h34 ribosomal protein, causing allosteric disruption of the primary tetracycline binding site(s), which releases the bound tetracycline. The ribosome returns to its normal conformational state and resumes protein synthesis. What is not clear is whether the *tet* ribosomal proteins actively prevent tetracycline from rebinding once it has been released or if the released tetracycline is able to rebind to the same or a different ribosome. Further details can be found in other publications [49, 50]. It has been assumed that all 13 ribosomal protection proteins in this group have similar mechanisms of action.

The majority of Gram-positive and Gram-negative genera carry either ribosomal protection genes alone or in combination with efflux/enzymatic genes as illustrated in Tables 5.2 and 5.3. The ribosomal protection genes predominate in Tc^r oral bacteria, Gram-positive bacteria anaerobic and urogenital Gram-negative bacteria, while they are less common among enteric Gram-negative bacteria. The *tet*(M) gene was identified in clinical *Enterococcus* spp. isolated between 1954 and 1955, which is approximately the same time as the first Gram-negative *tet* efflux genes were identified; however, the early *Enterococcus* spp. study was not published until 1997 although the Gram-negative studies were published 30 years earlier [13, 14]. Thus, both the *tet* efflux and *tet* ribosomal protection genes have been in the bacterial population for >50 years.

Nine of the ribosomal protection genes have a G + C% ranging between 30 and 40% and are thought to be of Gram-positive origin, while the *tet*(W) gene has a G + C% between 50 and 55% and its origin is unclear. The *Streptomycetes tet* and *otr*(A) genes have G + C% ranging from 68 to 78% and their origin is thought to be *Streptomycetes*. Despite the difference in G + C%, the Tet(M) protein shares 68–72% amino acid identity with the Tet(O), Tet(S), Tet(W), Tet(32), and Tet(36) proteins grouped together. The Tet(Q) proteins share 60% amino acid identity with the Tet(T) and are grouped together. The *tetB*(P), *otr*(A), and *tet* are grouped together and are unique to environmental bacteria [2]. The *tet*(32) gene has been found in two genera and the metagenome from the oral cavity of Northern European children [51, 52].

There have been more papers written about the *tet*(M) gene than any other ribosomal protection *tet* gene. The *tet*(M) gene is commonly found in oral, urogenital, aerobic and anaerobic Gram-positive and Gram-negative non-enteric bacteria, while it is less common in enteric genera (Tables 5.2 and 5.3). The *tet*(M) positive bacteria have been isolated from 80 genera including 41 Gram-positive and 39 Gram-negative genes found across the bacterial spectrum and from multiple different ecosystems. Some variability at the base pair level is found and different *tet*(M) genes may have a variation of their base pairs of at least 11%.

The *tet*(M) gene is usually part of a conjugative transposon that is often in the bacterial chromosome. In some *Clostridium perfringens* isolates, the *tet*(M) gene is found in the chromosome on an incomplete element and cannot move, while in other *C. perfringens* isolates the *tet*(M) gene is on complete transposons and can be conjugally transferred between isolates [32]. One exception of the chromosomal location for the *tet*(M) gene was found in the genus *Neisseria*. In *N. gonorrhoeae*, the *tet*(M) gene has normally been associated with 25.2 Mda conjugative plasmids that were isolated with the first *tet*(M) positive *N. gonorrhoeae* collected in 1983. These same *tet*(M) positive 25.2 Mda conjugative plasmids have since been found in Tc^r *N. meningitidis, Kingella denitrificans*, and *Eikenella corrodens* strains [53, 54]. The *tet*(M) plasmids confer high levels of Tc^r (MIC ≥16 μg ml^{-1}) and are closely related to the 24.5 Mda indigenous *N. gonorrhoeae* conjugative plasmids [55]. Only part of the *tet*(M) Tn*916*-like transposon is found on the 25.2 Mda Tc^r *Neisseria* plasmids, while the complete *tet*(M) transposon was integrated into a Tc^r *Haemophilus ducreyi* conjugative plasmid [55]. The *tet*(M) gene is also found naturally in commensal *Neisseria* spp. but here the *tet*(M) genes are located in the chromosome.

The *tetB*(P) gene has been found only in the genus *Clostridium*. It is unique among the ribosomal protection genes because all isolates that carry this gene also carry a *tetA*(P) gene which codes for an inducible efflux protein. The two genes are transcribed from a single promoter which is located 529 bp upstream of the *tetA*(P) start codon and the *tetB*(P) gene overlaps the *tetA*(P) gene by 17 nucleotides [56]. The *tetA*(P) gene has been found alone where it does confer Tc^r to the bacterial host, while the *tetB*(P) gene has not. When the *tetB*(P) gene was cloned away from the *tetA*(P) gene and introduced into *C. perfringens* (the natural host) and *E. coli* recipients, the resultant transformants had low-level Tc^r. Thus, it is not clear if the *tetB*(P) gene contributes to the natural host's Tc^r phenotype. Both the *tetA*(P) and *tetB*(P) genes are often associated with conjugative and nonconjugative plasmids.

5.7 Mosaic

Mosaic *tet* genes consist of regions from five of the known ribosomal *tet* genes with a descriptive designation such as *tet*(O/W) representing a hybrid between the *tet*(O) at one end and *tet*(W) at the other end of the gene. A *tet*(W/O/W) designation would represent a hybrid between the *tet*(O) and *tet*(W) genes with a partial *tet*(O) sequence between the ends of the *tet*(W) gene. Currently there have been 11 different mosaic combinations identified and the genes shown to still confer tetracycline resistance (Table 5.1). Mosaic genes can only be determined by sequencing the complete gene and at this time have been found in six different genera, four Gram-positive and two Gram-negative. One recent mosaic *tet*(W/N/W) has been identified from a Chinese pig manure sample and thus the host is unknown. The "N" sequence is a novel sequence not previously identified [57]. In *Clostridium* spp. and *Lactobacillus* spp. only a single mosaic has been identified while in the other species multiple different mosaic genes have been identified which carry variable amounts of any one gene and of the combination of ribosomal protections genes. Apparent mosaics for efflux gene combinations can be found in GenBank especially as part of whole genome sequences but it has not been determined whether these mosaics confer tetracycline resistance. This is one of the issues that have occurred as more sequences are being directly submitted from either complex samples or directly to GenBank without much available information, such as antibiotic susceptibility data (unpublished observations).

The gene originally designated *tet*(32) from a *Clostridium*-like strain has been sequenced and from bp 0 to 243 it had 100% identity with the same region in the *tet*(O) genes. The 158 bp non-coding region upstream of the structural gene showed 98% sequence homology with the upstream regions of the *tet*(O) genes from *S. mutans* and *Campylobacter jejuni*, GenBank # M20925 and M18896, respectively. The sequences at the end of the gene (1262–1782 bp) had a 98.8% sequence homology with the *tet*(O) gene. However, the sequences between bp 244–1263 share <70% similarity with any other known *tet* gene and since the overall DNA homology of the gene was <80% to the *tet*(O) or any other *tet* gene, it was given a new designation *tet*(O/32/O) [58]. The original work of Melville et al. [59] found by PCR that 6 of 9 rumen sheep samples and 8 of 11 pig fecal samples were positive for *tet*(32). However, from more recent data regarding its hybrid nature, it is unclear if these positive samples were actually detecting the *tet*(O/32/O) sequence or different genes or mosaic genes (Table 5.1).

5.8 Enzymatic

Thirteen genes now code for inactivating enzymes of which nine, *tet*(47)–*tet*(55), were identified from soil samples and have unknown bacterial hosts [41], and one, *tet*(37), was identified from a metagenomic analysis of an oral sample and also has an unknown bacterial host [40]. The recent identification of these new genes coding for inactivating enzymes suggests that this mechanism of tetracycline resistance may be very common in a variety of microbiomes and that as more complex samples are analyzed more inactivating enzymes are likely to be identified. The remaining three genes *tet*(X) (14 genera), *tet*(34) (4 genera), and *tet*(56) (1 genera) have only been identified in Gram-negative genera (Table 5.2). Why no Gram-positive bacteria have been found to carry this mechanism of tetracycline resistance is unknown at this time.

The *tet*(X) gene encodes for a flavin NADP-dependent monooxygenase that inactivates by regioselectively adding a hydroxyl group to the C-11a position of the antibiotic. This action requires oxygen and confers resistance to tetracycline, doxycycline, minocycline, and tigecycline [60, 61]. The *tet*(X) gene was originally found in a strict anaerobe, *Bacteroides* spp. where it was linked to an rRNA methylase gene, *erm*(F), that confers resistance to macrolides, lincosamides, and streptogramin B, and are part of the conjugative transposons CTnDOT, Tn*4351* and Tn*4400* [62]. The TetX protein requires oxygen to degrade the tetracycline; the *tet*(X) gene does not confer Tc^r in *Bacteroides* spp. where it was first identified [20]. The *erm*(F) and *tet*(X) genes have a G + C% content of 36 and 37%, respectively, suggesting that these genes did not originate in the *Bacteroides* spp. identified for which the G + C% content ranged between 40 and 48%. It was previously hypothesized that a functional *tet*(X) gene might be found in an environmental species. This hypothesis has since been proven to be correct when it was identified and shown to function in *Sphingobacterium* spp. [9, 63]. The *tet*(X) has been found in metagenomic analysis of environmental samples and more recently in other bacteria from clinical samples in Sierra Leone [64].

The *tet*(37) gene codes for a second NADP-dependent monooxygenase that is unrelated to the *tet*(X) gene but has a similar G + C% content of 37.9% and shares homology with other flavoproteins, oxidoreductases, and NADP-requiring enzymes [40]. The way this protein inactivates tetracycline is not clear. The *tet*(37) gene has only been cloned from the oral metagenome and no specific bacteria have been identified that carry this gene.

The *tet*(34) gene was first described in *Vibrio* spp. and codes for an enzyme that is similar to a xanthine-guanine phosphoribosyl transferase rather than a NADP-dependent monooxygenase [65]. Now tet(34) has been found in four Gram-negative genera. Little work has been done with this gene.

The other genes have been identified from functional genomics and the host sources are unknown [41].

5.9   Unknown

The *tet*(U) gene produces a small protein (105 amino acids) which confers low level tetracycline resistance [18]. The TetU protein has 21% similarity over its length to the TetM protein, but it does not include the consensus GTP-binding sequences, which are thought to be important for tetracycline resistance in these proteins. The *tet*(U) gene has been identified in a vancomycin and tetracycline resistant *S. aureus* strain that did not carry the *tet*(K), *tet*(L), *tet*(M), or *tet*(O) genes. From the same patient, vancomycin resistant enterococci were cultured that carried both the *tet*(U) and *tet*(L) genes and a few isolates also carried the *tet*(K) and/or *tet*(M) genes [66]. The *tet*(U) gene has also been identified in *Enterococcus* spp. The importance of the *tet*(U) gene is unclear since both *Enterococcus* and *Staphylococcus* isolates are able to carry a variety of efflux and ribosomal protection *tet* genes and whether it confers resistance has also been brought into question.

The tetracycline genes originally identified in the genus *Streptomyces* (*otr*[A], *otr*[B], *otr*[C], *tcr3*, *tet*) account for 29% of the genes listed in Table 5.1. However, the *otr*(A) and *otr*(B) are now found in *Bacillus* and *Mycobacterium* and may be associated with other related environmental genera suggesting gene exchange is occurring.

5.10   Conclusion

Tc^r bacteria are widely distributed throughout the world. They have been isolated from deep subsurface trenches, in waste water, surface and ground water, sediments and soils, and places that are relatively untouched by human civilization, such as penguins in Antarctica and seals from the Arctic [67–71]. To a great extent, what is in the environmental bacterial population remains largely unexplored. It has been estimated that 5–6% of the soil bacteria are actinobacteria and one out of every 1000 actinomycetes produces tetracycline. Thus it is not surprising that in recent years the number of tetracycline resistant environmental bacteria identified has greatly increased and as new ecosystems are examined and metagenomic DNA from a variety sources are analyzed, other tetracycline resistance genes will be identified and many other genera will be shown to carry these genes. Unfortunately, it is likely that human activities will lead to increasing bacterial resistance to tigecycline and that pathogens not currently resistant to tetracycline will acquire resistance over time, which will influence the continued use of tetracyclines as therapeutic agents in humans, animals, and plants and studies addressing this issue are needed.

Beside the characterized genes listed here and the variety of genera listed in the tables, there are other tetracycline resistance genes that can be found in GenBank, antibiotic resistant data bases such as Antibiotic Resistance Genes Databases (ARDB; https:// ardb.cbcb.umd.edu), the Comprehensive Antibiotic Resistant Database (CARD; http:// arpcard.mcmaster.ca), ResFinder (https://omictools.com/resfinder-tool) and~30 other data bases that are now available. All of these databases have many of same genes as listed in Table 5.1 but also list other genes not adequately characterized, identified only from microbiomes or functional analyses or given names that have now been changed (http://faculty.washington.edu/marilynr). Some of the newer data bases are being developed including ResFinder 3.1 (https://cge.cbs.dtu.dk/services/ResFinder) are now using the accepted names as found in http://faculty.washington.edu/marilynr and not just what is in GenBank files. These newer databases should help reduce the confusion of what a particular gene actually is, especially as these newer databases become more widely used. Using the accepted names of specific genes will help to unify the nomenclature within the literature as we go forward. However, what to do with genes that show homology from the whole genome sequence of specific isolates or metagenomic analysis of complex samples still needs to be adequately reconciled. Simply listing everything that has homology with a known gene even if it is unclear if that gene is functional has some major issues that will need to be addressed in the future.

References

1 Griffin, M.O., Fricovsky, E., Ceballos, G., and Villarreal, F. (2010). Tetracyclines: a pleitropic family of compounds with promising therapeutic properties, review of the literature. *Am. J. Phys. Cell Physiol.* 299: C539–C548.

2 Thaker, M., Spanogiannopoulos, P., and Wright, G.D. (2010). The tetracycline resistome. *Cell. Mol. Life Sci.* 67: 419–431.

3 Giamerellou, H. and Poulakou, G. (2011). Pharmacokinetic and pharmacodynamic evaluation of tigecycline. *Expert Opin. Drug Metab. Toxicol.* 7: 1459–1470.

4 Ghedin, E., Hailemariam, T., DePasse, J.V. et al. (2009). *Brugia malayi* gene expression in response to the targeting of the *Wolbachia* endosymbiont by tetracycline treatment. *PLoS Negl. Trop. Dis.* 3 (10): e525. https://doi.org/10.1371/journal.pntd.0000525.

5 Garrido-Mesa, N., Zarzuelo, A., and Galvez, J. (2013). Minocycline: far beyond an antibiotic. *Br. J. Pharmacol.* 169: 337–352.

6 Leigh, M.J.S., Nguyen, D.V., My, Y. et al. (2013). A randomized double-blind, placebo-controlled trial of minocycline in children and adolescents with Fragile X syndrome. *J. Dev. Behav. Pediatr.* 34: 147–155.

7 Sun, C., Wang, Q., Brubaker, J.D. et al. (2008). A robust platform for the synthesis of new tetracycline antibiotics. *J. Am. Chem. Soc.* 130: 17912–17927.

8 Dezube, B.J., Krown, S.E., Lee, J.Y. et al. (2006). Randomized phase II trial of matrix metalloproteinase inhibitor COL-3 in AIDS-related Kaposi's sarcoma: an AIDS malignancy consortium study. *J. Clin. Oncol.* 24: 1389–1394.

9 Ghosh, S. and LaPara, T.M. (2007). The effects of subtherapeutic antibiotic use in farm animals on the proliferation and persistence of antibiotic resistance among soil bacteria. *ISME J.* 1: 191–203.

10 Hong, P.-Y., Yannarell, A.C., Dai, Q. et al. (2013). Monitoring the perturbation of soil and ground water microbial communities due to pig production activities. *Appl. Environ. Microbiol.* 79: 2620–2629.

11 Aaretrup, F. (2012). Get pigs off antibiotics. *Nature* 486: 465–466.

12 Hughes, V.M. and Datta, N. (1983). Conjugative plasmids in bacteria of the "pre-antibiotic" era. *Nature* 301: 725–726.

13 Watanabe, T. (1963). Infective heredity of multiple drug resistance in bacteria. *Bacteriol. Rev.* 27: 87–115.

14 Atkinson, B.A., Abu-Al-Jaibat, A., and LeBlanc, D.J. (1997). Antibiotic resistance among enterococci isolated from clinical specimens between 1953 and 1954. *Antimicrob. Agents Chemother.* 41: 1598–1600.

15 Cousin, S.L. Jr., Whittington, W.L., and Roberts, M.R. (2003). Acquired macrolide resistance genes in pathogenic *Neisseria* spp. isolated between 1940 and 1987. *Antimicrob. Agents Chemother.* 47: 3877–3880.

16 Di Luca, M.C., D'Ercole, S., Petrelli, D. et al. (2010). Lysogenic transfer of *mef*(A) and *tet*(O) genes carried by Φm46.1 among group A streptococci. *Antimicrob. Agents Chemother.* 54: 4464–4466.

17 Waters, V.L. (2001). Conjugation between bacterial and mammalian cells. *Nat. Genet.* 29: 375–376.

18 Chopra, I. and Roberts, M.C. (2001). Tetracycline antibiotics: mode of action, applications, molecular biology and epidemiology of bacterial resistance. *Microbiol. Mol. Biol. Rev.* 65: 232–260.

19 Roberts, M.C. (1997). Genetic mobility and distribution of tetracycline resistance determinants. In: *Antibiotic Resistance: Origins, Evolution, Selection and Spread*, Ciba Foundation symposium 207 (ed. D.J. Chadwick and J.A. Goode), 206–218. Chichester UK: John Wiley & Sons.

20 Speer, B., Bedzyk, S., and Salyers, A.A. (1991). Evidence that a novel tetracycline resistance gene found on two *Bacteroides* transposons encodes an NADP-requiring oxidoreductase. *J. Bacteriol.* 173: 176–183.

21 Grimdmann, H., Aires-de-Sousa, M., Boyce, J., and Tiemersma, E. (2006). Emergence and resurgence of methicillin-resistant *Staphylococcus aureus* as a public-health threat. *Lancet* 368: 874–885.

22 Roberts, M.C. (2011). Environmental macrolide-lincosamide-streptogramin and tetracycline resistant bacteria. *Front. Microbiol.* 2: 1–8. https://doi.org/10.3389/fmicb.2011.00040.

23 DePaola, A. and Roberts, M.C. (1995). Class D and E tetracycline resistance determinants in Gram-negative catfish pond bacteria. *Mol. Cell. Probes* 9: 311–313.

24 Sorum, H., Roberts, M.C., and Crosa, J.H. (1992). Identification and cloning of a tetracycline resistance gene from the fish pathogen *Vibrio salmonicida*. *Antimicrob. Agents Chemother.* 36: 611–615.

25 Lawley, T.C., Burland, V., and Taylor, D.E. (2000). Analysis of the complete nucleotide sequence of the tetracycline-resistant transposon Tn*10*. *Plasmid* 43: 235–239.

26 Marshall, B., Roberts, M., Smith, A., and Levy, S.B. (1984). Homogeneity of tetracycline-resistance determinants in *Haemophilus* species. *J. Infect. Dis.* 149: 1028–1029.

27 Norgren, M. and Scott, J.R. (1991). The presence of conjugative transposon Tn*916* in the recipient strain does not impede transfer of a second copy of the element. *J. Bacteriol.* 173: 319–324.

28 Rice, L.B. (2007). Conjugative transposons. In: *Enzyme-Mediated Resistance to Antibiotics: Mechanisms, Dissemination, and Prospects for Inhibition* (ed. R.A. Bonomo and M. Tolmasky), 271–284. Washington DC: ASM.

29 Facinelli, B., Roberts, M.C., Giovanetti, E. et al. (1993). Genetic basis of tetracycline resistance in food borne isolates of *Listeria innocua. Appl. Environ. Microbiol.* 59: 614–616.

30 Giovanetti, E., Brenciani, A., Lupidi, R. et al. (2003). The presence of the *tet*(O) gene in erythromycin and tetracycline-resistant strains of *Streptococcus pyogenes. Antimicrob. Agents Chemother.* 47: 2844–2849.

31 Lancaster, H., Roberts, A.P., Dedi, R. et al. (2004). Characterization of Tn*916S*, a Tn*916*-like element containing the tetracycline resistance determinant *tet*(S). *J. Bacteriol.* 186: 4395–4398.

32 Soge, O.O., Beck, N., White, T.M., and Roberts, M.C. (2008). A novel transposon, Tn*6009*, composed of a Tn*916*-like element linked to *Staphylococcus aureus*-like *mer* operon. *J. Antimicrob. Chemother.* 62: 674–680.

33 Billington, S.J., Songer, J.G., and Jost, B.H. (2002). Widespread distribution of a Tet W determinant among tetracycline-resistant isolates of the animal pathogen *Arcanobacterium pyogenes. Antimicrob. Agents Chemother.* 46: 1281–1287.

34 Billington, S.J. and Jost, B.H. (2002). Multiple genetic elements carry the tetracycline resistance gene *tet*(W) in the animal pathogen *Arcanobacterium pyogenes. Antimicrob. Agents Chemother.* 50: 3580–3587.

35 Chung, W.O., Werckenthin, C., Schwarz, S., and Roberts, M.C. (1999). Host range of the *ermF* rRNA methylase gene in human and animal bacteria. *J. Antimicrob. Chemother.* 43: 5–14.

36 Wood, M.M. and Gardner, J.F. (2015). The integration and excision of CTnDOT. *Microbiol. Spectr.* 3 (2): https://doi.org/10.1128/microbiolspec.MDNA3-0020-2014.

37 Recchia, G.D. and Hall, R.M. (1995). Gene cassettes: a new class of mobile element. *Microbiology* 141: 3015–3027.

38 Fournier, P.-E., Vallenet, D., Barber, V. et al. (2006). Comparative genomics of multidrug resistance in *Acinetobacter baumannii. PLoS Genet.* 2 (1): e7.

39 Mendez, B., Tachibana, C., and Levy, S.B. (1980). Heterogeneity of tetracycline resistance determinants. *Plasmid* 3: 99–108.

40 Diaz-Torres, M.L., McNab, R., Spratt, D.A. et al. (2003). Novel tetracycline resistance determinate from the oral metagenome. *Antimicrob. Agents Chemother.* 47: 1430–1432.

41 Forsberg, K.J., Patel, S., Wencewicz, T.A., and Dantas, G. (2015). The tetracycline destructases: a novel family of tetracycline-inactivation enzymes. *Chem. Biol.* 22: 888–897.

42 Palm, G.J., Lederer, T., Orth, P. et al. (2008). Specific binding of divalent metal ions to tetracycline and to the Tet repressor/tetracycline complex. *J. Biol. Inorg. Chem.* 13: 1097–1110.

43 Warburton, P.J., Ciric, L., Lerner, A. et al. (2013). TetAB(46), a predicted heterodimeric ABC transporter conferring tetracycline resistance in *Streptococcus australis* isolated from the oral cavity. *J. Antimicrob. Chemother.* 68: 17–22.

44 Dugan, J., Rockey, D.D., Jones, L., and Andersen, A.A. (2004). Tetracycline resistance in *Chlamydia suis* mediated by genomic isolated inserted into the chlamydial *inv*-like gene. *Antimicrob. Agents Chemother.* 48: 3989–3995.

45 Suchland, R.J., Sandoz, K.M., Jeffrey, B.M. et al. (2009). Horizontal transfer of tetracycline resistance among *Chlamydia* spp. *in vitro. Antimicrob. Agents Chemother.* 53: 4604–4611.

46 L'Abbe-Lund, T.M. and Sorum, H. (2002). A global non-conjugative Tet C plasmid, pRAS3, from *Aeromonas salmonicida. Plasmid* 47: 172–178.

47 Roberts, M.C. (2005). MiniReview: update on acquired tetracycline resistance genes. *FEMS Microbiol. Lett.* 245: 195–203.

48 Leipe, D.D., Wolf, Y.I., Koonin, E.V., and Aravind, L. (2002). Classification and evolution of P-loop GTPases and related ATPases. *J. Mol. Biol.* 317: 41–72.

49 Connell, S.R., Tracz, D.M., Nierhaus, K.H., and Taylor, D.E. (2003). Ribosomal protection proteins and their mechanism of tetracycline resistance. *Antimicrob. Agents Chemother.* 47: 3675–3681.

50 Connell, S.R., Trieber, C.A., Einfeldt, E. et al. (2003). Mechanism of Tet(O), perturbs the conformation of the ribosomal decoding center. *Mol. Microbiol.* 45: 1463–1472.

51 Lancaster, H., Bedi, R., Wilson, M., and Mullany, P. (2005). The maintenance in the oral cavity of children of tetracycline-resistant bacteria and the genes encoding such resistance. *J. Antimicrob. Chemother.* 56: 524–531.

52 Warburton, P., Roberts, A.P., Alan, E. et al. (2009). Characterization of *tet*(32) genes from the oral metagenome. *Antimicrob. Agents Chemother.* 53: 273–276.

53 Roberts, M.C. and Knapp, J.S. (1988). Host range of the conjugative 25.2 Mdal tetracycline resistance plasmid from *Neisseria gonorrhoeae. Antimicrob. Agents Chemother.* 32: 488–491.

54 Roberts, M.C. and Knapp, J.S. (1988). Transfer of β-lactamase plasmids from *Neisseria gonorrhoeae* to *Neisseria meningitidis* and commensal *Neisseria* species by the 25.2-Megadalton conjugative plasmid. *Antimicrob. Agents Chemother.* 32: 1430–1432.

55 Roberts, M.C. (1989). Plasmids of *Neisseria gonorrhoeae* and other *Neisseria* species. *Clin. Microbiol. Rev.* 2: S18–S23.

56 Johanesen, P.A., Lyras, D., Bannam, T.L., and Rood, J.I. (2001). Transcriptional analysis of the *tet*(P) operon from *Clostridium perfringens. J. Bacteriol.* 183: 7110–7119.

57 Leclercq, S.O., Wang, C., Zhu, Y. et al. (2016). Diversity of the tetracycline mobilome within a Chinese pig manure sample. *Appl. Environ. Microbiol.* 82: 6454–6462.

58 Stanton, T.B., Humphrey, S.B., Scott, K.P., and Flint, H.J. (2005). Hybrid *tet* genes and *tet* nomenclature: request for opinions. *Antimicrob. Agents Chemother.* 49: 1265–1266.

59 Melville, C.M., Scott, K.P., Mercer, D.K., and Flint, H.J. (2001). Novel tetracycline resistance gene, *tet*(32), in the *Clostridium*-related human colonic anaerobe K10 and its transmission *in vitro* to the rumen anaerobe *Butyrivibrio fibrisolvens. Antimicrob. Agents Chemother.* 45: 3246–3249.

60 Moore, I.F., Hughes, D.W., and Wright, G.D. (2005). Tigecycline is modified by the flavin-dependent monooxygenase TetX. *Biochemist* 44: 11829–11835.

61 Yang, W.R., Moore, I.F., Koteva, K.P. et al. (2004). TetX is a flavin-dependent monooxygenase conferring resistance to tetracycline antibiotics. *J. Biol. Chem.* 279: 52346–52352.

62 Whittle, G., Whitehead, T.R., Hamburger, N. et al. (2003). Identification of a new ribosomal protection type of tetracycline resistance gene, *tet*(36), from swine manure pits. *Appl. Environ. Microbiol.* 69: 4151–4158.

63 Ghosh, S., Gralnick, J., Roberts, M.C. et al. (2009). *Sphingobacterium* sp. strain PM2-P1-29 harbors a functional *tet*(X) gene encoding for the degradation of tetracycline. *J. Appl. Microbiol.* 106: 1336–1342.

64 Leski, T.A., Bangura, U., Jimmy, D.H. et al. (2013). Multidrug-resistant *tet*(X)-containing hospital isolates in Sierra Leone. *Int. J. Antimicrob. Agents* 42: 83–86.

65 Nonaka, L. and Suzuki, S. (2002). New Mg^{2+} –dependent oxytetracycline resistance determinant Tet 34 in *Vibrio* isolates from marine fish intestinal contents. *Antimicrob. Agents Chemother.* 476: 1550–1552.

66 Weigel, L.M., Donlan, M., Shin, D.H. et al. (2004). High-level vancomycin-resistant *Staphylococcus aureus* isolates associated with a polymicrobial biofilm. *Antimicrob. Agents Chemother.* 51: 231–238.

67 Brown, M.G., Mitchell, E.H., and Balkwill, D.L. (2008). Tet 42, a novel tetracycline resistance determinant isolated form deep terrestrial subsurface bacteria. *Antimicrob. Agents Chemother.* 52: 4518–4521.

68 Donato, J.J., Moe, L.A., Converse, B.J. et al. (2010). Metagenomic analysis of apple orchard soil reveals antibiotic resistance genes encoding predicted bifunctional proteins. *Appl. Environ. Microbiol.* 76: 4396–4401.

69 Glad, T., Fristiansen, V.F., Nielsen, K.M. et al. (2010). Ecological characterization of the colonic microbiota in arctic and sub-arctic seals. *Microb. Ecol.* 60: 320–330.

70 Kümmerer, K. (2004). Resistance in the environment. *J. Antimicrob. Chemother.* 54: 311–320.

71 Rahman, M.H., Sakamoto, K.Q., Nonaka, L., and Suzuki, S. (2008). Occurrence and diversity of tetracycline *tet*(M) in enteric bacteria of Antarctic Adelie penguins. *J. Antimicrob. Chemother.* 62: 627–628.

6

Fluoroquinolone Resistance

Karl Drlica[1], Xilin Zhao[1,2], Muhammad Malik[1], Hiroshi Hiasa[3], Arkady Mustaev[1], and Robert Kerns[4]

[1] New Jersey Medical School, Rutgers Biomedical and Health Sciences, Public Health Research Institute, Newark, NJ, USA
[2] State Key Laboratory of Molecular Vaccinology and Molecular Diagnostics, School of Public Health, Xiamen University, Xiamen, China
[3] Department of Pharmacology, University of Minnesota Medical School, Minneapolis, MN, USA
[4] Division of Medicinal and Natural Products Chemistry, University of Iowa, Iowa City, IA, USA

6.1 Introduction

6.1.1 Overview

Fluoroquinolones are broad-spectrum antibacterials that have received considerable attention due to the development of resistance to other agents. Unfortunately, increased use is leading to widespread fluoroquinolone resistance. However, resistance has stimulated the development of new quinolone-like derivatives and studies to obtain a deeper understanding of quinolone action. One product of the work is a framework for describing how resistance emerges. That framework, called the mutant selection window hypothesis, provides a rationale for adjusting doses to restrict the selective amplification of mutant subpopulations. The present chapter describes our current understanding of quinolone action and resistance. We also update work on the mutant selection window and on the emerging problem of plasmid-borne fluoroquinolone resistance. We apologize in advance for omitting references to many fine papers – for editorial reasons the number of citations was limited.

Since the term resistance is often used loosely, it is necessary to define terms. Resistance is defined clinically by empirical breakpoints. An isolate is either resistant or susceptible. To express an increase in clinical resistance, we mean an increase in the prevalence of resistance, not a decrease in the susceptibility (minimal inhibitory concentration, MIC) of an individual isolate. With respect to resistant mutants and resistant subpopulations, we mean cells having a higher MIC than the relevant parental strain or majority population. Resistant mutants and subpopulations are not necessarily resistant in a clinical sense. A resistance mutation is a genetic change that results in a resistant mutant or subpopulation.

Bacterial Resistance to Antibiotics – From Molecules to Man, First Edition.
Edited by Boyan B. Bonev and Nicholas M. Brown.
© 2020 John Wiley & Sons Ltd. Published 2020 by John Wiley & Sons Ltd.

6.1.2   The Resistance Problem

Antimicrobial resistance begins as a local phenomenon that, if not controlled, expands to become a national and eventually an international problem. The first step is selective enrichment of resistant subpopulations, often through a stepwise process that gradually reduces susceptibility and increases the probability that the next step will occur. Once susceptibility to quinolones is lost, the resistant strains can spread if fitness remains high or if compensatory changes raise fitness sufficiently. In some cases, chromosomal resistance genes are mobilized by plasmids. Because plasmids often carry genes for resistance to several antimicrobials, they can promote the selection of quinolone resistance even when members of non-quinolone drug families are used. Many fluoroquinolone-resistant pathogens have now been reported (Table 6.1).

Hospital outbreaks of drug-resistant pathogens are the most visible resistance problem. Often these infections are caused by clonal spread of the pathogens. An early example emerged from the use of ciprofloxacin with *Staphylococcus aureus*. By the 1990s, more than 80% of the isolates from hospitals in many countries were considered fluoroquinolone resistant. This widespread dissemination may be connected to methicillin resistance, since fluoroquinolone-resistant, methicillin-sensitive strains have not spread. More recently, *Klebsiella pneumoniae* has made headlines. When a multidrug-resistant strain in North Carolina, USA, acquired resistance to carbapenems, it became very difficult to control. The strain, which is now resistant to fluoroquinolones and most other agents, challenges even the best infection control teams. We expect outbreak problems involving quinolones to expand, in part due to widespread, routine fluoroquinolone use in long-term care facilities. Resistant mutants enriched in those facilities can enter hospital systems as seriously ill patients move between long-term and acute-care institutions. The broad-spectrum nature of quinolones also contributes to nosocomial resistance problems: quinolone use disturbs digestive flora and allows the expansion of *Clostridium difficile*. This organism, which causes severe diarrhea, is difficult to control, especially when it becomes multidrug resistant. Outbreaks of these and other pathogens make local surveillance an important part of hospital infection control programs.

Community-acquired resistance tends to increase gradually and generally receives less media coverage than nosocomial resistance. However, resistance in the community can eventually eliminate the usefulness of an antibiotic class. Gradual loss of susceptibility is seen as an upward creep in MIC and increasing prevalence of resistant isolates. Controlling community resistance problems can be complex, as illustrated by the treatment of tuberculosis. On a global basis, *Mycobacterium tuberculosis* is becoming increasingly multidrug resistant (MDR, defined as resistance to isoniazid and rifampicin). For many years efforts to slow the spread of MDR tuberculosis relied on marginally effective fluoroquinolones, agents that were most effective with Gram-negative infections. In about 2000, new fluoroquinolones having good activity with Gram-positive bacteria were introduced into clinical practice, and two of those compounds, gatifloxacin and moxifloxacin, are being examined as replacements for older quinolones. Since tuberculosis treatment requires many months of drug administration, use of broad-spectrum quinolones could accelerate the emergence of resistance for many other pathogenic bacteria. Thus, what may be good for tuberculosis patients, may not be good for the overall community.

Table 6.1 Examples of fluoroquinolone resistance.

Bacterial species	Compound tested[a]	Geographic location	Prevalence of resistance[b]	No. isolates tested
Acinetobacter baumannii	Cip	Saudi Arabia, Adult ICU unit, King Fahad National Guard Hospital, Riyadh.	Increased from 79% (2004) to 90% (2009)	886
	Quinolones[c]	USA. Nashville General Hospital, Nashville TN	46% (also pan-resistant) (2006–2008)	247
Campylobacter sp	Cip, Nal	Poland	74%, 73% (2009–2010)	498[c]
	Cip, Nal	UK	22% (2001–2005)	1002[d]
Clostridium difficile	Cip, Lev, Mox	Spain, Hospital General Universitario Gregorio Marañón, Universidad Complutense, Madrid	92%, 90%, 73% (2000–2006)	101
Escherichia coli	Cip	Canada	8% (2007)	1702
	Cip	USA	Increased from 3% (2000) to 17% (2010)	1 836 598
	Cip	UK	Increased from 6% (2001) to 19% (2011)	228 138
Klebsiella pneumoniae	Quinolones[c]	South Korea, Samsung Medical Centre, Seoul,	10% (2009–2010)	414[e]
	Cip	Canada	12% (2007–2009)	274[f]
Mycobacterium tuberculosis	Ofl	China, Shanghai Pulmonary Hospital, Shanghai	72% (2007–2009)	368 MDR+ 126 XDR
	Lev, Ofl	India, various tertiary care hospitals treating TB in Delhi	11% (2007–2010)	465 MDR +18 XDR
Neisseria gonorrhoeae	Cip	USA STD clinics in 30 US cities	12% (30% in MSM)[g] (2005–2010)	34 600 (8117 MSM)[g]
	Cip	Australia	30% (53% in Victoria) (2012)	1212 (312 Victoria)
	Cip	Latin America (nine countries)	Increased from 2% (2000) to 31% (2009)	11 400

(Continued)

Table 6.1 (Continued)

Bacterial species	Compound tested[a]	Geographic location	Prevalence of resistance[b]	No. isolates tested
	Cip	Canada	Increased from 1% (2000) to 25% (2009)	>40 000
Pseudomonas aeruginosa	Cip, Lev	France Centre Hospitalier Universitaire de Lille, Lille	24%, 26% (2009)	115
	Cip	USA	Increased from 20% (1997) to 25% (2009)	924, 740
	Cip	Saudi Arabia, Adult ICU Unit, King Fahad National Guard Hospital, Ryadh.	Increased from 33% (2004) to 51% (2009)	855
Salmonella enterica	Nal	Nepal, Dhulikhel Hospital Kathmandu	77% (2009–2010)	114
	Cip, Nal	Senegal, Institut Pasteur, Dakar	4%, 27% (2009)	45
Salmonella typhi	Cip	Cambodia, Sotr Nikom District, Siem Reap Province	90% intermediate (2007–2011)	148
	Nal	India, Pondicherry	78% (2005–2009)	337
Staphylococcus aureus	Lev	USA	40% (2009)	4210
	Cip	Pakistan, Karachi	35% (2010–2011)	465
Streptococcus pneumoniae	Cip, Lev, Mox	Canada	1.7% (1998) to 2.4% (2009), 0.73% (1998) to 1.4% (2009) 0.31% (1998) to 0.48% (2009)	26 081

[a] Abbreviations: Nal (nalidixic acid), Cip (ciprofloxacin), Lev (levofloxacin), Ofl (ofloxacin), and Mox (moxifloxacin).
[b] Years in parenthesis indicate time of sampling.
[c] Specific names of the drugs were not reported; commas separate data for individual drugs.
[d] Poultry isolates.
[e] Community acquired bacteremia.
[f] Urinary tract infections.
[g] MSM: men who have sex with men.

6.1.3 Quinolone-class Compounds

Thousands of quinolones have been synthesized, with more than a dozen reaching the clinical market. Another half-dozen are in veterinary use. Since the various derivatives differ significantly in activity, we briefly introduce the major members of the class.

We define quinolone-class compounds loosely as including a variety of related agents that poison bacteria by trapping gyrase and topoisomerase IV on DNA as drug-enzyme-DNA complexes. The compounds, which include true quinolones, naphthyridones, and quinazolinediones, all have a pair of core ring structures (see nalidixic acid, Figure 6.1). To provide a framework for comparison, the compounds are often grouped into "generations." We prefer a generation scheme governed by relationships between action mechanism and drug structure, with a focus on lethal activity (Figure 6.1, Table 6.2). We note that clinicians tend to define quinolone generations according to clinical characteristics and indications (http://www.aafp.org/afp/2002/0201/p455.html); the two schemes only partially overlap.

All generation schemes acknowledge nalidixic acid as the prototype quinolone, although it is formally a naphthyridone (nitrogen at position 8). We consider nalidixic and oxolinic acids as representative first-generation compounds. Their lethal activity is characteristically blocked by co-treatment with inhibitors of protein synthesis, by anaerobic growth, and by suspension of bacterial cultures in saline. Lethal activity is also blocked by agents that interfere with the accumulation of hydroxyl radical, a highly toxic reactive oxygen species (ROS). Interfering with ROS accumulation has little effect on the ability of quinolones to block bacterial growth, indicating that blocking growth is mechanistically distinct from killing cells (for general relationships between blocking growth and killing bacteria see [1]).

The addition of fluorine at position C-6, a piperazinyl ring to C-7, and an exchange of carbon for nitrogen at position-8 produced norfloxacin (for numbering system, see ciprofloxacin, Figure 6.1). This second-generation compound has substantially improved activity and pharmacokinetic properties, but lethal activity is still blocked by the inhibition of protein synthesis. Norfloxacin can kill *Escherichia coli* under anaerobic conditions, but only at elevated drug concentrations [2], and it kills *E. coli* suspended in saline. How each of the structural changes creating norfloxacin enhance lethal activity is unknown.

Ciprofloxacin emerged from norfloxacin by replacement of the N-1 ethyl with a cyclopropyl group. That single change broadened the spectrum of infections covered and conferred lethal activity with non-growing (chloramphenicol-treated) cells. The latter observation defines ciprofloxacin as a member of the third generation. Ciprofloxacin remains the most active of the commercially available fluoroquinolones with Gram-negative bacteria.

Fourth-generation compounds, such as sparfloxacin, moxifloxacin, and gatifloxacin, are distinguished from ciprofloxacin by having a halogen or methoxy group attached to position C-8. The C-8 methoxy moiety is associated with improved activity with Gram-positive bacteria, and it increases activity with resistant mutants. With *E. coli*, the lethal activity of the fourth-generation compound PD161144 is unaffected by anaerobiosis, inhibitors of protein synthesis, or agents that interfere with the accumulation of reactive oxygen species.

We emphasize that our generation scheme is imperfect, because it may or may not apply when considering activity with organisms other than *E. coli*. For example, two C-8 methoxy compounds, gatifloxacin and moxifloxacin, kill growing *M. tuberculosis* equally. But when protein synthesis is inhibited by chloramphenicol, the lethal activity of gatifloxacin is blocked, while that of moxifloxacin is not [3]. By this criterion, moxifloxacin would be a fourth-generation quinolone and gatifloxacin would be in the first or second generation. Moreover, several important quinolones lie outside our generational scheme.

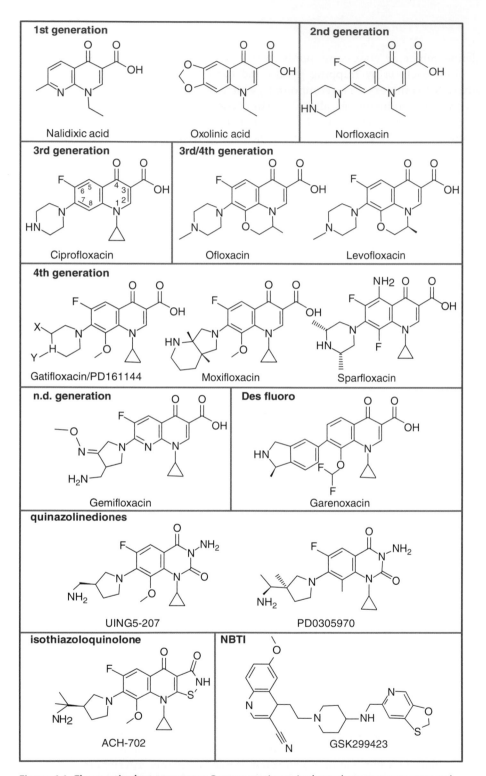

Figure 6.1 Fluoroquinolone structures. Representative quinolone-class agents are arranged according to categories. The numbering system for positions in quinolones is shown with ciprofloxacin. For gatifloxacin, substituents are X = methyl and Y = H; for PD161144 X = H and Y = ethyl.

Table 6.2 Major fluoroquinolone and quinolone-like compounds.

Type	Representative compound	Distinctive features
First-generation quinolone	Nalidixic acid, Oxolinic acid	No C-6 fluorine or C-7 ring system; lethal activity blocked by inhibition of protein synthesis or anaerobic conditions; narrow spectrum of treatable bacterial diseases.
Second-generation quinolone	Norfloxacin	C-6 fluorine, C-7 ring; lethal activity blocked by inhibition of protein synthesis, reduced by anaerobic conditions
Third-generation quinolone	Ciprofloxacin	N-1 cyclopropyl group; lethal activity only partially blocked by inhibition of protein synthesis
Third-fourth generation quinolone	Levofloxacin, Ofloxacin	Fused ring for N-1 and C-8 groups; most properties similar to ciprofloxacin
Fourth-generation quinolone	Moxifloxacin, Gatifloxacin, Sparfloxacin	C-8 substituent (*e.g.*, methoxy); elevated activity with GyrA resistance
Quinolone type not determined	Gemifloxacin	With some Gram-positive bacteria it has similar activity with gyrase and topoisomerase IV
Des-fluoro quinolone	Garenoxacin	Lacks C-6 fluorine; good activity with *Staphylococcus aureus*
Quinazolinedione	UING5–207, PD0305970	Lacks C-3 carboxyl, insensitive to GyrA resistance mutations
Isothiazoloquinolone	ACH-702	Lacks C-3 carboxyl, broad spectrum, good against Gram-positive bacteria and anaerobes

One is levofloxacin, a popular compound that shows activity with both Gram-negative and Gram-positive bacteria. The distinctive feature of levofloxacin is the fusion between the N-1 and C-8 substituents. This fused ring structure places levofloxacin between the third and fourth generations. Two newer agents, garenoxacin and gemifloxacin, also differ from the ciprofloxacin-based lineage. Garenoxacin is noteworthy because it lacks the C-6 fluorine characteristic of fluoroquinolones.

A variety of related compounds also form drug–topoisomerase–DNA complexes (Figure 6.1, Table 6.2). We briefly comment on the quinazolinediones, because they have been used in resistance studies. Two general classes emerged from initial work, the 2,4-diones and the 1,3-diones. Both dione types lack the carboxyl group that is characteristic of the quinolones. The absence of the carboxyl moiety renders the diones insensitive to the major resistance mutations in *gyrA* and *parC* [4, 5].

6.2 Mechanism of Action

6.2.1 Overview

Knowing how the quinolones act provides a context for understanding resistance. A key observation is that the compounds are "concentration-dependent killers." Thus, raising their concentration improves the ability to eradicate an infecting pathogen population, to reduce the chance that new mutants will arise, and to kill resistant mutants that might already be present. Below we describe several steps in quinolone action that occur at different drug concentrations.

At low quinolone concentrations, the drugs form bacteriostatic ternary complexes with DNA and either gyrase or topoisomerase IV; at higher concentrations, roughly above five-times the MIC, the compounds kill rapidly growing cells within an hour or two. Rapid killing occurs by two lethal pathways (Figure 6.2). In one, bacteria respond to lethal lesions by generating ROS that accelerate cell death [7, 8]. Killing by this pathway requires ongoing protein synthesis, which is required for ROS accumulation. Protein synthesis is also required for the chromosome fragmentation associated with nalidixic acid treatment. The second mode of killing, which is most prominent with the newest quinolones (*e.g.*, PD161144 with *E. coli*), occurs in the absence of growth, protein synthesis, or oxygen. Factors affecting rapid lethality are summarized in Table 6.3.

A slow form of cell death occurs when growing cultures of *E. coli* are incubated overnight with quinolone. This slow death, which is commonly expressed as minimal bactericidal concentration (MBC), is used to evaluate compounds for pharmaceutical purposes. MBC is usually lower than quinolone concentrations needed to rapidly kill, sometimes only two- or four-times MIC. Although we know that ROS do not contribute to MBC, little else is known about the lethal mechanism of slow death; thus, it is not discussed here.

6.2.2 Bacteriostatic Activity: Formation of Cleaved Complexes

DNA gyrase and DNA topoisomerase IV each has four subunits, two with ATPase activity (GyrB or ParE in gyrase and topoisomerase IV, respectively) and two with DNA-cleaving activity (GyrA and ParC). The enzymes act by passing one region of duplex DNA through another: a pair of nicks separated by four base-pairs are introduced into DNA to create a gate for DNA strand passage. The quinolones trap the enzymes in the gate-open form. Since the drug–enzyme–DNA complexes contain broken DNA, they are called "cleaved complexes." Use of other terms, such as "cleavage" or "cleavable" complexes, is based on the early idea that DNA cleavage occurs only after protein denaturation. This idea is probably incorrect. The ends of the DNA in the complexes are covalently bound to enzyme, which allows the topoisomerases to reseal breaks when the drug is removed by dilution. DNA resealing also occurs when cleaved complexes are treated with the metal chelator EDTA or when they are incubated at temperatures of approximately 65 °C.

Current work is guided by X-ray crystal structures of cleaved complexes (Figure 6.3). In these structures two drug molecules are seen intercalated into DNA at the nick sites such that the C-7 ring system faces GyrB (ParE) and the 3-carboxy end extends into GyrA (ParC). In GyrA (ParC), the drug carboxyl group forms a magnesium-water bridge with GyrA-83 and GyrA-87 (*E. coli* gyrase nomenclature). This bridge stabilizes quinolone-containing complexes and is an important feature of activity. Preventing the formation of the bridge is the major source of target-based resistance [4, 5, 9].

We recently examined drug binding interactions using a derivative of ciprofloxacin that can cross-link to cysteines if located near the drug binding site [10]. As predicted by published crystal structures, *E. coli* GyrB-466C formed cross-links with the distal end of the quinolone C-7 ring. Surprisingly, cross-link formation was also observed with GyrA-81C. Since GyrB-466 and GyrA-81 are far apart in the crystal structure

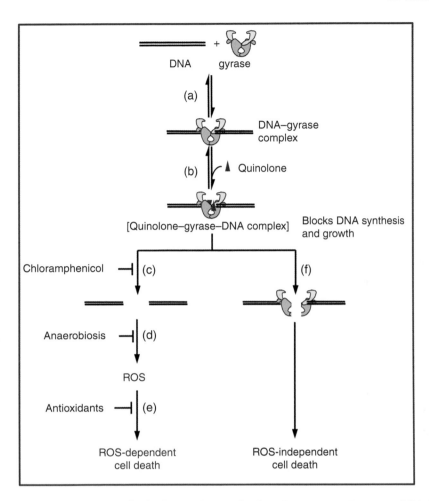

Figure 6.2 **Overview of quinolone action mechanism.** Gyrase or topoisomerase IV binds to DNA (a). Quinolone binds to the complex, DNA breaks are introduced by the enzyme, and a cleaved complex forms (b). The cleaved complex reversibly blocks DNA replication, transcription, and bacterial growth. Quinolone concentrations that generate cleaved complexes correspond to MIC. DNA breaks released from the cleaved complexes stimulate two lethal pathways whose relative usage depends on quinolone structure. The ROS-dependent pathway requires ongoing protein synthesis to release DNA breaks (c) and aerobic conditions to allow a cascade of ROS (d). Antioxidants can interfere with the ROS cascade (e) and block cell death. With some highly active fluoroquinolones, cell death is insensitive to agents that block protein synthesis or the accumulation of ROS (f), probably because death occurs before ROS can act (quinolones that kill by the ROS-independent pathway stimulate accumulation of ROS). Compounds such as norfloxacin and ciprofloxacin diplay features of both pathways. *Source:* The figure is modified from Drlica et al. [6]. © American Society for Microbiology.

(17 Å), two quinolone-binding modes must exist. Additional support for a secondary binding mode comes from work on derivatives of ciprofloxacin in which an aryl group is added to the C7 piperazinyl ring [11]. When the aryl moiety is dinitrophenyl, the fluoroquinolone is particularly insensitive to GyrA resistance substitutions, and it selectively enriches an uncommon *gyrA* resistance mutation. Both observations are

Table 6.3 Factors that affect quinolone lethality with little effect on cleaved complex formation.

Treatment/ mutation	Quinolone tested	Effect on lethality	Proposed explanation
Anaerobic growth	Nalidixic acid	Blocks	Oxygen required for ROS cascade
	Ciprofloxacin	Intermediate	Functions by both pathways
	PD161144	No effect	Death occurs without the need of ROS, ROS are still produced but have little contribution to cell death
Chloramphenicol	Nalidixic and Oxolinic acids	Blocks	Exposure of DNA breaks to fragment chromosomes requires a newly synthesized, unstable protein
	Ciprofloxacin	Intermediate	Functions by both pathways
	PD161144	No effect	Chromosome fragmentation occurs without the need for new protein
Plumbagin	Ciprofloxacin	Reduces	Raises superoxide to protective concentrations, induction of SoxRS
Thiourea	Norfloxacin, Nalidixic acid	Blocks	Radical scavenger
2,2′ bipyridyl	Norfloxacin, Nalidixic acid	Blocks	Iron chelator, blocks Fenton reaction and production of hydroxyl radical
Δ*katG*	Norfloxacin	Enhances	Elimination of catalase, a safety valve for ROS cascade
Δ*lepA*	Oxolinic acid	Reduces	Loss of EF4-mediated inhibition of tmRNA
Δ*sodA*Δ*sodB*	Norfloxacin	Reduces	Raises superoxide to protective concentrations, induction of SoxRS
Δ*yihE*	Nalidixic acid	Enhances	Loss of inhibition of MazF toxin

inconsistent with existing X-ray structures of cleaved complexes. Determining the precise orientation of quinolones in the second mode is the subject of ongoing studies.

Cleaved complexes block growth by interfering with the movement of replication and transcription complexes. Cleaved-complex formation and the inhibition of DNA replication are reversible [10]; consequently, they do not by themselves account for quinolone-mediated cell death. Indeed, blocking replication by other means, such as by treatment with hydroxyurea or by shifting a *dnaB* (Ts) mutant to restrictive temperature in minimal medium, has no effect on rapid, quinolone-mediated killing.

A common misconception is that blocking growth, revealed as MIC or a drop in efficiency of plating, is equivalent to cell death. In these two tests, the assay system measures culture growth or colony formation in the presence of drug – when growth is blocked by a lower drug concentration than required to kill cells, the assay measures bacteriostasis, not cell death. Killing is measured by incubation in the presence of drug followed by colony formation in the absence of drug. We emphasize the distinction between bacteriostatic and bactericidal tests because recent work with a variety of antimicrobials has been misleading.

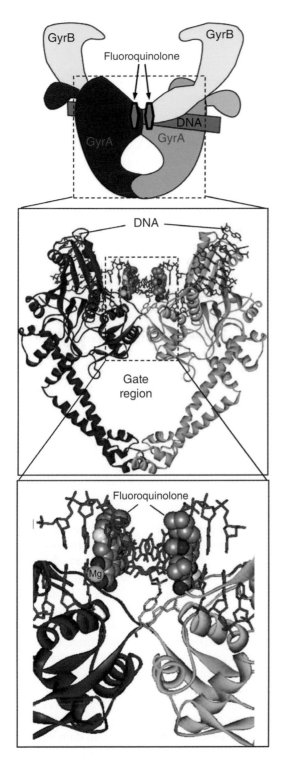

Figure 6.3 Structure of cleaved complexes. The top part of the figure shows a low-resolution cartoon of the two GyrA and two GyrB subunits in gyrase (topoisomerse IV shows a similar structure with its ParC and ParE subunits). The center of the figure shows detail from X-ray crystallography with GyrB (ParE) omitted for clarity. The two quinolone molecules (moxifloxacin) bound per enzyme are seen to be in parallel binding sites intercalated into DNA at the nicks. The quinolone C-7 ring system faces GyrB (ParE), while the carboxyl end of the drug extends to helix IV of GyrA (ParC). The bottom part of the figure shows an enlargement of the DNA gate region of GyrA. *Source:* The figure was prepared from data in Protein Data Base, entry PDB 2XKK.

With fluoroquinolones, the mechanistic distinction between blocking growth and lethal activity is clear. One line of support for this statement concerns the observation that a variety of conditions block rapid lethal activity without affecting MIC [2]. Among these are anaerobic growth and treatment with antioxidants. Another line reveals perturbations that preferentially enhance lethal action. These include: (i) deficiencies in *yihE* and *katG* [8, 12], two genes that in the wild-type form protect from death arising from quinolone-mediated accumulation of ROS, (ii) deficiencies in some repair proteins, and (iii) deficiencies in a variety of genes that are still uncharacterized [13]. The opposite is also true. For example, deletion of *lepA*, which encodes a translation elongation factor, *protects* from cell death without affecting MIC. Finally, quinolone concentrations required to cause chromosome fragmentation and rapid cell death are higher than required to block growth [14, 15]. In the following two sections we consider events associated with rapid death.

6.2.3 Bactericidal Activity: Protein-Synthesis-Dependent Pathway

For first- and second-generation quinolones, an unidentified factor seems to release DNA ends from the constraint exerted by gyrase and topoisomerase IV. That release, which requires ongoing protein synthesis [14, 15], fragments chromosomes and stimulates a lethal pathway involving several genes and the production of toxic ROS (Figure 6.4). The bactericidal action of third- and fourth-generation compounds, which is less dependent on ongoing protein synthesis, is discussed in the next section.

An early step in the genetic pathway to death is likely to be generation of truncated proteins that insert into cell membranes. One contributing factor is the MazF toxin, an endoribonuclease that cleaves mRNA and blocks translation during periods of stress. At non-lethal concentrations of quinolone, MazF is protective; its action allows cells time to efflux noxious molecules and repair damage. But at high lethal levels of quinolone, toxic levels of truncated proteins are expected to accumulate from MazF action.

Another contributing factor is EF4, a ribosomal elongation protein normally sequestered in the cell membrane. During stress, EF4 enters the cytosol, binds to stress-stalled ribosomes, and stimulates ribosomal back-translation. That allows protein synthesis to recover from moderate stress. But EF4 also blocks tagging of truncated proteins for degradation, which would allow the accumulation of toxic peptides. Indeed, the absence of EF4 *protects E. coli* from being killed by quinolones (wild-type protein would be destructive) [16]. Thus, EF4 and MazEF appear to be bifunctional proteins that protect from moderate stress but stimulate death when stress is severe. How their action is connected to quinolone-mediated chromosome fragmentation is currently unknown.

Insertion of truncated proteins into the cell membrane activates a two-component membrane stress-response system called Cpx. When stress is high, activation of Cpx stimulates another two-component system (Arc) that then leads to a cascade of reactive oxygen species [17]. (The destructive feature of Cpx is revealed by a CpxR deficiency being protective.) The Cpx system also serves to repair membrane protein damage [18] and to mitigate MazF toxicity [12] when stress is moderate. Thus, Cpx has a destructive and a protective action; it represents a third bifunctional system involved in determining whether stress-mediated damage is repairable.

Several safety valves exist to keep moderate stress from killing cells. One is a protein kinase called YihE. The absence of YihE increases the lethal action of nalidixic acid by

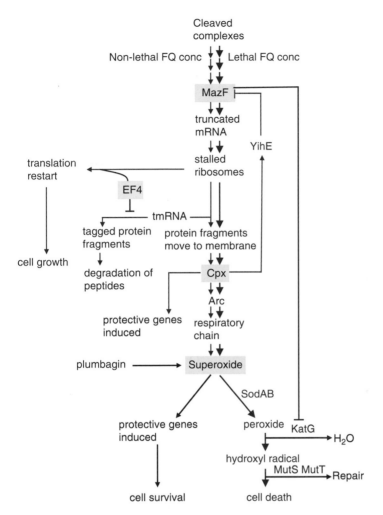

Figure 6.4 Scheme showing the live-or-die stress response pathway envisioned for quinolones. Stress caused by moderate fluoroquinolone concentration (blue type and lines) results in a protective stress response. The quinolones form cleaved complexes that in an unknown way stimulate the MazF toxin to cleave mRNA, thereby halting translation and allowing cells time to repair damage. The protein fragments resulting from MazF-mediated mRNA cleavage are tagged for degradation by tmRNA and EF4, a ribosomal protein that facilitates ribosomal restart. Peptides that enter cell membranes stimulate the Cpx two-component system to induce genes involved in membrane protein repair. Safety valves YihE kinase and KatG catalase negatively regulate MazF and detoxify peroxide, respectively. At high quinolone concentrations, chromosome fragmentation requiring ongoing protein synthesis initiates a lethal pathway that overwhelms the protective response and results in cell death (red lines). mRNA cleavage by MazF is extensive, and EF4 blocks the tagging of peptides by tmRNA. Thus, high levels of protein fragments accumulate and enter cell membranes. That causes the Cpx system to induce the Arc two-component system, which in turn leads to high-level production of superoxide. Superoxide dismutates to peroxide; the Fenton reaction then converts peroxide to hydroxyl radical. Hydroxyl radical damages nucleotides and many macromolecule types, causing mutations and cell death. To assure death, MazF cleaves *katG* mRNA, which lowers the level of KatG, a protein that would otherwise reduce peroxide levels. Other protective functions, such as induction of membrane repair by Cpx and induction of the SoxRS regulon by superoxide, are overwhelmed. Bifunctional factors are highlighted in yellow.

100-fold, but only if the MazEF toxin is present [12]. Thus, YihE appears to be a negative regulator of MazF. Another safety valve is the *katG* catalase/peroxidase, which detoxifies peroxide by converting it to water. Deletion of *katG* increases norfloxacin lethality by 20-fold without affecting MIC [8].

Superoxide occupies a key position in the lethal stress response. At moderate stress levels, superoxide is thought to accumulate and stimulate protective responses, such as induction of the SoxRS regulon. During periods of harsh stress, superoxide appears to accumulate to high levels that dismutate to peroxide. Peroxide is then converted to hydroxyl radical whose toxic effects overwhelm the protective functions of superoxide. The bifunctional nature of superoxide explains how low concentrations of metabolic generators of superoxide, such as plumbagin and paraquat, reduce the lethal action of bleomycin and the quinolones, even though high concentrations of plumbagin kill bacteria.

The final steps in ROS-mediated killing are largely due to DNA damage caused by the production of a hydroxyl radical and 8-oxy-guanine [19]. Other contributions to cell death derive from protein carbonylation and lipid peroxidation, both arising from hydroxyl radical action. Chromosome fragmentation associated with first-generation quinolones seems to be repaired during the overnight incubation required to measure cell death, as blocking the ROS cascade blocks death unless cells are deficient in repair functions.

In summary, quinolones kill in part through a surge in reactive oxygen species: mutations and treatments expected to enhance the surge increase quinolone lethality, and those that suppress the surge reduce lethality. Our current speculation is that chromosomal breaks stimulate the accumulation of ROS, which is self-propagating (production of hydroxyl radical promotes additional macromolecular damage that then promotes more accumulation of hydroxyl radical). Bifunctional proteins and superoxide collectively make a live-or-die decision based on the level of stress [12].

We note that several laboratories recently challenged the general idea that an ROS cascade is associated with the lethal action of antimicrobials [20–23]. The challenging arguments, which ignored several important supportive results, stimulated a new round of refinement that has now solidified a role for ROS [24]. The challenges also revealed a need to clarify the experimental definition of lethal stress response. One issue is that the ROS contribution often derives from a cellular *response* to antimicrobial-mediated damage and thus occurs after the primary lesion forms; consequently, studies that focus on factors acting at or before lesion formation, such as drug uptake, efflux, and target interactions, are largely uninformative. Their effects can be removed from consideration by normalizing lethal activity to growth inhibition (MIC) [8]. Other issues are the choice of test compound and the degree of exposure. Both are important to observe a ROS contribution, as is timing (ROS accelerate killing rather than increase the extent). Quinolone studies are also complicated by the existence of two killing pathways. As we point out later, one pathway seems to be independent of ROS, presumably because killing occurs before ROS can act. When a quinolone acts using both pathways, as is the case with norfloxacin, drug concentration becomes a crucial parameter [2].

Two aspects of the ROS-lethality hypothesis remain unresolved. One concerns the widespread use of an iron chelator (2,2′-bipyridyl) to attribute lethal action to the conversion of peroxide to hydroxyl radical. The possibility of off-target effects has not been eliminated. A second issue is anaerobic killing. Some antimicrobials are lethal in the absence of oxygen. We postulate that under anaerobic conditions, non-oxygen radicals perform the same function as ROS during aerobic growth. This is an unexplored area of antimicrobial lethality.

6.2.4 Bactericidal Activity: Protein Synthesis-Independent Pathway

At least two lethal pathways must exist, because some of the newer fluoroquinolones, such as PD161144, kill *E. coli* when protein synthesis is blocked by chloramphenicol [2], when the ROS cascade is prevented by anaerobiosis [2], and when hydroxyl radical accumulation is blocked by treating cells with the radical scavenger thiourea and the iron chelator 2,2′-bipyridyl [25] (iron is important for conversion of peroxide to hydroxyl radical). Since some fluoroquinolones destabilize drug–gyrase–DNA complexes formed on isolated bacterial nucleoids [15], we postulate that protein-synthesis-independent cell death arises from destabilization of the gyrase subunits. Destabilization then leads to bacterial chromosome fragmentation. The hypothesis is supported by the observation that an alanine-to-serine substitution located on the GyrA-GyrA interface, which is expected to destabilize the GyrA-GyrA interaction, allows first-generation quinolones, such as nalidixic acid, to kill *E. coli* in the presence of chloramphenicol [15]; this activity is not observed with wild-type cells.

Although drug-structure studies have not produced a clear mechanism for lethal chromosome fragmentation, they do support the existence of the second lethal pathway. One important aspect of drug structure is the C-7 ring system. As pointed out earlier, gatifloxacin and moxifloxacin, which differ only in their C-7 ring systems, differ in their ability to kill mycobacteria treated with chloramphenicol. Another is the 3-carboxyl group. The lethal action of a quinazolinedione, which lacks the carboxyl, shows reduced inhibition by chloramphenicol with mycobacteria [26]. A third is the N-1 cyclopropyl group: ciprofloxacin, which exhibits some lethal activity in the presence of chloramphenicol, differs from norfloxacin, which does not, only at the N-1 position. Also important is the C-8 substituent. When this group is fused to the N-1 moiety, as is the case with levofloxacin, lethal activity in the presence of chloramphenicol is reduced. Understanding how these structural features contribute to lethal activity is hindered by the absence of a simple biochemical surrogate for cell death.

6.2.5 Special Features of Topoisomerase IV

Formation of drug–enzyme–DNA complexes with topoisomerase IV may lead to a third death scenario. With this enzyme, killing by the older quinolones is blocked by chloramphenicol. However, inhibition of replication occurs slowly and parallels cell death ([27]; when gyrase is the primary target, DNA replication is inhibited much faster than cells are killed). Moreover, with topoisomerase IV, resistance is genetically codominant [27]; with gyrase, it is recessive. These differences between gyrase and topoisomerase IV probably reflect the cellular location of the enzymes more than the drug mechanism, since complexes formed with topoisomerase IV are capable of rapidly blocking replication in *recA, seqA, or gyrB* mutants [27], and X-ray structures of cleaved complexes are similar for the two enzymes. Gyrase is likely located close to replication forks that are blocked rapidly, while topoiosmerase IV may normally be positioned behind replication forks.

Differences between gyrase and topoisomerase IV are likely to be important for the emergence of resistance. With Gram-positive organisms, many fluoroquinolones have topoisomerase IV as the primary target; therefore, these compounds are not expected to have rapid killing power. Resistance with *S. pneumoniae* emerges at lower drug concentrations for levofloxacin, which has topoisomerase IV as the primary target, than for moxifloxacin, which has gyrase as the primary target.

6.3 Quinolone Resistance: General Aspects

6.3.1 Sources of Resistance

Although resistance is traditionally addressed by developing new compounds, efforts to understand the sources of resistance may lead to more durable solutions to the resistance problem. To provide an overview, we briefly sketch the major sources, which range from molecular events to our consumption practices.

Mutations are currently the major genetic source of quinolone resistance. They derive largely from errors in DNA replication and repair – currently we can do nothing to stop these natural processes. For many pathogens, mutations occur spontaneously at a frequency of about one in 10^7 cells. This basal level is raised orders of magnitude by the mutagenic action of the quinolones themselves, largely through induction of error-prone repair pathways. Some pathogen strains also contain faulty replication proteins called mutators (an example is the *E. coli dnaQ-49* mutator, which creates a deficiency in the proofreading activity of DNA polymerase). Mutators usually represent only a small fraction of a bacterial population, probably because they tend to reduce bacterial fitness; however, they can be significant when weak selective pressure allows them to increase in abundance.

A second source of resistance derives from plasmids that carry resistance genes. Plasmid-mediated resistance stems from mobilization of chromosomal genes encoding efflux pumps, enzymes that alter quinolone structure, and proteins that interfere with quinolone–topoisomerase interactions. Broad-spectrum plasmids disseminate these types of resistance factors to a variety of bacterial species.

Our use of quinolones constitutes a different type of source, one that we can influence. Indeed, if we never used quinolones, resistance would be rare. A major "treatment-based" source of resistance is widely acknowledged to be overuse. Many examples show a correlation between the use of antibiotics and the prevalence of resistance. Over the last two decades quinolone use has increased dramatically. In one U.S. study, fluoroquinolone prescriptions rose about 9% per year between 1990 and 1998; in 1998 they were almost 13 million annually [28]. In another study, quinolone prescriptions tripled from 7 million in 1995 to 22 million in 2002 [29]. Thus, it is not surprising to see examples in which resistance also increased (Table 6.1). A more recent survey of fluoroquinolone prescriptions indicates that they may have peaked in the USA: quinolone prescriptions were 120 per 1000 outpatients in 2007; by 2010 that number dropped to 94/1000 (http://www.cddep.org/resistancemap/use/quinolones). Whether decreased use will actually translate into a decrease in the prevalence of resistance is not known.

Less widely appreciated is the role played by antimicrobial concentration. Resistant mutants are selectively enriched when quinolone concentrations fall inside a specific concentration range. For many pathogens, the quinolone doses currently used place drug concentrations inside that range for much of the dosing interval. For these situations, we expect fluoroquinolone dose to affect the number of prescriptions required to create a resistant infection: lower doses should generate resistant infections more often than higher doses.

Agricultural use also contributes to resistance. For example, enrofloxacin was licensed for use in Danish agriculture in 1991. Within two years, fluoroquinolone resistance appeared in *E. coli* isolated from cattle and in *Staphylococcus hyicus* obtained from pigs [30]. By 1998, 30% of the *S. aureus* isolates obtained from poultry in Denmark were ciprofloxacin-resistant. Increased prevalence of resistance among isolates of *Salmonella* and *Campylobacter* is also attributed to agricultural use of fluoroquinolones. In 2005 the US Food and Drug Administration won a court ruling that restricted some agricultural use of fluoroquinolones, but unlimited use is still common in many parts of the world.

A consequence of the clinical and agricultural use of quinolones is environmental contamination, largely because the agents are very stable. For example, in laboratory experiments we found that ciprofloxacin placed in agar retains full activity for at least 60 days of incubation at 37 °C. Excretion and disposal place the compounds in groundwater where they may contribute to the emergence of plasmid-mediated resistance.

6.3.2 Effect of Quinolone Concentration: The Mutant Selection Window

6.3.2.1 Definition of the Window

When bacterial cultures are plated on agar containing various concentrations of antibiotic, the number of colonies recovered drops sharply as drug concentration increases. In some cases, mutant subpopulations are evident as shoulders in the mutant recovery curve as it drops. This type of measurement, which is commonly applied with *S. aureus*, is called population analysis. When we examined the behavior of mycobacteria to fluoroquinolones, we noticed a distinct response. Increasing the drug concentration shows no effect on colony formation until MIC is approached; then colony recovery drops sharply. At higher drug concentrations, a broad plateau in colony recovery is seen, and at very high concentrations mutant recovery drops sharply again. This second drop occurs at the MIC of the least susceptible, first-step (next-step) mutant subpopulation. For convenience we call this mutant MIC the mutant prevention concentration (MPC), because it severely limits the recovery of mutants (at concentrations above MPC, bacterial growth requires the acquisition of two or more concurrent resistance mutations, which is a rare event). At drug concentrations slightly below MIC, selection pressure is low, and little selective amplification occurs. The concentration range between MIC and MPC allows selective growth of resistant mutants – that range is called the mutant selection window. The selection window concept provides a framework for describing the emergence of antimicrobial resistance.

Selective pressure can extend to concentrations much lower than MIC when very sensitive competition methods are used for detection [31]. Consequently, low-level environmental contamination with antimicrobials and the use of low antimicrobial concentrations as growth promoters with food animals are not innocuous.

6.3.2.2 Effects of Quinolone Species, Resistance Mutations, and Plasmids on the Selection Window

The selection window idea explains major features of resistant mutant enrichment. For example, fluoroquinolone concentration (position in the window) affects the identity of the mutants enriched. With *Mycobacterium smegmatis*, non-gyrase mutants are selected at low quinolone concentrations (slightly above the MIC for 99% of the population). As fluoroquinolone concentration is increased, the frequency of non-gyrase variants drops. In contrast, *gyrA* mutants, which were initially a small fraction of the mutant population, become increasingly prevalent. At very high fluoroquinolone concentration, colonies arise only from the GyrA variant having the highest MIC. A similar response is seen with *S. pneumoniae*, although both gyrase and topoisomerase IV are targets. Non-topoisomerase mutants are recovered at low fluoroquinolone concentrations; target mutants are recovered at higher concentrations [32]. Whether GyrA or ParC variants of *S. pneumoniae* are obtained depends on the particular fluoroquinolone examined. Thus, mutants with low MIC tend to be more abundant than those that are less susceptible; use of low doses and weak derivatives is more likely than more stringent conditions to begin a stepwise climb to resistance.

A second feature is the shape of the mutant selection curve. The presence of a broad plateau in mutant recovery is most clearly observed when only one intracellular target (gyrase) is present. When the pathogen has two targets (gyrase and topoisomerase IV), the curve becomes more complex, often with inflection points. A plateau can be seen with two-target situations if susceptibility of the two targets differs sufficiently, as when *S. aureus* is exposed to norfloxacin [33]. This fluoroquinolone strongly prefers topoisomerase IV over gyrase. The plateau is reduced, sometimes to an inflection point, when a compound attacks the two targets more equally, as when ciprofloxacin is administered to *S. aureus* or when gemifloxacin is applied to *S. pneumoniae*. A steep selection curve (narrow selection window) is also seen when inhibitor action is unaffected by target mutations. For example, some quinazolinediones inhibit growth of wild-type cells and GyrA mutants with equal effectiveness (MIC and MPC are similar). Having a steep selection curve should reduce the probability that treatment will enrich mutant subpopulations, because normal fluctuating drug concentrations will be inside the window for shorter times than when the window is broad.

Mutations that perturb intracellular drug concentration are expected to shift the position of the window relative to the extracellular drug concentration, the concentration that is readily measured. For example, up-regulation of efflux is expected to lower the intracellular drug concentration, which would make the window higher when based on extracellular concentration: MIC and MPC would be higher. An upward shift in the window boundaries is also expected for factors that: (i) reduce drug uptake (*e.g.*, mutation of *ompF*), (ii) interfere with cleaved complex formation (*e.g.*, presence of *qnr*), and (iii) degrade quinolones (e. g presence of *aac (6')-Ib-cr*). Conversely, efflux inhibitors are expected to shift the selection curve to lower concentrations, because they would elevate intracellular drug concentration. That explains why an efflux inhibitor can appear to reduce mutation frequency by orders of magnitude when based on extracellular concentration: extracellular concentration may be above the intracellular window.

A different type of change in the selection curve is seen in the plateau level (inflection point with some drug–pathogen combinations). This value represents the relative size of the resistant mutant subpopulation. For example, the level is expected to be higher when mutators are present. A dramatic effect is expected from plasmid-mediated resistance, because plasmids enter a bacterial population at high frequency, sometimes as high as 10^{-2}. That makes the plateau level high. Conditions that raise the plateau level increase the probability that the entire bacterial population will become resistant.

The selection window hypothesis also addresses combination therapy (administration of two or more antimicrobials lacking cross-resistance). The simultaneous presence of two antimicrobials should close the selection window, because the probability is low that a double mutant will arise. A key to successful combination therapy is to keep the concentration of both drugs above their respective MICs – if that is not achieved, periods will exist in which one drug is alone inside the selection window, a situation that is equivalent to monotherapy. Attaining good pharmacokinetic overlap is difficult, because pharmacokinetic profiles rarely coincide. One solution is to chemically link fluoroquinolones with another antimicrobial, as has been done with rifampicin. Then the two will have identical pharmacokinetic profiles.

6.3.2.3   Use of the Selection Window Hypothesis versus Traditional Dosing Strategies
The selection window hypothesis differs qualitatively from the older idea in which the danger zone for selection of resistant mutants lies below MIC rather than between the MIC and MPC. The two ideas make different predictions about the emergence of resistance. According to the older view, eradication of the susceptible population suppresses acquisition of resistance ("Dead bugs don't mutate"). This idea applies when resistance arises largely from newly formed resistant mutants, *i.e.*, from mutations occurring after administration of a drug. Such would be the case when quinolone treatment induces the mutagenic SOS response. In contrast, the selection window hypothesis maintains that resistance can emerge from mutations present before *and* during treatment with quinolone. Examination of many pathogen-antimicrobial combinations reveals that recommended doses often place drug concentrations inside the selection window where the drugs enrich resistant mutant subpopulations.

An important feature of the window idea is that mutant subpopulations present before treatment can lead to resistance even if the susceptible population is quickly eradicated. Evidence for the presence of pre-existing mutants derives from an experiment in which noses of tuberculosis patients were sampled for *S. aureus*. This bacterium is very susceptible to rifampicin, a drug commonly used to treat tuberculosis. Subpopulations of rifampicin-resistant *S. aureus* present before treatment were readily recovered after a treatment that eradicated the susceptible population.

6.3.2.4   Pharmacodynamics and the Selection Window
Pharmacokinetic-pharmacodynamic analyses seek to relate dose, fluctuating drug concentration, and clinical outcome to identify effective doses. Antimicrobial efficacy is associated with two parameters: (i) drug activity with a particular pathogen, generally measured as MIC, and (ii) the drug concentration achieved at the infection site. For fluoroquinolones, the two parameters are combined by dividing the area under the time–concentration curve in a 24-hour period (AUC_{24}) by MIC. With some infections, a threshold value of this pharmacodynamic index (AUC_{24}/MIC) correlates empirically with favorable patient and microbiological outcome (the maximal concentration divided by MIC also correlates with outcome).

The dose that would provide a favorable clinical outcome for large numbers of patients can be estimated by combining the drug concentration obtained for patient populations (AUC determinations) with bacterial susceptibility (MIC) from large collections of clinical isolates. Conceptually, this empirical approach can be applied to the resistance problem by substituting MPC for MIC, since MPC is the MIC of the least susceptible mutant subpopulation. However, we do not currently have enough measurements of MPC for accurate calculations. A key advance would be an assay for MPC that could be readily automated by clinical laboratories, since that would generate the MPC data set needed to determine mutant-restricting doses.

As an alternative to MPC-based calculations, animal studies are performed to determine a multiple of MIC that blocks the growth of resistant subpopulations. Those data are then used to bridge to human populations. Unfortunately, MIC is not always proportional to MPC. Consequently, the values of AUC_{24}/MIC that restrict growth of mutant subpopulations can vary significantly among bacterial isolates. Moreover, the animal studies may not be large enough to encompass the full range of resistant mutant subpopulations. Thus, the empirical MIC-based approach may underestimate the dose needed to restrict the enrichment of resistant mutants. Nevertheless, MIC-based methods are useful for identifying doses that will *fail* to suppress the emergence of resistance.

6.3.2.5 Experimental Support for the Selection Window Hypothesis

The mutant selection window is defined with static drug concentrations, using growth on agar plates [34] or in large volumes of liquid medium [35]; in contrast, infections are usually treated with fluctuating concentrations. Consequently, it has been important to determine how well the static boundaries apply to the dynamic situation. Measurements with *in vitro* dynamic models show that the window can be observed with fluctuating antimicrobial concentrations for quinolones, vancomycin, and daptomycin. It is also readily seen in rabbits infected with *S. aureus* and treated with levofloxacin or vancomycin. In all cases, static data fit well with dynamic measurements. Testing the selection window idea with patient populations is difficult, because very large clinical trials are needed to see the emergence of resistance.

6.3.2.6 Lethal Action and Resistant Mutant Selection

As emphasized in a previous section, bacteriostatic and bactericidal events associated with fluoroquinolone treatment are mechanistically distinct. They are also distinct in their effects on resistance. Restricting resistance by keeping concentrations above the mutant selection window is based on blocking mutant *growth*. Killing the bulk, susceptible population is an added effect that directly reduces pathogen numbers. That reduction should reduce treatment times, which in turn should lower costs, toxic side effects, and the chance that new resistance will develop. Removal of susceptible cells from an infection should also increase the probability that host defense systems eliminate resistant mutants.

Direct killing of resistant mutants is a separate issue. Some fluoroquinolones, such as the C-8-methoxy derivatives, more avidly kill resistant mutants than their C-8-H counterparts. This mutant-active feature could contribute to the C-8-methoxy compounds being better at restricting the selection of resistant mutants. Enhanced killing of resistant mutants should also contribute to the ability of a compound to control a larger bacterial population.

The contribution of lethal effects is addressed by pharmacodynamic considerations. When values of AUC_{24}/MIC are related to reduction of pathogen number, they take into account the lethal activity of the antimicrobial. Likewise, AUC_{24}/MPC includes the effects of lethal activities on resistant mutant subpopulations. As long as a threshold of AUC_{24}/MPC is exceeded, lethal compounds should not require maintenance of fluoroquinolone concentrations above the MPC throughout therapy to restrict the emergence of resistance. Examination of experimental results (rabbits infected with *S. aureus* and treated with levofloxacin) show restricted amplification of resistant mutant subpopulations when fluoroquinolone concentrations are above MPC for only 20% of the dosing interval [36]. In this case, MPC overestimates the concentration needed to suppress the emergence of resistance.

6.3.2.7 Application of the Selection Window Hypothesis
The selection window provides principles for comparing compounds and doses for the ability to restrict the emergence of resistance during treatment. Thus, measurement of MPC can be added to traditional analysis of new compounds to develop those most likely to restrict the emergence of resistance. Measurement of MPC can also be used to compare existing compounds and doses that have already been approved. Both applications can be implemented with information in the published literature.

Restricting the emergence of resistance by changing dosing guidelines from an MIC basis to an MPC basis is more complex. One problem is that measuring MPC is labor intensive and not easily adapted to the laboratory setting of clinical microbiology. Thus, obtaining the necessary pharmacodynamic data requires technical improvement. A second issue is that dosing to exceed MPC and restrict the emergence of resistance generally requires doses that are higher than needed to cure susceptible disease, and raising the dose may increase the frequency of toxic side effects.

Other potential problems directed at MPC-based dosing include: (i) effects of body compartments on susceptibility and drug concentration, and (ii) effects of mixed infections with pathogens having very different values of MPC. Such problems are not unique to MPC; they are also encountered with current MIC-based dosing. In general, fluoroquinolones are so potent that raising the dose is unlikely to apply selective pressure that was not already present with MIC-based dosing. We conclude that a major issue is whether restricting the emergence of resistance is worth the added cost of larger drug amounts and potential side effects.

6.3.2.8 Consideration from Evolutionary Biology
A recent discussion of the evolutionary aspects of dosing by Day and Read [37] introduces effects on competing organisms. The report concludes that the optimal dosing strategy is to use either the highest tolerable dose or the lowest clinically effective dose. The highest tolerable dose may or may not place drug concentrations above the upper boundary of the selection window (*e.g.*, the MPC). When the highest tolerable dose produces concentrations inside the selection window, it will apply selective pressure. For most antimicrobial-pathogen combinations examined, FDA-approved doses place drug concentrations inside the selection window. Thus, dosing as high as tolerable is not necessarily a way to severely restrict the emergence of resistance.

Another idea proposed by Day and Read, to use the lowest concentration that is clinically effective, is not supported by experimental evidence; low concentrations that are just above MIC enrich mutant subpopulations faster and more extensively than higher doses. The key idea is that low doses allow a larger fraction of the pathogen population to begin the "hill climb" to resistance.

A third issue requiring clarification is the idea that dosing above MPC cannot be used as a rule of thumb, because horizontal transfer occurs at a high frequency. MIC and MPC are points on the x axis of population analysis curves that define the selection window. Frequency is the y axis; it differs according to the source. For example, spontaneous mutations occur at a frequency of less than 10^{-7}, mutator mutations increase spontaneous frequency to about 10^{-4}, horizontal transfer is about 10^{-2}, and the value is 1 when the entire population is resistant. Thus, the chance that resistance will emerge depends on both the frequency at which resistance arises and the concentration relative to the selection window. Whether keeping concentrations above MPC is sufficient to restrict the emergence of resistance transferred horizontally has not been tested experimentally.

6.3.3 Stepwise Accumulation of Resistance Mutations

Unless pathogens are exposed to high quinolone doses, resistance emerges from the gradual accumulation of mutations that lower susceptibility in small steps. In molecular terms, the mutations lower the intracellular drug concentration and/or the sensitivity of the DNA topoisomerases to the drugs. Repeated cycles of fluoroquinolone treatment, punctuated by periodic outgrowth of pathogen populations, lead to the stepwise accumulation of mutations. The order in which target and non-target alleles arise probably depends on the incremental increase in quinolone concentration. If the initial concentration is low, non-target alleles will be selected, as seen with mycobacteria and *S. pneumoniae*. If the initial concentration is moderately high, target mutations are selected. Then small increases in concentration lead to the recovery of additional non-target alleles. The principle also applies to resistance substitutions in the target genes. For example, in laboratory experiments with *Haemophilus influenzae*, a first-step mutation occurs in *gyrA*, the second step in *parC*, a third is at a second position in *gyrA*, and the fourth is at a second position in *parC* (Figure 6.5). Clinical resistance among *E. coli* isolates often maps to these four genes [39].

Stepwise accumulation of resistance mutations has important implications for surveillance work, because resistance is defined by interpretive breakpoints (breakpoints are values of MIC that empirically discriminate between resistant and susceptible isolates). Isolates having MICs below the breakpoint can carry resistance mutations but still be scored as susceptible. Those mutations can increase the probability that a mutant with a second resistance allele will be selectively amplified [40, 41]. The result is expected to be a sudden increase in the prevalence of resistance due to a "susceptible" bacterial population passing the resistance threshold with higher probability than if all cells were wild-type. Thus, the problem of fluoroquinolone resistance may be larger than revealed by surveillance studies.

While gradual accumulation of resistance mutations is a common phenomenon among antimicrobials, it is not seen with every agent. For example, resistance to rifampicin in *S. aureus* derives from a resistance mutation in RNA polymerase that is highly protective: resistance arises in a single step, and no achievable drug

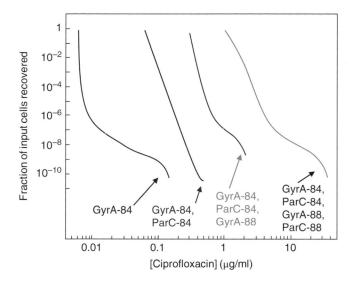

Figure 6.5 Stepwise accumulation of fluoroquinolone-resistant topoisomerase mutations. The left-most curve (black) represents the fraction of input cells recovered as colonies when applied to agar containing the indicated concentrations of ciprofloxacin. Blue arrow points to concentration at which a first-step GyrA-84 mutation was recovered. When cultures of the GyrA-84 mutant were grown and applied to ciprofloxacin-containing agar, mutants were recovered at various drug concentrations as indicated by the blue line. Red arrow points to drug concentration at which a GyrA-84 ParC-84 double mutant was recovered. When cultures of the double mutant were grown and applied to ciprofloxacin-containing agar, mutants were recovered at various drug concentrations as indicated by the red line. Green arrow points to drug concentration at which a GyrA-84 ParC-84 GyrA-88 triple mutant was recovered. When cultures of the triple mutant were grown and applied to ciprofloxacin-containing agar, mutants were recovered at various drug concentrations as indicated by the green line. Black arrow points to drug concentration at which a GyrA-84 ParC-84 GyrA-88 ParC-88 quadruple mutant was recovered. *Source:* Figure adapted from Li et al. [38]. © American Society for Microbiology.

concentration can exert the selective pressure needed for the strain to acquire an additional resistance mutation. This example bolsters the conclusion that quinolone resistance arises stepwise by showing that single-step resistance can be detected.

6.3.4 Quinolone-induced Quinolone Resistance

As mentioned above, the quinolones are themselves mutagenic. Many years ago it was suggested that they induce mutations through the generation of free radicals, since the mutagenic effect of nalidixic acid, pipemidic acid, and norfloxacin was blocked by β-carotene, a scavenger of free radicals. We now understand that hydroxyl radical production is a part of the cellular response to quinolone treatment and that free radicals produce 8-oxo-guanine, a highly mutagenic compound. The quinolones also induce the SOS response, which contains genes whose products perform error-prone DNA repair. A third type of induced mutagenesis comes from misrepair of quinolone–topoisomerase–DNA complexes. Thus, it is likely to be important to keep drug concentrations high enough to kill mutants as well as the wild-type cells that give rise to the mutants.

6.3.5 Cross-resistance

In general, mutations that confer resistance to one quinolone also lower susceptibility to other members of the class. Nevertheless, cross-resistance among quinolones can be low with organisms that contain two targets (gyrase and DNA topoisomerase IV) if the primary target differs for the two compounds. Such is the case with *S. pneumoniae*. Some quinolone derivatives, such as ciprofloxacin and levofloxacin, have topoisomerase IV as the primary target, while others, such as sparfloxacin, moxifloxacin, and gatifloxacin, have gyrase as the preferred target. Consequently, a ciprofloxacin-resistant mutation may have little effect on the MIC for gatifloxacin and *vice versa*. While these target differences allow isolates of *S. pneumoniae* having a resistance allele for one quinolone to be considered susceptible to another, the first-step mutation is not silent: it can profoundly increase the frequency at which the isolate acquires a second quinolone-resistant mutation. In these situations patients must be carefully monitored for the emergence of resistance, especially with pathogens likely to spread through hospitals and other institutions.

Cross-resistance between quinolones and other antimicrobials generally occurs through efflux mutations and plasmid-mediated resistance, both of which can affect multiple drug classes. However, an example of target-based cross-resistance has been noted: chloroquine treatment of malaria seems to lead to the emergence of ciprofloxacin-resistant *E. coli* [42]. In this case, the resistant mutants exhibited amino acid substitutions that are typical of resistance selected by the quinolones. In retrospect, cross-resistance with chloroquine is not surprising, because the quinolone class emerged from studies with chloroquine. Since chloroquine is very widely used, it may be an important source of quinolone resistance in countries having a high burden of malaria.

6.3.6 Drug Tolerance and Persisters

Antibiotic tolerance occurs when a lethal antibiotic blocks bacterial growth but fails to kill cells. Tolerance is thought to be an important precursor of clinical resistance, because it creates a pathogen subpopulation that survives antibiotic treatment (members of tolerant subpopulations are called persisters). When quinolone treatment stops, the residual, surviving bacteria, derived from persister subpopulations, can grow and cause relapse. Additional resistant mutants arise, and when the population is subjected to a second round of antimicrobial treatment, the resistant fraction of the population is enriched. Stepwise accumulation of resistant mutants is expected from repeated cycles of treatment and outgrowth, either in the same patient or following dissemination to other patients. The presence of persister cells can render some infections, such as those caused by *M. tuberculosis*, particularly difficult to treat.

Single-cell analysis performed with *E. coli* suggests that persistence derives from a small fraction of the population stochastically residing in a non-growing state [43], which, as pointed out above, restricts the ability of most quinolones to kill cells. The non-growing state is generated by the action of toxins that degrade RNA, and persister frequency is normally kept low by the action of antitoxins. The Lon protease, activated by polyphosphate, degrades antitoxins, thereby activating the toxins that then block growth. Since polyphosphate levels are regulated by (p)ppGpp, the latter ultimately serves as one way in which persister levels may be controlled.

With quinolones, the level of persisters has been assessed as plateau values when time-kill or concentration-kill experiments are performed – persisters are cells that remain alive after the susceptible population is killed, provided that care is taken to avoid scoring resistant mutant subpopulations. For *E. coli* growing in culture, persister cells constitute about 10^{-4} to 10^{-5} of the population [43]. Mutations that alter the plateau level of concentration-kill curves are used to identify factors involved in persister formation. For example, an *E. coli* mutant deficient in superoxide dismutase (*sodA sodB*) exhibits a higher survival plateau with norfloxacin [8] (in the absence of superoxide dismutase, superoxide concentrations are expected to accumulate and be protective, as illustrated in Figure 6.4). Moreover, subinhibitory concentrations of plumbagin or paraquat, metabolic generators of superoxide, also protect *E. coli* cells from being killed by quinolones and other antibiotics. How superoxide is related to toxin-mediated persistence is unknown.

6.4 Quinolone Resistance: Molecular Basis

6.4.1 Permeability-based Resistance

The quinolones must enter cells to act; consequently, mutations that reduce drug uptake lower susceptibility. Gram-negative bacteria have two membranes in the cell envelope. In enterobacteria, the outer membrane is itself composed of two layers: an outer surface of lipopolysaccharide and an inner phospholipid bilayer. Underlying the phospholipid is a peptidoglycan network and then the plasma membrane. The quinolones enter cells through the surface membranes and through porins, which are water-filled, barrel-like proteins that serve as passageways. The relative use of the two entry modes depends on the hydrophobicity of the drug molecules, with hydrophilic quinolones entering largely through porins. Reduced quinolone susceptibility correlates with a deficiency in particular porin proteins (OmpF in *E. coli*; D2 in *Pseudomonas aeruginosa*) and in amino acid substitutions that narrow the porin passageway [44].

The expression of porins is under complex regulation, part of which involves a master regulator called multiple antibiotic resistance (Mar). One consequence of *mar* mutations is that selection of resistance to other antibiotics, such as chloramphenicol or tetracycline, can lead to membrane-associated quinolone resistance through down-regulation of *ompF*. Thus, antibiotic resistance determinants involved in drug uptake can affect several drug classes. That makes quinolone use vulnerable to the heavy agricultural use of other antibiotics.

Impaired drug uptake due to resistance mutations has only a modest effect on susceptibility, especially for the newer quinolones. However, reduced uptake is a widespread phenomenon, and clinically resistant isolates often exhibit low porin expression [45]. Reduced fluoroquinolone uptake is probably an important part of the hill-climb to resistance.

6.4.2 Efflux-based Fluoroquinolone Resistance

Up-regulation of genes involved in transport-mediated bacterial processes results in a modest loss of fluoroquinolone susceptibility [46, 47]. These processes include efflux systems that pump out metabolic end products, signaling molecules, and noxious materials, such as antibiotics. Efforts to improve quinolone-class antibacterials show that many quinolones are substrates for one or more bacterial efflux pump.

Efflux systems are grouped into five types: ATP-binding cassette (ABC) superfamily, major facilitator (MF) superfamily, multidrug and toxic compound extrusion (MATE) family, resistance-nodulation-division (RND) family, and small multidrug resistance (SMR) family. Bacteria have genes for multiple members of each efflux gene family (the *E. coli* genome contains genes for at least 37 efflux transporters). Virtually every type of bacterium expresses one or more efflux pump that exports some, if not many quinolones. Since fluoroquinolones are structurally diverse, surprises occasionally emerge. For example, gatifloxacin inhibits some efflux pumps in *P. aeruginosa*, which allows this quinolone to retain activity with efflux-expressing, ciprofloxacin-resistant strains [48].

The AcrAB-TolC system of *E. coli* is one of the better characterized systems [49]. This system recognizes many fluoroquinolones and exports a variety of agents that include tetracycline, β-lactams, chloramphenicol, erythromycin, rifampicin, dyes, disinfectants, and organic solvents. The TolC part of the pump forms a long channel spanning both the outer membrane and the periplasmic space. AcrA is a lipoprotein located in the periplasm; the inner membrane protein, AcrB, is a proton-motive-force transporter of the RND type. Strains lacking the AcrAB proteins are hypersusceptible to many quinolones; conversely, fluoroquinolone-resistant efflux mutants often overproduce the periplasmic protein AcrA. Such mutants show decreased accumulation of ciprofloxacin and ethidium bromide; they also display decreased susceptibility to tetracycline, ampicillin, and chloramphenicol. Deletion or inactivation of proteins that repress expression of *acrAB* (AcrR and AcrS) reduces fluoroquinolone susceptibility [50, 51]. Thus, efflux can be perturbed at many points.

Efflux can be a significant contributor to fluoroquinolone resistance. For example, at least six RND-type transporters pump quinolones out of *P. aeruginosa* [52, 53]. Three have been extensively examined: (i) MexAB-OprM (regulated by *nalB* [*mexR*]), (ii) MecCD-OprJ (regulated by *nfxB*), and (iii) MexEF-OprN (regulated by *nfxC* [*mexT*]). High-level clinical resistance of *P. aeruginosa* to fluoroquinolones has been attributed largely to over-expression of efflux pumps. Since the use of quinolones selects strains expressing high levels of efflux that also remove members of other antimicrobial classes, treatment of *P. aeruginosa* with fluoroquinolones can facilitate the emergence of multidrug resistance. We note that resistant mutants of *P. aeruginosa* also contain mutations in genes encoding gyrase and DNA topoisomerase IV (with cultured cells, gyrase mutants are recovered when fluoroquinolone concentrations are high enough to suppress efflux-mediated growth). Thus, *P. aeruginosa* behaves like other bacteria by gradually accumulating both target- and non-target-based resistance determinants.

Gram-positive bacteria are also subject to efflux-mediated resistance. In these organisms, members of the MF and SMR families are most often associated with resistance. The *S. aureus* NorA pump system, which exports hydrophilic fluoroquinolones, was among the earliest to be recognized as interfering with the clinical utility of quinolones. In *S. pneumoniae*, the major NorA-type pump is called PmrA. This pump recognizes older fluoroquinolones, such as ciprofloxacin and norfloxacin, but it is less effective with newer agents that have bulky groups attached to the C-7 position [54]. In

Gram-positive bacteria, variants of NorA-type efflux are frequently defined by the ability of reserpine, a pump inhibitor, to lower fluoroquinolone MIC. However, the modest increase in susceptibility associated with reserpine treatment fails to account for the contribution of efflux to the emergence of mutation-based resistance in *S. pneumoniae*. It seems that PmrA effects are supplemented by over-expression of the ABC transport proteins PatA and PatB, which can be fluoroquinolone induced [55, 56].

6.4.3 Topoisomerase-Protecting Proteins: Qnr

In 1998 a strain of *K. pneumoniae* was described that exhibited plasmid-mediated resistance to fluoroquinolones and 13 other agents [57]. The fluoroquinolone-resistance factor was called Qnr, an abbreviation for quinolone resistance. Several different types of Qnr protein were subsequently discovered, and examination of nucleotide/ amino acid sequences revealed that Qnr proteins are part of a family characterized by pentapeptide repeats. Every fifth amino acid of these repeats is either leucine or phenylalanine [58], which suggests that the proteins may be involved in protein–protein interactions. The pentapeptide protein family includes roughly 500 members that display a wide variety of properties.

Structural analysis of MfpA, a member of the family found in mycobacteria, showed that the protein was similar in structure to B-form DNA. That led to the proposal that Qnr proteins act as DNA mimics, competing with DNA for binding to gyrase and thereby interfering with supercoiling and the formation of cleaved complexes. Subsequent work revealed that MfpA may be atypical: another member of the Qnr class, QnrB1, has little ability to block the supercoiling activity of gyrase, but it avidly destabilizes cleaved complexes [59]. Even a 1000-fold excess of ciprofloxacin over Qnr fails to overcome the Qnr-gyrase interaction [60]. How Qnr-mediated complex destabilization occurs is unknown.

While the MIC of *qnr*-containing strains is often below the resistance breakpoint, the effect of *qnr* is substantial, sometimes raising MIC by more than 100-fold. Since quinolone resistance arises in a stepwise fashion, reduced susceptibility due to the presence of *qnr* is expected to be an important factor in the emergence of resistance, either because it adds to the effect of an existing resistance allele to render a strain clinically resistant or it serves as an early step in the path to resistance. Thus, the occurrence of *qnr* genes on plasmids, as discussed in a subsequent section, is a significant threat to fluoroquinolone efficacy.

6.4.4 Quinolone-modifying Enzymes: Aac(6′)-1b-cr

In 2006 another type of low-level fluoroquinolone resistance was observed when two *qnr*-containing plasmids conferred different levels of protection from quinolones [61]. A mutant library, prepared by insertion into the more protective plasmid, was examined for mutations that lowered MIC to the level attributed to the presence of *qnr*. Insertions mapping within or immediately upstream of *aac(6′)-Ib* were associated with increased quinolone susceptibility [61]. The *aac(6′)-Ib* gene encodes an aminoglycoside acetyltransferase that confers resistance to tobramycin, kanamycin, and amikacin [62]. Changes in two codons, Trp-102 to Arg and Asp-179 to Tyr, create a ciprofloxacin-resistance phenotype [63]. The variant enzyme was named Aac(6′)-Ib-cr to indicate its ability to acetylate the unsubstituted nitrogen of the ciprofloxacin C-7 piperazinyl ring [61]. Aac(6')-Ib-cr also lowers susceptibility to norfloxacin, which has the same C-7 ring as ciprofloxacin. However, it has no effect on quinolones, such as enrofloxacin, pefloxacin, levofloxacin, and gemifloxacin, that lack an unsubstituted C-7 ring nitrogen [63]. Thus, for some quinolones, *aac (6')-Ib-cr* action serves as an early step in the gradual climb to resistance.

6.4.5 Topoisomerase-based Resistance

Gyrase and topoisomerase IV are the intracellular targets of the quinolones; mutations in *gyrA, gyrB, parC,* and *parE* contribute to resistance. For both the *gyrA* and *parC* genes, most resistance mutations map within a narrow stretch of nucleotides termed the quinolone-resistance-determining region (QRDR). The mutations cause substitutions in gyrase and topoisomerase IV that are close to the binding site of the quinolones, as defined by X-ray crystallography of cleaved complexes. In *E. coli,* the GyrA QRDR comprises amino acids 51 through 106; a similar region exists in ParC. Amino acid changes that lower susceptibility the most correspond to alterations at positions 81, 83, and 87 (*E. coli* GyrA numbering system); fluoroquinolone-resistant clinical isolates usually have changes at positions 83 and 87 [39]. The narrow spectrum of resistance alleles facilitates the development of DNA-based diagnostic methods, which are likely to be particularly important with *M. tuberculosis* due to the long culture times required for susceptibility testing.

A combination of X-ray crystallography and biochemistry explain resistance due to the major GyrA mutations [4, 5, 64]. In cleaved complexes, the 3-carboxyl substituent of quinolones forms a magnesium–water bridge with amino acids 83 and 87 that contributes to quinolone activity. Amino acid substitutions at those positions eliminate the bridge and the activity it confers to the drugs. A GyrA-81Cys substitution has a similar effect, presumably due to steric interference with formation of the bridge. Since quinazolinediones lack the carboxyl group and cannot form the bridge [65], their activity is unaffected by these resistance mutations. A current challenge is to find ways to increase the stability of dione-containing cleaved complexes such that traditional GyrA-mediated quinolone resistance is bypassed without also inhibiting human topoisomerase II [5].

As with other resistance determinants, target-based resistance builds in a stepwise manner, generally at positions equivalent to *E. coli* GyrA amino acids 83 and 87. For example, with clinical isolates of *E. coli,* resistant mutants having low values of MIC tend to contain amino acid changes at either GyrA-83 or GyrA-87 [39]. As MIC increases, additional changes are found in ParC, and at even higher values of MIC, a third mutation is found in GyrA. At very high values, isolates tend to have two GyrA and two ParC changes. These are the results predicted by studies with cultured cells (Figure 6.5).

We conclude that bacterial cells can lower quinolone susceptibility in many ways: reduced drug uptake, increased drug efflux, drug modification, and target modification (Figure 6.6). The presence of multiple resistance factors allows a variety of paths in which small steps are taken during the climb to high-level resistance. If quinolone concentrations are high enough (above MPC), the cells are forced jump up a steep cliff.

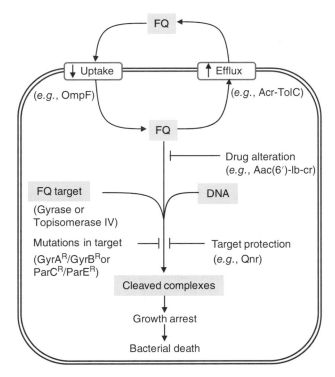

Figure 6.6 **Factors contributing fluoroquinolone resistance.** A bacterial cell is sketched with the major resistance factors shown red.

6.5 Plasmid-mediated Resistance

Fluoroquinolone-resistance genes carried by plasmids pose a serious threat to quinolone efficacy for two reasons. First, plasmids can carry genes for resistance to multiple antibiotics, which means that fluoroquinolone resistance can be acquired through the use of other antimicrobial classes. Second, plasmids can enter a bacterial population at a much higher frequency than spontaneous resistance mutations. That makes plasmid-based resistant cells a larger fraction of the bulk population and more likely to become the dominant class when fluoroquinolones are applied. We discuss three types of plasmid-mediated quinolone resistance: interference with cleaved complexes, inactivation of quinolones by chemical modification, and elevated drug efflux.

The best-studied family of plasmid-mediated fluoroquinolone-resistance genes is called *qnr* [57], which was briefly discussed earlier. Diversity among the Qnr determinants is substantial. For example, five major family groups, termed *qnrA, qnrB, qnrC, qnrD,* and *qnrS,* have been identified (reviewed in [66]), and some family groups have many members (seven forms of *qnrA* and 42 forms of *qnrB* have been found). The surveys that revealed Qnr diversity also show that Qnr-expressing plasmids are globally distributed (Table 6.4) and that they occur in a wide variety of bacterial species (Table 6.5). Thus, Qnr is likely to be a major contributor to

Table 6.4 Distribution of plasmid-borne resistance.

Resistance gene	Location of isolate
aac	Italy
aac	China
aac	France
aac	Japan
aac	Spain
aac	United States
aac	Uruguay
qnr	Bangladesh
qnr	China
qnr	France
qnr	Germany
qnr	Japan
qnr	Korea
qnr	Thailand
qnr	UK
qnr	United States
qnr	Taiwan
qnr	Turkey
qnr	Italy
qepA	Belgium
qepA	Japan
qepA	China
qepA	France

quinolone resistance. We note that the effect of Qnr varies considerably from one quinolone to another (Table 6.6); these differences are currently unexplained.

A second type of plasmid-borne resistance gene encodes the quinolone-acetylating enzyme Aac (6′)-Ib-cr. As pointed out in a previous section, this factor lowers the activity of compounds, such as ciprofloxacin and norfloxacin, by about fourfold. Although the geographical distribution of *aac (6′)-Ib-cr* has not been studied as thoroughly as that of *qnr*, it is clear that drug-modification-based resistance is widely distributed (Table 6.4) As with *qnr*, *aac (6′)-Ib-cr*-containing plasmids are found in a variety of Enterobacteriaceae (Table 6.5), and the prevalence of *aac (6′)-Ib-cr* can be substantial (Table 6.7). Since *aac (6′)-Ib-cr* is part of an integron cassette, it is likely to move readily among plasmids. Indeed, fluoroquinolone-resistance plasmids are conjugative and tend to have a broad host range.

The most recently discovered type of plasmid-mediated quinolone resistance involves efflux pumps. The QepA efflux pump, which was first found in Japan in 2006

Table 6.5 Bacterial species carrying fluoroquinolone-resistance plasmids.

Resistance gene	Bacterial species
aac	*Aeromonas species*
aac	*Citrobacter freundi*
aac	*Enterobacter cloacae*
aac	*Escherichia coli*
aac	*Klebsiella pneumoniae*
aac	*Salmonella species*
qnr	*Aeromonas* species
qnr	*C. freundii*
qnr	*Citrobacter koseri*
qnr	*Enterobacter aerogenes*
qnr	*Enterobacter amnigenus*
qnr	*Enterobacter sakazakii*
qnr	*Enterobacter cloacae*
qnr	*Escherichia coli*
qnr	*Klebsiella oxytoca*
qnr	*K. pneumoniae*
qnr	*Providencia stuartii*
qnr	*Proteus mirabilis*
qnr	*Pseudomonas fluorescens*
qnr	*Salmonella enterica*
qnr	*Serratia marcescens*
qnr	*Shigella dysenteriae*
qnr	*Shigella flexneri*
qnr	*Vibrio species*
qepA	*E. coli*

in an *E. coli* isolate, increases MIC by about 10-fold for hydrophilic fluoroquinolones, such as norfloxacin and ciprofloxacin. So far, the prevalence of QepA-mediated resistance in humans is low (0.3% among *E. coli* isolates collected from 140 Japanese hospitals between 2002 and 2006 [68], 0.8% of extended spectrum beta lactamase (ESBL)-producing enterobacterial isolates collected in France during 2007). Another *E. coli* efflux system, *oqxAB*, has recently been found in food animals in China and Denmark, and plasmid-mediated efflux has been reported in *S. aureus*. Such reports are likely to increase now that laboratories are looking for efflux determinants.

Plasmid-mediated resistance seems to derive from the mobilization of chromosomal genes followed by enrichment due to environmental contamination with quinolones. Integrons in conjugative plasmids mobilize other chromosomal genes; thus, a potential mechanism is available. Moreover, the aquatic organisms *Shewanella algae* and *Vibrio*

Table 6.6 Effect of Qnr on fluoroquinolone susceptibility.

Fluoroquinolone	Transconjugant MIC_{90} with qnr^a	MIC of recipient lacking qnr^a	Ratio
AM-1121	0.5	0.008	63
Bayy3118	0.125	0.004	31
Ciprofloxacin	1	0.008	125
Garenoxacin	2	0.008	250
Gatifloxacin	0.5	0.008	63
Gemifloxacin	1	0.004	250
Levofloxacin	0.5	0.015	33
Moxifloxacin	1	0.03	33
Nalidixic acid	32	4	8
Premafloxacin	0.25	0.03	8
Sitafloxacin	0.125	0.008	16
Sparfloxacin	1	0.008	125

[a] MIC in ug/ml for *Escherichia coli* receiving *qnr*-containing plasmids from 17 clinical isolates of *E. coli* and *Klebsiella pneumoniae* [67].

splendidus contain chromosomal genes for *qnrA* and *qnrS*, respectively, suggesting that mobilization occurred with these bacteria [66]. The selective pressure needed to amplify the resistant, plasmid-bearing strains could easily result from massive use of quinolones contaminating aquatic environments. It is reasonable to assume that resistance genes then move from the aquatic environment to food animals and eventually to humans. Indeed, clonal spread of plasmid-mediated fluoroquinolone resistance is now occurring in hospitals [69].

6.6 Concluding Remarks

As we enter an era in which antimicrobial resistance commands attention, two phenomena stand out: antimicrobial consumption remains high, and immunosuppression is widespread. They will continue to drive the emergence of resistance, perhaps faster than we can develop new compounds. One underlying flaw in the administration of antibiotics is the almost universal use of cure as a clinical endpoint. The emergence of resistance was largely ignored when therapeutic protocols were developed, and those protocols continue to result in administering drug concentrations that usually result in cure. But when applied with hundreds of millions of prescriptions, dosing-to-cure inevitably enriches resistant subpopulations and leads to resistance. The heart of this problem is the use of MIC, a bacteriotatic parameter, as the metric for bacterial susceptibility.

In the present chapter we have presented a case for using MPC as a susceptibility standard. MPC-based dosing uniquely aligns the interests of individual and public health, because it can potentially improve treatment outcome and reduce selection of

Table 6.7 Prevalence of plasmids with *qnr* or *aac (6')-Ib-cr* in bacterial strains with reduced susceptibility to ciprofloxacin[a].

Country	Year	Bacterial species	Prevalence of *qnr* plasmid	Prevalence of *aac (6')-Ib-cr* plasmid
China	2008	*Citrobacter freundii*	43% (17/40)	28% (11/40)
		Enterobacter cloacae	63% (27/43)	9% (4/43)
		Escherichia coli	5% (5/105)	13% (14/105)
		Klebsiella pneumoniae	50% (38/77)	21% (16/77)
Korea	2005	*C. freundi*	60% (39/65)	
		E. cloacae	57% (51/89)	
		Enterobacter aerogenes	19% (5/26)	
		Serratia marcescens	2% (2/110)	
USA	1999–2004	*Enterobacter*		7.5% (12/160)
		E. coli		32% (15/47)
		K. pneumoniae		16% (17/106)
		Enterobacter	31% (50/160)	
		E. coli	4% (2/47)	
		K. pneumoniae	20% (21/106)	

[a] All strains had MIC for ciprofloxacin that was greater than 0.25 µg/ml.

resistance – doses need to exceed a mutant-restricting threshold of AUC_{24}/MPC. No compound has been chosen for this type of dosing, and it is unlikely that the criterion will be applied until regulatory agencies change their general guidelines. The issue is a higher chance of toxic side effects with MPC rather than MIC as a standard.

Another approach focuses on improving lethal activity. Compounds that rapidly kill bacteria will quickly reduce pathogen population size, which for quinolones will suppress the mutagenic effects of the SOS response. A problem with lethal activity is that we do not understand the mechanism well enough to design more highly lethal compounds. Moreover, we do not have simple biochemical surrogates of killing that can be used for high throughput screening to find lethal compounds. Lethal action is fundamentally a basic research problem, not an optimization issue that can be handled easily by the pharmaceutical industry.

We conclude that ways exist for antimicrobials, in particular quinolones that bypass existing resistance issues, to maintain utility within the current environment of heavy consumption. Implementation will require a difficult industry and regulatory shift that can come only after research efforts lay a suitable foundation.

Acknowledgments

We thank Marila Gennaro, Richard Pine, and Bo Shopsin for critical comments on the manuscript. The work was supported by NIH grant AI073491.

References

1 Zhao, X., Hong, Y., and Drlica, K. (2015). Moving forward with ROS involvement in antimicrobial lethality. *J. Antimicrob. Chemother.* 70: 639–642.

2 Malik, M., Hussain, S., and Drlica, K. (2007). Effect of anaerobic growth on quinolone lethality with *Escherichia coli. Antimicrob. Agents Chemother.* 51: 28–34.

3 Malik, M. and Drlica, K. (2006). Moxifloxacin lethality with *Mycobacterium tuberculosis* in the presence and absence of chloramphenicol. *Antimicrob. Agents Chemother.* 50: 2842–2844.

4 Aldred, K., McPherson, S., Wang, P. et al. (2012). Drug interactions with *Bacillus anthracis* topoisomerase IV: biochemical basis for quinolone action and resistance. *Biochemistry* 51: 370–381.

5 Aldred, K., McPherson, S., Turnbough, C. et al. (2013). Topoisomerase IV-quinolone interactions are mediated through a water-metal ion bridge: mechanistic basis of quinolone resistance. *Nucleic Acids Res.* 41: 4628–4639.

6 Drlica, K., Malik, M., Kerns, R., and Zhao, X. (2008). Quinolone-mediated bacterial death. *Antimicrob. Agents Chemother.* 52: 385–392. https://doi.org/10.1128/AAC.01617-06.

7 Kohanski, M., Dwyer, D., Hayete, B. et al. (2007). A common mechanism of cellular death induced by bactericidal antibiotics. *Cell* 130: 797–810.

8 Wang, X. and Zhao, X. (2009). Contribution of oxidative damage to antimicrobial lethality. *Antimicrob. Agents Chemother.* 53: 1395–1402.

9 Aldred, K., Schwanz, H., Li, G. et al. (2013). Overcoming target-mediated quinolone resistance in topoisomerase IV by introducing metal ion-independent drug-enzyme interactions: implications for drug design. *ACS Chem. Biol.* 8: 2660–2668.

10 Mustaev, A., Malik, M., Zhao, X. et al. (2014). Fluoroquinolone-gyrase-DNA complexes: two modes of drug binding. *J. Biol. Chem.* 289: 12300–12312.

11 Malik, M., Mustaev, A., Schwanz, H. et al. (2016). Suppression of gyrase-mediated resistance by C7 aryl fluoroquinolones. *Nucleic Acids Res.* 44 (7): 3304–3316.

12 Dorsey-Oresto, A., Lu, T., Mosel, M. et al. (2013). YihE kinase is a novel regulator of programmed cell death in bacteria. *Cell Rep.* 3: 528–537.

13 Han, X., Dorsey-Oresto, A., Malik, M. et al. (2010). *Escherichia coli* genes that reduce the lethal effects of stress. *BMC Microbiol.* 10: 35.

14 Chen, C.-R., Malik, M., Snyder, M., and Drlica, K. (1996). DNA gyrase and topoisomerase IV on the bacterial chromosome: quinolone-induced DNA cleavage. *J. Mol. Biol.* 258: 627–637.

15 Malik, M., Zhao, X., and Drlica, K. (2006). Lethal fragmentation of bacterial chromosomes mediated by DNA gyrase and quinolones. *Mol. Microbiol.* 61: 810–825.

16 Li, L., Hong, Y., Luan, G. et al. (2014). Ribosomal elongation factor 4 promotes cell death associated with lethal stress. *MBio* 5: e01708.

17 Kohanski, M., Dwyer, D., Wierzbowski, J. et al. (2008). Mistranslation of membrane proteins and two-component system activation trigger antibiotic-mediated cell death. *Cell* 135: 679–690.

18 Raivio, T. and Silhavy, T. (2001). Periplasmic stress and ECF sigma factors. *Annu. Rev. Microbiol.* 55: 591–624.

19 Foti, J., Devadoss, B., Winkler, J. et al. (2012). Oxidation of the guanine nucleotide pool underlies cell death by bactericidal antibiotics. *Science* 336: 315–319.

20 Liu, Y. and Imlay, J. (2013). Cell death from antibiotics without the involvement of reactive oxygen species. *Science* 339: 1210–1213.

21 Mahoney, T. and Silhavy, T.J. (2013). The Cpx stress response confers resistance to some, but not all, bactericidal antibiotics. *J. Bacteriol.* 195: 1869–1874.

22 Ezraty, B., Vergnes, A., Banzhaf, M. et al. (2013). Fe-S cluster biosynthesis controls uptake of aminoglycosides in a ROS-less death pathway. *Science* 340: 1583–1587.

23 Keren, I., Wu, Y., Inocencio, J. et al. (2013). Killing by bactericidal antibiotics does not depend on reactive oxygen species. *Science* 339: 1213–1216.

24 Dwyer, D., Belenky, P., Yang, J. et al. (2014). Antibiotics induce redox-related physiological alterations as part of their lethality. *Proc. Natl. Acad. Sci. USA* 111: E2100–E2109.

25 Wang, X., Zhao, X., Malik, M., and Drlica, K. (2010). Contribution of reactive oxygen species to pathways of quinolone-mediated bacterial cell death. *J. Antimicrob. Chemother.* 65: 520–524.

26 Malik, M., Marks, K., Mustaev, A. et al. (2011). Fluoroquinolone and quinazolinedione activities against wild-type and gyrase mutant strains of *Mycobacterium smegmatis*. *Antimicrob. Agents Chemother.* 55: 2335–2343.

27 Khodursky, A. and Cozzarelli, N. (1998). The mechanism of inhibition of topoisomerase IV by quinolone antibacterials. *J. Biol. Chem.* 273: 27668–27677.

28 Sahm, D., Peterson, D., Critchley, I., and Thronsberry, C. (2000). Analysis of ciprofloxacin activity against *Streptococcus pneumoniae* after 10 years of use in the United States. *Antimicrob. Agents Chemother.* 44: 2521–2524.

29 Linder, J., Huang, E., Steinman, M. et al. (2006). Fluoroquinolone prescribing in the United States: 1995 to 2002. *Am. J. Med.* 118: 259–268.

30 Aarestrup, F., Jensen, N., Jorsal, S., and Nielsen, T. (2000). Emergence of resistance to fluoroquinolones among bacteria causing infections in food animals in Denmark. *Vet. Rec.* 146: 76–78.

31 Hughes, D. and Andersson, D. (2012). Selection of resistance at lethal and non-lethal antibiotic concentrations. *Curr. Opin. Microbiol.* 15: 555–560.

32 Pan, X.-S., Ambler, J., Mehtar, S., and Fisher, L.M. (1996). Involvement of topoisomerase IV and DNA gyrase as ciprofloxacin targets in *Streptococcus pneumoniae*. *Antimicrob. Agents Chemother.* 40: 2321–2326.

33 Zhao, X. and Drlica, K. (2002). Restricting the selection of antibiotic-resistant mutants: measurement and potential uses of the mutant selection window. *J. Infect. Dis.* 185: 561–565.

34 Dong, Y., Zhao, X., Domagala, J., and Drlica, K. (1999). Effect of fluoroquinolone concentration on selection of resistant mutants of *Mycobacterium bovis* BCG and *Staphylococcus aureus*. *Antimicrob. Agents Chemother.* 43: 1756–1758.

35 Quinn, B., Hussain, S., Malik, M. et al. (2007). Daptomycin inoculum effects and mutant prevention concentration with *Staphylococcus aureus*. *J. Antimicrob. Chemother.* 60: 1380–1383.

36 Cui, J., Liu, Y., Wang, R. et al. (2006). The mutant selection window demonstrated in rabbits infected with *Staphylococcus aureus*. *J. Infect. Dis.* 194: 1601–1608.

37 Day, T. and Read, A. (2016). Does high-dose antimicrobial chemotherapy prevent the evolution of resistance? *PLoS Comput. Biol.* 12: e1004689.

38 Li, X., Mariano, N., Rahal, J.J. et al. (2004). Quinolone-resistant *Haemophilus influenzae*: Determination of mutant selection window for ciprofloxacin, garenoxacin, levofloxacin, and moxifloxacin. *Antimicrob. Agents Chemother.* 48: 4460–4462. https://doi.org/10.1128/AAC.48.11.4460-4462.2004.

39 Baudry-Simner, P., Singh, A., Karlowsky, J. et al. (2012). Mechanisms of reduced susceptibility to ciprofloxacin in *Escherichia coli* isolates from Canadian hospitals. *Can. J. Infect. Dis. Med. Microbiol.* 23: e60–e64.

40 Zhao, X., Xu, C., Domagala, J., and Drlica, K. (1997). DNA topoisomerase targets of the fluoroquinolones: a strategy for avoiding bacterial resistance. *Proc. Natl. Acad. Sci. U. S. A.* 94: 13991–13996.

41 Li, X., Zhao, X., and Drlica, K. (2002). Selection of *Streptococcus pneumoniae* mutants having reduced susceptibility to levofloxacin and moxifloxacin. *Antimicrob. Agents Chemother.* 46: 522–524.

42 Davidson, R., Davis, I., Willey, B. et al. (2008). Antimalarial therapy selection for quinolone resistance among *Escherichia coli* in the absence of quinolone exposure, in tropical South America. *PLoS One* 3: e2727.

43 Maisonneuve, E., Castro-Camargo, M., and Gerdes, K. (2013). (p)ppGpp controls bacterial persistence by stochastic induction of toxin-antitoxin activity. *Cell* 154: 1140–1150.

44 Masi, M. and Pages, J.-M. (2013). Structure, function, and regulation of outer membrane proteins involved in drug transport in Enterobacteriaceae: the OmpF/C-TolC case. *Open Microbiol. J.* 7 (suppl 1M2): 22–33.

45 Karczmarczyk, M., Martins, M., Quinn, T. et al. (2011). Mechanisms of fluoroquinolone resistance in *Escherichia coli* isolates from food-producing animals. *Appl. Environ. Microbiol.* 77: 7113–7120.

46 Poole, K. (2008). Bacterial multidrug efflux pumps serve other functions. *Microbe* 3: 179–185.

47 Piddock, L. (2006). Multidrug-resistance efflux pumps – not just for resistance. *Nat. Rev. Microbiol.* 4: 629–636.

48 Pankey, G. and Ashcraft, D. (2005). In vitro synergy of ciprofloxacin and gatifloxacin against ciprofloxacin-resistant *Pseudomonas aeruginosa*. *Antimicrob. Agents Chemother.* 49: 2959–2964.

49 Cloeckaert, A. and Chaslus-Dancla, E. (2001). Mechanisms of quinolone resistance in *Salmonella*. *Vet. Res.* 32: 291–300.

50 Hirakawa, H., Takumi-Kobayashi, A., Theisen, U. et al. (2008). AcrS/EnvR represses expression of the *acrAB* multidrug efflux genes in *Escherichia coli*. *J. Bacteriol.* 190: 6276–6279.

51 Wang, H., Dzink-Fox, J., Chen, M., and Levy, S.B. (2001). Genetic characterization of highly fluoroquinolone-resistant clinical *Escherichia coli* strains from China: role of *acrR* mutations. *Antimicrob. Agents Chemother.* 45: 1515–1521.

52 Masuda, N., Sakagawa, E., Ohya, S. et al. (2000). Substrate specificities of MexAB-OprM, MexCD-OprJ, and MexXY-oprM efflux pumps in *Pseudomonas aeruginosa*. *Antimicrob. Agents Chemother.* 44: 3322–3327.

53 Poole, K. (2000). Efflux-mediated resistance to fluoroquinolones in Gram-negative bacteria. *Antimicrob. Agents Chemother.* 44: 2233–2241.

54 Beyer, R., Pestova, E., Millichap, J.J. et al. (2000). A convenient assay for estimating the possible involvement of efflux of fluoroquinolones by *Streptococcus pneumoniae* and *Staphylococcus aureus*: evidence for diminished moxifloxacin, sparfloxacin, and trovafloxacin efflux. *Antimicrob. Agents Chemother.* 44: 798–801.

55 Marrer, E., Schad, K., Satoh, A.T. et al. (2006). Involvement of the putative ATP-dependent efflux proteins PatA and PatB in. *Antimicrob. Agents Chemother.* 50: 685–693.

56 Avrain, L., Garvey, M., Mesaros, N. et al. (2007). Selection of quinolone resistance in *Streptococcus pneumoniae* exposed *in vitro* to subinhibitory drug concentrations. *J. Antimicrob. Chemother.* 60: 965–972.

57 Martinez-Martinez, L., Pascual, A., and Jacoby, G. (1998). Quinolone resistance from a transferrable plasmid. *Lancet* 351: 797–799.

58 Bateman, A., Murzin, A., and Teichmann, S. (1998). Structure and distribution of pentapeptide repeats in bacteria. *Protein Sci.* 7: 1477–1480.

59 Vetting, M., Hegde, S., Wang, M. et al. (2011). Structure of QnrB1, a plasmid-mediated fluoroquinolone resistance factor. *J. Biol. Chem.* 286: 25265–25273.

60 Tran, J., Jacoby, G., and Hooper, D. (2005). Interaction of the plasmid-encoded quinolone resistance protein Qnr with *Escherichia coli* DNA gyrase. *Antimicrob. Agents Chemother.* 49: 118–125.

61 Robicsek, A., Strahilevitz, J., Jacoby, G. et al. (2006). Fluoroquinolone-modifying enzyme: a new adaptation of a common aminoglycoside acetyltransferase. *Nat. Med.* 12: 83–88.

62 Tolmasky, M., Roberts, M., Woloj, M., and Crosa, J. (1986). Molecular cloning of amikacin resistance determinants from a *Klebsiella pneumoniae* plasmid. *Antimicrob. Agents Chemother.* 30: 315–320.

63 Robicsek, A., Jacoby, G., and Hooper, D. (2006). The worldwide emergence of plasmid-mediated quinolone resistance. *Lancet Infect. Dis.* 6: 629–640.

64 Wohlkonig, A., Chan, P., Fosberry, A. et al. (2010). Structural basis of quinolone inhibition of type IIA topoisomerases and target-mediated resistance. *Nat. Struct. Mol. Biol.* 17: 1152–1153.

65 Laponogov, I., Pan, X., Veselkov, D. et al. (2010). Structural basis of gate-DNA breakage and resealing by type II topoisomerases. *PLoS One* 5: e11338.

66 Poirel, L., Cattoir, V., and Nordmann, P. (2012). Plasmid-mediated quinolone resistance interactions between human, animal, and environmental ecologies. *Front. Microbiol.* 3: 1–7.

67 Wang, M., Sahm, D., Jacoby, G. et al. (2004). Activities of newer quinolones against *Escherichia coli* and *Klebsiella pneumoniae* containing the plasmid-mediated quinolone resistance determinant *qnr. Antimicrob. Agents Chemother.* 48: 1400–1401.

68 Yamane, K., Wachino, J., Suzuki, S., and Arakawa, Y. (2008). Plasmid-mediated *qepA* gene among *Escherichia coli* clinical isolates from Japan. *Antimicrob. Agents Chemother.* 52: 1564–1566.

69 Frasson, I., Cavallaro, A., Bergo, C. et al. (2011). Prevalence of aac(6′)-1b-cr plasmid-mediated and chromosome-encoded fluoroquinolone resistance in Enterobacteriaceae in Italy. *Gut Pathog.* 3: 12.

7

Dihydropteroate Synthase (Sulfonamides) and Dihydrofolate Reductase Inhibitors

Clemente Capasso[1] and Claudiu T. Supuran[2]

[1] *Istituto di Bioscienze e Biorisorse - CNR, Napoli, Italy*
[2] *Dipartimento di Scienze Farmaceutiche, Università degli Studi di Firenze, Polo Scientifico, Florence, Italy*

7.1 Introduction

7.1.1 Structure and Mechanism of Action of Dihydropteroate Synthase and Dihydrofolate Reductase

Prokaryotes and unicellular eukaryotes such as yeasts or *Plasmodium* can synthesize folate *de novo via* the folic acid synthesis pathway (Figure 7.1). The steps of this pathway include: (i) the condensation of dihydropteridine pyrophosphate (DHPP) **1** with *p*-aminobenzoic acid (pABA) **2** by dihydropteroate synthase (DHPS, EC 2.5.1.15) to form dihydropteroate (DHP) **3**; (ii) the addition of glutamate to DHP by dihydrofolate synthase (DHFS, EC 6.3.2.12) to form dihydrofolate (DHF) **4**; and (iii) the reduction of DHF to form tetrahydrofolate (THF) **5**, catalyzed by dihydrofolate reductase (DHFR, EC 1.5.1.3) [1].

DHPS is thus an essential enzyme responsible for the *de novo* synthesis of the folate molecule. Tetrahydrofolate produced by this pathway is required for one carbon transfer reactions in the biosynthesis of a range of biomolecules, such as nucleotides and amino acids [2]. DHFR is the enzyme responsible for the maintenance of folate pools in their physiologic, reduced states. DHFR, in fact, is needed for the intracellular conversion of synthetic folic acid into the THF forms that can participate in folate/homocysteine metabolism. Reduction of DHFR enzymatic activity diminishes the THF pool inside the cells, affecting the level of folate coenzymes and thus purine and pyrimidine biosynthesis [3]. This may as well influence homocysteine levels and methylation processes, because methyl-THF is needed for the re-methylation of homocysteine to form methionine, thereby ensuring the provision of *S*-adenosylmethionine (SAM) necessary for most biological methylation processes [4].

In contrast to prokaryotes and lower eukaryotes, higher eukaryotic cells cannot synthesize folates *de novo*, being thus totally dependent on exogenous folic acid. Thus, the folic acid synthesis pathway represents a convenient target for obtaining drugs for the treatment of infectious diseases, caused by bacteria and protozoan parasites. Indeed, at

Bacterial Resistance to Antibiotics – From Molecules to Man, First Edition.
Edited by Boyan B. Bonev and Nicholas M. Brown.
© 2020 John Wiley & Sons Ltd. Published 2020 by John Wiley & Sons Ltd.

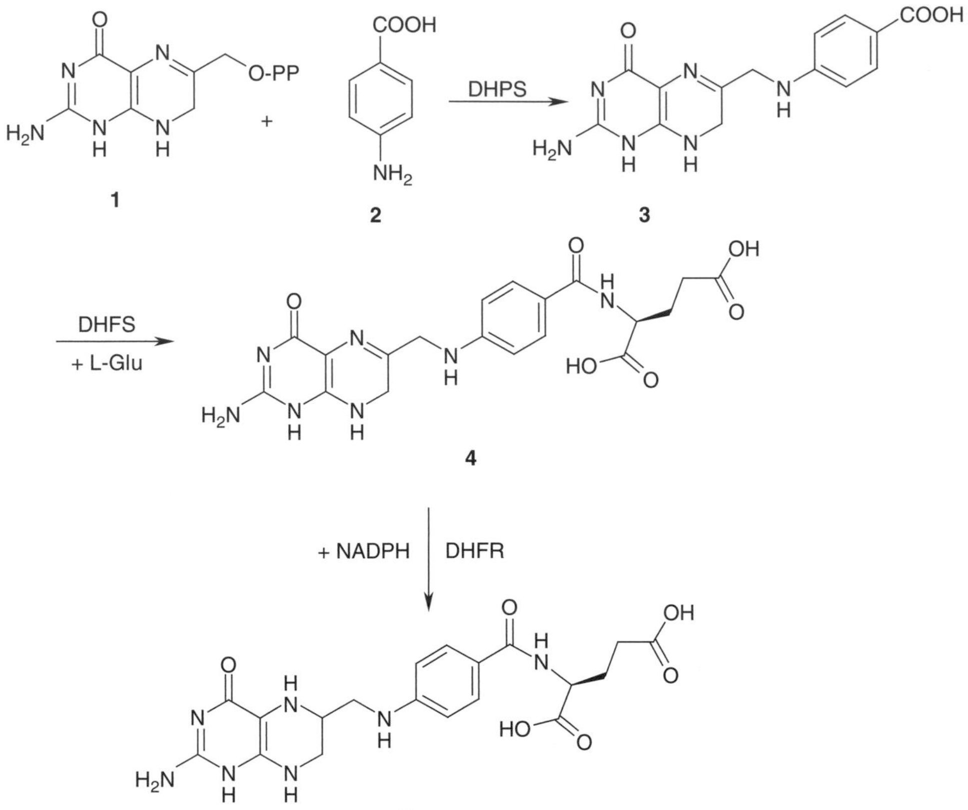

Figure 7.1 Biosynthesis of THF **5** from dihydropteridine pyrophosphate **1** (see text for details).

least two steps in this pathway are targeted by antibacterial drugs in clinical use for decades: DHPS is the target of sulfa drugs, whereas DHFR is targeted by trimethoprim (and its congeners) [5–7]. These enzymes have been validated as antimicrobial therapeutic targets by chemical and genetic means at the beginning of research in the field of antibacterials [8, 9].

7.1.2 Dihydropteroate Synthase

DHPS activity was identified 1962 in *Escherichia coli* by Brown et al. [8] and later in crude cell extracts of several organisms by Shiota and coworkers [9, 10]. Humans lack an equivalent to dihydropteroate synthase, which is an advantage from the drug design viewpoint, since no toxicity to the host should emerge by using compounds targeting the bacterial/protozoan enzyme. DHPSs are inhibited by sulfonamides, which function as competitive inhibitors of one of the DHPS substrates, pABA **2**. The DHPS encoded by the folP gene in *Bacillus anthracis* is also inhibited by sulfonamides [11]. The protein was shown to be a homodimer of two 30 kDa subunits. The chromosomal gene that encodes for DHPS in *Streptococcus pneumoniae* was cloned, sequenced, and shown to code for a protein of 34 kDa [12]. A similar sequence was also identified in a *Bacillus subtilis* folic acid biosynthetic operon [13]. Two other genes (sul*I* and sul*II*) that code for plasmid-borne sulfonamide-resistant DHPSs have also been sequenced [14, 15]. DHPS possesses a classic (β/α) TIM barrel structure, and the pABA binding site is at the edge of the barrel, being comprised of loop regions where the mutations that confer sulfonamide resistance are found [16]. In contrast, the other substrate of DHPS, DHPP **1**, binds in a deep, structured pocket within the DHPS β-barrel. This pocket is highly conserved among various bacterial species [16].

7.1.3 Dihydrofolate Reductase

DHFR is the enzyme responsible for the NADPH-dependent reduction of 5,6-dihydrofolate to 5,6,7,8-tetrahydrofolate, an essential cofactor used in the biosynthetic pathways of purines, thymidylate, methionine, glycine, pantothenic acid, and *N*-formyl-methionyl tRNA [17, 18]. DHFR is ubiquitously expressed in all organisms, being found in all dividing cells of prokaryotes and eukaryotes. The mammalian enzymes are all highly similar in sequence, while each bacterial form is distinct. The DHFR sequence in humans is 30% similar to that of *E. coli* and 70% similar to other mammalian DHFRs [19]. Human DHFR is a small enzyme with a primary structure of 186 amino acid residues and a molecular weight of about 20 kDa [19]. Structural studies show that DHFR is a monomeric molecule with many secondary structural elements. The protein is divided into two subdomains, the adenosine-binding subdomain and the loop subdomain [20]. The adenosine-binding subdomain is the larger of the two and binds the adenosine moiety of NADPH. The loop subdomain contains three loops. Between the two subdomains is the active site, where folate and NADPH bind. The size of the active site is regulated by the movements of the two subdomains [20–22]. At the transcriptional level, DHFR is governed by a TATA-less promoter that is controlled by numerous transcription factors, including Sp1 and E2F, which are important for its regulation throughout the cell cycle [3, 23]. In addition to key roles in folate metabolism, DHFR can reduce 7,8-dihydrobiopterin (BH2) to 5,6,7,8-tetrahydrobiopterin (BH4) [24], the natural cofactor required for the hydroxylation of aromatic amino acids, such as phenylalanine, tyrosine, and tryptophan in higher animals [25]. Also, BH4 is an essential cofactor for nitric oxide synthesis involved in oxidative stress as a regulator of nitric oxide synthase or as a direct radical scavenger [26]. In bacteria, there are DHFR encoded within mobile elements (*e.g.*, plasmids), which are rapidly spread within a bacterial community [27]. They are grouped into two main families, A and B [27, 28]. The A gene (*dfrA*) family encodes proteins of 152–189 amino acids with identity levels of 20–90% and some structural and sequence similarities to the chromosomal enzyme. At least 20 different *dfrA* sequences have been reported so far [29, 30]. The B gene (*dfrB*) family, by contrast, encodes a unique group of enzymes referred to as DfrB, which are completely unrelated to the chromosomal DHFRs or other DHFRs in terms of sequence and structure. The *dfrB* genes encode similar and much shorter proteins (78 residues) with identity levels of 75% or above, which are extremely resistant to trimethoprim (TMP) [31–33].

7.2 Clinically Used Inhibitors

7.2.1 Sulfa Drugs Inhibitors

The primary sulfonamides were the first antimicrobial drugs, discovered in 1935 by Domagk, and they paved the way for the antibiotic revolution in medicine [34]. The first sulfonamide showing effective antibacterial activity, Prontosil 6, was a prodrug, with the real antibacterial agent being sulfanilamide 7, a compound isosteric/isostructural with pABA 2. Sulfanilamide is generated by *in vivo* reduction of prontosil [34, 35]. In fact, the aniline moiety is common in both compounds 2 and 7, whereas the carboxyl moiety of pABA is mimicked by the primary sulfamoyl moiety of sulfanilamide. In the following years after the discovery of sulfanilamide as a bacteriostatic agent [35], a range of analogs have entered into clinical use (constituting the so-called sulfa drug class of antibacterials), and many of these compounds are still widely used (mainly in combination with DHFR inhibitors) [36].

The clinically used sulfa drugs (including prontosil which is no longer used) are shown in Figure 7.2, and they include sulfamethoxazole 8, sulfisomidine 9, sulfacetamide 10, and sulfathiazole 11 among others [36, 37]. All these derivatives are competitive inhibitors (with pABA) of DHPS, the first enzyme in the biosynthesis of tetrahydrofolate; however, significant resistance developed over the years to these drugs, which drastically limit their use in the clinic (see later in the text).

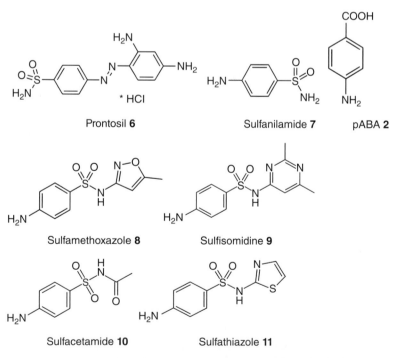

Figure 7.2 Sulfa drugs in clinical use (**7–11**) and prontosil **6**, the lead molecule generating this class of pharmacological agents. These compounds are structurally similar to pABA **2** and compete with this compound for the biosynthesis of DHP **3** [37]. Many other analogs are known [37].

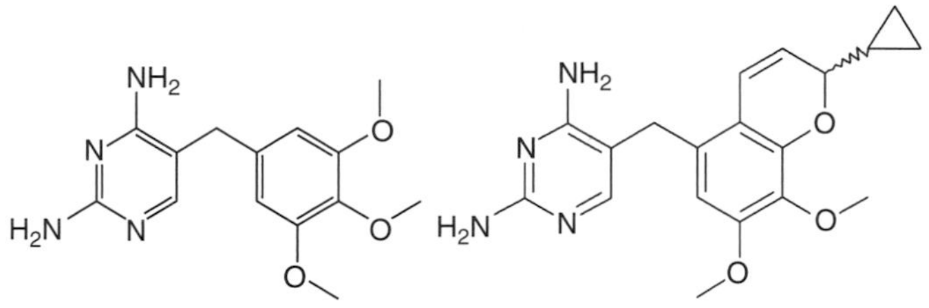

Trimethoprim **12**   Iclaprim **13**

Figure 7.3 Chemical structure of the DHFR inhibitor trimethoprim **12**. Its congeners (brodimoprim, tetroxoprim, *etc.*) have slightly different moieties replacing the 4-methoxy group of 12 [36]. The new DHFR inhibitor iclaprim **13** is also shown [36].

7.2.2 DHFR Inhibitors

DHFR inhibitors such as trimethoprim **12**, are in clinical use since the 1960s [32], whereas newer analogs (*e.g.*, brodimoprim, tetroxoprim, iclaprim) were constantly discovered and developed [36] (Figure 7.3). Such compounds compete with the pteridine moiety of DHF for binding to the enzyme (see Figure 7.1) and thus impair the generation of THF, a compound crucial in many biosynthetic processes (as shown in the "Introduction" section). Sulfa drugs and DHFR inhibitors are usually used in combination as antibacterial drugs, just to avoid the drug resistance problems typical of these pharmacologic agents [36, 37].

7.3 Molecular Mechanisms of Drug Resistance

The molecular mechanisms of drug resistance both to sulfa drugs [37] and trimetho-prim-like DHFR inhibitors [36, 38] are understood in detail because several X-ray crystal structures of DHPS and/or DHFR (both wild type and mutated proteins) are available, in complex with substrates and/or inhibitors.

7.3.1 Resistance to Sulfa Drugs

Crystallographic work with *B. anthracis* (BaDHPS) and *Yersinia pestis* (YpDHPS) DHPS, complexed with DHPP **1**, pABA **2**, and the pABA analog *p*-hydroxybenzoic acid, afforded a deep understanding of the catalytic, inhibition, and resistance mechanisms to sulfa drugs of this enzyme [37]. First, the reaction between **1** and **2** occurs not *via* an S_N-2 like but *via* an S_N-1 reaction mechanism with formation of a DHP carbocation (at C-9) which then attacks the weakly nucleophilic amino group of pABA (or its analog *p*-hydroxybenzoic acid), with formation of DHP **3** (the adduct of **3** and BaDHPS has been isolated and characterized at the atomic level). Furthermore, when BaDHPS was soaked in DHPP **1** and sulfathiazole **11**, the covalent compound **14** (Figure 7.4) was observed in the X-ray crystal structure, bound within the active site of the enzyme and stabilized by the so-called loop 1–loop 2 substructure of the protein [37].

The same crystallographic study from White's group [37] showed that in the YpDHPS–sulfa drug complexes, the sulfa drug perfectly fits the pABA binding pocket,

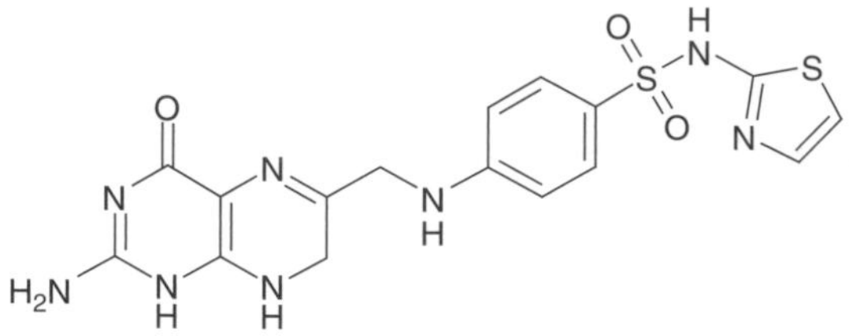

14

Figure 7.4 Covalent adduct of DHPP with the sulfa drug sulfathiazole as determined by X-ray crystallography [37].

with the SO_2 oxygens matching the COOH group of pABA, and the phenylene moieties of the two types of compounds being superposable, and engaging the same hydrophobic pocket in the loop 1–loop 2 substructure of the enzyme [37]. The resistance mutations in DHPS all cluster around this substructure, and make up residues Phe33, Thr67, and Pro69. These amino acids form key elements for the pABA-binding site. The sulfa drugs have been observed positioned (with their heterocyclic ring, which is attached to the SO_2NH moiety) outside the DHPS substrate envelope, being located such that mutations at Phe33 and pro69 can impede the binding of the drug [37]. Thus, the presence of the heterocyclic ring (or the acetamido moiety in **10**) in the sulfa drugs (which has no counterpart in the substrate molecule, pABA) leads to the generation of mutations within the DHPS active site, and the consequent drug resistance problems.

7.3.2 Resistance to Trimethoprim-like DHFR Inhibitors

TMP is a synthetic drug that was introduced for clinical use in Western Europe in the early 1960s [32]. Although its structural similarity to dihydrofolate makes it a competitive inhibitor of the ubiquitous chromosomal DHFR in bacteria, fungi, and protozoa, mammalian DHFRs are resistant to TMP [32]. Therefore, the specificity and selectivity of this antifolate drug have led to its widespread use in the treatment of human infections. However, bacteria have developed resistance mechanisms, including the acquisition of plasmid-derived versions of DHFR, leading to high-level resistance or point mutations of the chromosomal DHFR, which have been demonstrated to confer intermediate resistance [33].

Just a single amino acid substitution, Phe98 to Tyr98, in DHFR was shown to be the molecular origin of trimethoprim **12** resistance (in *Staphylococcus aureus* [38]). This amino acid substitution at the active site was found in all *S. aureus* TMP-resistant clinical isolates tested so far. Ternary complexes of the chromosomal *S. aureus* DHFR (SaDHFR) with methotrexate and TMP, in the presence of nicotinamide adenine dinucleotide phosphate (NADPH), as well as that of mutant Phe98Tyr DHFR SaDHFR (F98Y) ternary folate–NADPH complex, have been obtained and their structure determined by high-resolution X-ray crystallography [38]. Critical evidence concerning the resistance mechanism has also been provided by NMR spectral analyses of ^{15}N-labeled TMP in the ternary complexes of both wild-type and mutant enzyme. These studies show that the mutation results in the loss of a hydrogen bond between the 4-amino group of TMP and the carbonyl oxygen of Leu5. This mechanism of resistance is predominant in both transferable plasmid-encoded and non-transferable chromosomally encoded resistance [38]. Knowledge of the resistance mechanism at a molecular level could help in the design of antibacterials active against multi-resistant *S. aureus* (MRSA), one of today's most serious problems in clinical infectology, but this serious clinical problem has not yet been resolved.

7.4 Clinical Impact of Resistance to Sulfa Drugs and DHFR Inhibitors

The wide use (and misuse) of antibacterial treatment led to the selection for bacterial strains with physiologically or genetically enhanced capacity to survive high doses of antibacterials of both these pharmacological classes [37–40]. The biochemical mechanisms by which these processes occur were outlined above.

Co-trimoxazole, a combination of trimethoprim **12** and sulfamethoxazole **8** (in the ratio of 1:5) is still widely clinically used, although significant resistance problems to both drugs emerged [39, 40]. The synergistic effects of the two drugs are due to the fact they inhibit successive steps in folate synthesis (Figure 7.1).

Presently, sulfa drugs/DHFR inhibitors are primarily used for the treatment of uncomplicated urinary tract infections (cystitis) (mainly trimethoprim-sulfa-methoxazole), this regimen being preferred to the quinolones (even though approximately 20% of *E. coli* are resistant to co-trimoxazole) [41]. Skin infections are sometimes treated with sulfacetamide or a sulfacetamide-silver(I) salt [42]. With its greater efficacy against a limited number of bacteria, co-trimoxazole remains indicated for some infections; for example, it is used as prophylaxis in patients at risk for *Pneumocystis jirovecii* pneumonia and as therapy in Whipple's disease [39, 40, 43]. Gram-positive bacteria are generally moderately susceptible to this combination therapy [43].

7.5 Conclusions

Since the introduction of the sulfa drugs into clinical use in the 1940s, many antibiotics have been developed, possessing a wide range of mechanisms of action. However, their extensive use and misuse raised a serious public health problem due to the worldwide emergence of bacterial pathogens resistant to multiple antibiotics. There is thus a stringent need to develop new antibiotics to keep pace with this bacterial resistance. Recent advances in microbial genomics, synthetic organic chemistry, and X-ray crystallography have provided opportunities to identify novel antibacterial targets for the development of new classes of antibiotics and to design more potent antimicrobial compounds derived from existing antibiotics in clinical use for decades, such as the sulfa drugs or the trimethoprim-like DHFR inhibitors. Although resistance to these pharmacological agents constitutes a serious problem, sulfa drugs and DHFR inhibitors still have a firm place in the armamentarium of antibacterials and may lead to new agents in which the resistance problem may be overcome by structure-based drug design.

References

1 Djapa, L.Y., Zelikson, R., Delahodde, A. et al. (2006). *Plasmodium vivax* dihydrofolate reductase as a target of sulpha drugs. *FEMS Microbiol. Lett.* 256 (1): 105–111. Epub 2006/02/21.

2 Hartman, P.G. (1993). Molecular aspects and mechanism of action of dihydrofolate reductase inhibitors. *J. Chemother.* 5 (6): 369–376. Epub 1993/12/01.

3 Chen, M.J., Shimada, T., Moulton, A.D. et al. (1984). The functional human dihydrofolate reductase gene. *J. Biol. Chem.* 259 (6): 3933–3943. Epub 1984/03/25.

4 Sohn, K.J., Jang, H., Campan, M. et al. (2009). The methylenetetrahydrofolate reductase C677T mutation induces cell-specific changes in genomic DNA methylation and uracil misincorporation: a possible molecular basis for the site-specific cancer risk modification. *Int. J. Cancer* 124 (9): 1999–2005. Epub 2009/01/07.

5 Khalil, I., Ronn, A.M., Alifrangis, M. et al. (2003). Dihydrofolate reductase and dihydropteroate synthase genotypes associated with *in vitro* resistance of *Plasmodium falciparum* to pyrimethamine, trimethoprim, sulfadoxine, and sulfamethoxazole. *Am. J. Trop. Med. Hyg.* 68 (5): 586–589. Epub 2003/06/19.

6 Mockenhaupt, F.P., Teun Bousema, J., Eggelte, T.A. et al. (2005). *Plasmodium falciparum* DHFR but not DHPS mutations associated with sulphadoxine-pyrimethamine treatment failure and gametocyte carriage in northern Ghana. *Tropical Med. Int. Health* 10 (9): 901–908. Epub 2005/09/02.

7 Benkovic, S.J., Fierke, C.A., and Naylor, A.M. (1988). Insights into enzyme function from studies on mutants of dihydrofolate reductase. *Science* 239 (4844): 1105–1110. Epub 1988/03/04.

8 Brown, G.M. (1962). The biosynthesis of folic acid. II. Inhibition by sulfonamides. *J. Biol. Chem.* 237: 536–540. Epub 1962/02/01.

9 Shiota, T., Disraely, M.N., and McCann, M.P. (1964). The enzymatic synthesis of Folate-like compounds from hydroxymethyldihydropteridine pyrophosphate. *J. Biol. Chem.* 239: 2259–2266. Epub 1964/07/01.

10 Shiota, T., Baugh, C.M., Jackson, R., and Dillard, R. (1969). The enzymatic synthesis of hydroxymethyldihydropteridine pyrophosphate and dihydrofolate. *Biochemistry* 8 (12): 5022–5028. Epub 1969/12/01.

11 Valderas, M.W., Andi, B., Barrow, W.W., and Cook, P.F. (2008). Examination of intrinsic sulfonamide resistance in *Bacillus anthracis*: a novel assay for dihydropteroate synthase. *Biochim. Biophys. Acta* 1780 (5): 848–853. Epub 2008/03/18.

12 Lopez, P., Espinosa, M., Greenberg, B., and Lacks, S.A. (1987). Sulfonamide resistance in *Streptococcus pneumoniae*: DNA sequence of the gene encoding dihydropteroate synthase and characterization of the enzyme. *J. Bacteriol.* 169 (9): 4320–4326. Epub 1987/09/01.

13 Slock, J., Stahly, D.P., Han, C.Y. et al. (1990). An apparent *Bacillus subtilis* folic acid biosynthetic operon containing pab, an amphibolic trpG gene, a third gene required for synthesis of para-aminobenzoic acid, and the dihydropteroate synthase gene. *J. Bacteriol.* 172 (12): 7211–7226. Epub 1990/12/01.

14 Radstrom, P. and Swedberg, G. (1988). RSF1010 and a conjugative plasmid contain *sulII*, one of two known genes for plasmid-borne sulfonamide resistance dihydropteroate synthase. *Antimicrob. Agents Chemother.* 32 (11): 1684–1692. Epub 1988/11/01.

15 Sundstrom, L., Radstrom, P., Swedberg, G., and Skold, O. (1988). Site-specific recombination promotes linkage between trimethoprim- and sulfonamide resistance genes. Sequence characterization of *dhfrV* and *sulI* and a recombination active locus of Tn21. *Mol Gen Genet* 213 (2–3): 191–201. Epub 1988/08/01.

16 Morgan, R.E., Batot, G.O., Dement, J.M. et al. (2011). Crystal structures of *Burkholderia cenocepacia* dihydropteroate synthase in the apo-form and complexed with the product 7,8-dihydropteroate. *BMC Struct. Biol.* 11: 21. Epub 2011/05/11.

17 Blakley, R.L. (1995). Eukaryotic dihydrofolate reductase. *Adv. Enzymol. Relat. Areas Mol. Biol.* 70: 23–102. Epub 1995/01/01.

18 Heaslet, H., Harris, M., Fahnoe, K. et al. (2009). Structural comparison of chromosomal and exogenous dihydrofolate reductase from *Staphylococcus aureus* in complex with the potent inhibitor trimethoprim. *Proteins* 76 (3): 706–717. Epub 2009/03/13.

19 Pan, Y.C., Domin, B.A., Li, S.S., and Cheng, Y.C. (1983). Studies of amino-acid sequence in dihydrofolate reductase from a human methotrexate-resistant cell line KB/6b. Structural and kinetic comparison with mouse L1210 enzyme. *Eur. J. Biochem.* 132 (2): 351–359. Epub 1983/05/02.

20 Kovalevskaya, N.V., Smurnyy, Y.D., Polshakov, V.I. et al. (2005). Solution structure of human dihydrofolate reductase in its complex with trimethoprim and NADPH. *J. Biomol. NMR* 33 (1): 69–72. Epub 2005/10/14.

21 Cody, V., Luft, J.R., Ciszak, E. et al. (1992). Crystal structure determination at 2.3 A of recombinant human dihydrofolate reductase ternary complex with NADPH and methotrexate-gamma-tetrazole. *Anticancer Drug Des.* 7 (6): 483–491. Epub 1992/12/01.

22 Cody, V., Wojtczak, A., Kalman, T.I. et al. (1993). Conformational analysis of human dihydrofolate reductase inhibitor complexes: crystal structure determination of wild type and F31 mutant binary and ternary inhibitor complexes. *Adv. Exp. Med. Biol.* 338: 481–486. Epub 1993/01/01.

23 Jensen, D.E., Black, A.R., Swick, A.G., and Azizkhan, J.C. (1997). Distinct roles for Sp1 and E2F sites in the growth/cell cycle regulation of the DHFR promoter. *J. Cell. Biochem.* 67 (1): 24–31. Epub 1997/11/05.

24 Curtius, H.C., Heintel, D., Ghisla, S. et al. (1985). Tetrahydrobiopterin biosynthesis. Studies with specifically labeled (2H)NAD(P)H and 2H2O and of the enzymes involved. *Eur. J. Biochem.* 148 (3): 413–419. Epub 1985/05/02.

25 Kim, H.L., Choi, Y.K., Kim, D.H. et al. (2007). Tetrahydropteridine deficiency impairs mitochondrial function in *Dictyostelium discoideum* Ax2. *FEBS Lett.* 581 (28): 5430–5434. Epub 2007/11/03.

26 Channon, K.M. (2004). Tetrahydrobiopterin: regulator of endothelial nitric oxide synthase in vascular disease. *Trends Cardiovasc. Med.* 14 (8): 323–327. Epub 2004/12/15.

27 Huovinen, P., Sundstrom, L., Swedberg, G., and Skold, O. (1995). Trimethoprim and sulfonamide resistance. *Antimicrob. Agents Chemother.* 39 (2): 279–289. Epub 1995/02/01.

28 Huovinen, R.L., Alanen, K.A., and Collan, Y.U. (1995). Cell proliferation in dimethylbenz(A)anthracene(DMBA)-induced rat mammary carcinoma treated with antiestrogen toremifene. *Acta Oncol.* 34 (4): 479–485. Epub 1995/01/01.

29 Kehrenberg, C. and Schwarz, S. (2005). dfrA20, A novel trimethoprim resistance gene from Pasteurella multocida. *Antimicrob. Agents Chemother.* 49 (1): 414–417. Epub 2004/12/24.

30 Alonso, H. and Gready, J.E. (2006). Integron-sequestered dihydrofolate reductase: a recently redeployed enzyme. *Trends Microbiol.* 14 (5): 236–242. Epub 2006/04/06.

31 Howell, E.E. (2005). Searching sequence space: two different approaches to dihydrofolate reductase catalysis. *Chembiochem* 6 (4): 590–600. Epub 2005/04/07.

32 Skold, O. (2001). Resistance to trimethoprim and sulfonamides. *Vet. Res.* 32 (3–4): 261–273. Epub 2001/07/04.

33 Dale, G.E., Broger, C., D'Arcy, A. et al. (1997). A single amino acid substitution in *Staphylococcus aureus* dihydrofolate reductase determines trimethoprim resistance. *J. Mol. Biol.* 266 (1): 23–30. Epub 1997/02/14.

34 Domagk, G. (1950). Experimental bases in chemotherapy of bacterial infections with sulfonamides and related substances with special reference to its application in surgery. *Langenbecks Arch. Klin. Chir. Ver. Dtsch. Z. Chir.* 264: 102–123.

35 Tréfouël, J.T., Nitti, F., and Bovet, D. (1935). Activité du p.aminophénylsulfamide sur l'infection streptococcique expérimentale de la souris et du lapin. *C. R. Soc. Biol.* 120: 756–761.

36 Hawser, S., Lociuro, S., and Islam, K. (2006). Dihydrofolate reductase inhibitors as antibacterial agents. *Biochem. Pharmacol.* 71: 941–948.

37 Yun, M.K., Wu, Y., Li, Z. et al. (2012). Catalysis and sulfa drug resistance in dihydropteroate synthase. *Science* 335 (6072): 1110–1114.

38 Dale, G.E., Broger, C., D'Arcy, A. et al. (1997). A single amino acid substitution in *Staphylococcus aureus* dihydrofolate reductase determines trimethoprim resistance. *J. Mol. Biol.* 266: 23–30.

39 Tadesse, D.A., Zhao, S., Tong, E. et al. (2012). Antimicrobial drug resistance in *Escherichia coli* from humans and food animals, United States, 1950–2002. *Emerg. Infect. Dis.* 18: 741–749.

40 Dini, L., du Plessis, M., Frean, J., and Fernandez, V. (2010). High prevalence of dihydropteroate synthase mutations in *Pneumocystis jirovecii* isolated from patients with *Pneumocystis pneumonia* in South Africa. *J. Clin. Microbiol.* 48: 2016–2021.

41 Täuber, M.G. and Mühlemann, K. (2009). Antibiotic therapy in the outpatient setting: update 2009. *Praxis* 98: 877–883.

42 Mastrolorenzo, A. and Supuran, C.T. (2000). Antifungal activity of Ag(I) and Zn(II) complexes of Sulfacetamide derivatives. *Met. Based Drugs* 7: 49–54.

43 Yoneyama, H. and Katsumata, R. (2006). Antibiotic resistance in bacteria and its future for novel antibiotic development. *Biosci. Biotechnol. Biochem.* 70: 1060–1075.

8

Anti-tuberculosis Agents

Ying Zhang

Department of Molecular Microbiology and Immunology, Bloomberg School of Public Health, Johns Hopkins University, Baltimore, MD, USA

8.1   Introduction

Tuberculosis (TB) caused by *Mycobacterium tuberculosis* remains a leading infectious disease worldwide despite availability of chemotherapy and BCG vaccine. Each year there are about nine million new cases and close to two million deaths. *M. tuberculosis* is a highly successful pathogen and has latently infected one-third of the world population, about two billion people. In people with a latent tuberculosis infection, the risk of this becoming active tuberculosis is approximately 5–10% during their lifetime, but when the immune system is compromised, such as co-infections with HIV, malnutrition, or aging, latent infection can reactivate and develop into active disease [1]. Although TB has a cure rate of 85–95%, therapy is lengthy and takes at least six months. This prolonged treatment makes patient compliance to therapy difficult and is a frequent cause for selection of drug-resistant TB. Increasing emergence of drug-resistant TB [2] poses a significant threat to disease control globally. It is estimated that about 50 000 new cases of multidrug-resistant TB (MDR-TB) and 40 000 cases of XDR-TB (MDR-TB with additional bacillary resistance to a fluoroquinolone and a second-line injectable drug) occur globally each year [3]. Drug-resistant TB and HIV infection, which weakens the host immune system, present serious challenges for effective TB control.

8.1.1   A Brief History of Current TB Drugs and Therapy

The first TB drug streptomycin (SM), discovered from soil microbes in 1943 by Schatz and Waksman, marked the beginning of modern TB therapy [4]. This was followed by the discovery of a series of anti-TB drugs from synthetic compounds including para-aminosalicylic acid (PAS) in 1946 and isoniazid (INH) and pyrazinamide (PZA) in 1952 based on nicotinamide activity against tubercle bacilli in the animal model [5]. The nicotinamide lead also led to the discovery of ethionamide (ETH)/prothionamide (PTH) in 1956 [6]. Ethambutol (EMB) was discovered in 1961 at Lederle [7]. Further screening for antibiotics from soil microbes led to discovery of cycloserine (CS) [8], kanamycin [9], and its derivative amikacin, and viomycin [10], capreomycin ([11]), and rifamycins [12] and its derivative rifampin (RIF) [13]. The broad-spectrum quinolones developed in the 1980s have high activity against mycobacteria and are used as second-line drugs for the treatment of MDR-TB [14, 15].

The British Medical Research Council performed the first ever randomized clinical trial (RCT) with SM in 1946, proving SM efficacy [16], but resistance to SM was soon observed. It was noted that addition of PAS to SM prevented SM resistance, thus establishing the important principle of drug combination for the treatment of TB. The same principle of drug combination was subsequently applied to the treatment of various infectious diseases such as AIDS and those caused by *Helicobacter pylori,* and also cancer. Addition of INH to SM and PAS in the 1950s finally led to a consistent cure in 18–24 months. In the 1970s, RIF used in combination with INH further shortened treatment to 9 months [16]. In the 1970 and 1980s, re-evaluation of PZA in combination with INH and RIF led to the current six month TB chemotherapy. Since 1995, WHO recommends the six months therapy (part of the Directly Observed Therapy – short course [DOTS] strategy) as the preferred standard chemotherapy for treating drug-susceptible TB. Treatment of drug-resistant TB is more complex and requires 18–24 months with a cocktail of first-line and second-line agents.

8.1.2 Different Classes of TB Drugs

The current TB drugs can be divided into first-line drugs and second-line drugs. The first-line drugs include INH, RIF (and its derivative rifapentine), PZA and EMB (Figure 8.1). The second-line drugs include SM, kanamycin, amikacin, capreomycin,

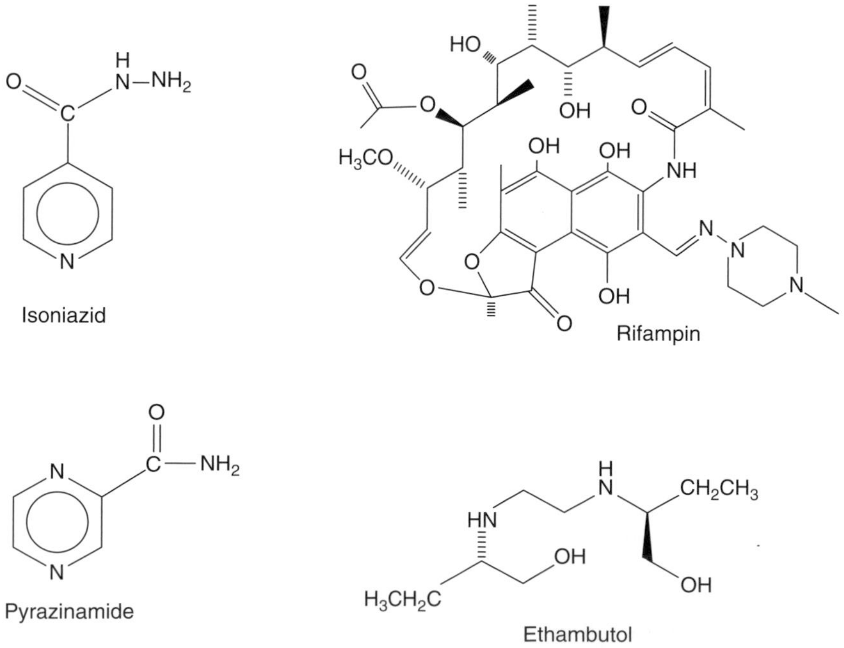

Figure 8.1 Structures of front-line TB drugs.

CS, PAS, ETH/PTH, and fluoroquinolones (FQ). First-line drugs are used in combination to prevent drug resistance and improve therapeutic efficacy for the treatment of all newly diagnosed TB patients, whereas second-line drugs are used in combination with first-line drugs PZA and EMB for the treatment of MDR-TB. According to the specificity of the drugs, TB drugs can also be grouped into TB or mycobacteria-specific drugs such as INH, PZA, EMB, PAS, ETH, and broad-spectrum antibiotics such as RIF, SM, kanamycin, amikacin, capreomycin, CS and FQ. The mechanisms of action and resistance to TB-specific drugs are specific to *M. tuberculosis*, whereas mechanisms of action and resistance of the broad-spectrum drugs in *M. tuberculosis* are mostly the same as in other bacterial species. The TB drugs can also be divided based on their mechanism of action as inhibitors of cell wall synthesis (INH, EMB, ETH, CS), inhibitors of nucleic acid synthesis (rifampin, quinolones), inhibitors of protein synthesis (aminoglycosides, capreomycin), and inhibitors of energy metabolism (PZA, bedaquiline). TB drugs can also divided into drugs active against only growing bacteria (INH, EMB, ETH, PAS, CS), drugs active against both growing and non-growing bacteria (RIF, FQ, bedaquiline), and drug only active against non-growing persisters (PZA).

8.2 Molecular Mechanisms of Drug Resistance

8.2.1 The Five Major Mechanisms of Drug Resistance

Antibiotic resistance in bacteria can occur by five major mechanisms [17]: (i) decreased uptake or impermeability, (ii) increased efflux, (iii) enzymatic inactivation, (iv) modification or over-expression of drug target, and (v) inactivation of prodrug activating enzymes or defective prodrug activation. It is worth mentioning inactivation of prodrug activation enzyme is a new mechanism of drug resistance that was first convincingly demonstrated in 1992 with INH resistance mediated by mutations in catalase-peroxidase enzyme (KatG), which activates INH [18]. Then in 1996, PZA was found to be a prodrug that is activated by nicotinamidase/pyrazinamidase (PncA) and mutations in PncA cause PZA resistance [19]. In 1998, a similar mechanism was demonstrated for metronidazole resistance in *H. pylori*, due to mutations in *rdxA* encoding NADPH nitroreductase [20] or in *frxA* encoding NAD(P)H flavin oxidoreductase involved in metronidazole activation [21]. Subsequently, resistance to various TB drugs including ethionamide, thioamide (thiacetazone), and PAS were all found to be caused by a similar mechanism of mutations in prodrug activation enzymes. However, this new mechanism of drug resistance is not mentioned in most published reviews or book chapters on antibiotic resistance. We would like to emphasize the importance of mutation in prodrug activating enzymes as a new mechanism of antibiotic resistance and prodrug as a strategy for developing new effective antibiotics in general.

8.2.2 Intrinsic Resistance

Mycobacterium tuberculosis is intrinsically resistant to many antibiotics, which can be caused by the following factors. For example, *M. tuberculosis* has a hydrophobic cell envelope, which provides a permeability barrier for some antibiotics [22]. *M. tuberculosis* also possesses some enzymes such as beta-lactamase that inactivate penicillin [23]. In addition, *M. tuberculosis erm37* encodes a 23S rRNA methyltransferase, which methylates its primary target at A2058 and could also attach additional methyl groups to the neighboring nucleotides A2057 and A2059 of the 23S rRNA [24]. The gene *erm37*, present in *M. tuberculosis* complex organisms but absent in many non-tuberculous mycobacteria (NTM), is responsible for the natural macrolide resistance in *M. tuberculosis* and for susceptibility to NTM [25]. *whiB7*, a transcriptional activator present only in Actinomycetes, was found to be responsible for intrinsic resistance to various antibiotics including chloramphenicol, clarithromycin, erythromycin, lincomycin, spectinomycin, streptomycin, and tetracycline *M. tuberculosis* [26]. Inactivation of *whiB7* in *M. tuberculosis* caused increased susceptibility to macrolides, a lincosamide, and streptomycin [26]. Antibiotics (erythromycin, tetracycline, streptomycin) and fatty acids (palmitic acid being most active), and other stimuli induced WhiB7 expression, causing an inducible antibiotic resistance [26]. *whiB7* induces a regulon of eight genes including the known efflux gene *tap*, encoding an efflux pump that confers low-level resistance to aminoglycosides and tetracycline, and *erm37 (Rv1988)*, encoding a ribosomal methyltransferase which confers macrolide, lincosamide, and streptogramin resistance by modification of 23S rRNA, in response to antibiotics [26]. There are seven *whiB* genes, *whiB1–7*, in *M. tuberculosis*, which are induced by different stress conditions and antibiotics [27]. It was found that the cell wall inhibitors like INH, EMB and cycloserine induced *whiB2*, whereas aminoglycosides induced *whiB7* primarily and also other *whiB* genes such as *whiB2*, *whiB3*, and *whiB6* to a lesser extent [27]. However, *whiB* genes have not been found to cause clinically relevant TB drug resistance.

8.2.3 Unique Features of Drug Resistance in *M. tuberculosis*

Several features are worth noting when discussing mechanisms of drug resistance in *M. tuberculosis*. First, mobile genetic elements such as plasmids and transposons, which are known to mediate drug resistance in various bacterial species, do not cause drug resistance in *M. tuberculosis*. Instead, drug resistance in *M. tuberculosis* is caused by spontaneous (and perhaps drug induced) mutations in chromosomal genes at a frequency of 10^{-5} to 10^{-8}. Because such mutations resulting in drug resistance are unlinked, the probability of developing resistance to three drugs used simultaneously occurs at 10^{-18} to 10^{-20}. Thus, in theory, the chance of drug resistance is virtually nonexistent when three effective drugs are used in combination, but in reality drug resistance occurs frequently clinically through poor patient compliance to therapy or sometimes physician errors, resulting in drug-resistant TB. Second, the MDR/XDR phenotype is caused by sequential accumulation of mutations in different drug resistance genes involved in individual drug resistance (Table 8.1). Thirdly, a significant number of TB drugs are prodrugs that require activation by *M. tuberculosis* enzymes to kill or inhibit the bacteria. Thus, inactivation of prodrug activating enzymes KatG, PncA, EthA, and ThyA, responsible for INH, PZA, ethionamide and PAS resistance, respectively, is a fairly common mechanism of resistance in *M. tuberculosis*. In addition, target alterations as in mutations in *inhA*, *rpoB*, *rpsL* or *rrs*, *embB*, and *gyrA*, responsible for INH, RIF, streptomycin, ethambutol and quinolone resistances, respectively, are also frequent mechanisms of drug resistance in *M. tuberculosis* [28]. It has been recently shown that various compensatory mutations are closely associated with MDR/XDR-TB in clinical isolates [29]. Some of these mutations may not be causal in drug resistance but may help to improve the fitness of the organism, while others may help to stabilize or increase the level of resistance [30, 31].

Table 8.1 Mechanisms of drug action and resistance in *Mycobacterium tuberculosis*.

Drugs (Year of discovery)	MIC (µg/ml)	Gene(s) involved in Resistance	Gene Function	Role	Mechanism of Action
Isoniazid (1952)	0.02–0.2	*katG*	Catalase-peroxidase	Prodrug conversion	Inhibition of mycolic acid biosynthesis and other multiple effects on DNA, lipids, carbohydrates, and NAD metabolism
		inhA	Enoyl ACP reductase	Drug target	
Rifampin (1966)	0.05–1	*rpoB*	β-subunit of RNA polymerase	Drug target	Inhibition of RNA synthesis
Pyrazinamide (1952)	16–50 (pH 5.5)	*pncA*	Nicotinamidase/ pyrazinamidase	Prodrug conversion	Depletion of membrane energy
		rpsA	Ribosomal protein S1	Drug target	Inhibition of trans-translation
		panD	Aspartate decarboxylase	Drug target?	Inhibition of pantothenate and coenzyme A synthesis
Ethambutol (1961)	1–5	*embB*	Arabinosyl transferase	Drug target	Inhibition of arabinogalactan synthesis
Streptomycin (1944)	2–8	*rpsL*	S12 ribosomal protein	Drug target	Inhibition of protein synthesis
		rrs	16S rRNA	Drug target	
		gidB	rRNA methyltransferase (G527 in 530 loop)	Drug target	
Amikacin Kanamycin (1957)	2–4 2–4	*rrs*	16S rRNA	Drug target	Inhibition of protein synthesis
		rrs	16S rRNA,		
		eis promoter	Aminoglycoside acetyltransferase 16S rRNA	Drug inactivation	
Capreomycin (1960)		*tlyA*	2'-O-methyltransferase	Target sensitization	
Quinolones (1963)	0.5–2.5	*gyrA*	DNA gyrase subunit A	Drug target	Inhibition of DNA gyrase
		gyrB	DNA gyrase subunit B		
Ethionamide (1956)	2.5–10	*etaA/ethA* *inhA*	Flavin monooxygenase	Prodrug conversion Drug target	Inhibition of mycolic acid synthesis
PAS (1946)	1–8	*thyA*	Thymidylate synthase	Drug activation	Inhibition of folic acid metabolism
		dhfr	Dihydrofolate reductase	Drug target	Inhibition of folic acid metabolism by incorporation into the folate pathway by dihydropteroate synthase and dihydrofolate synthase to generate a hydroxyl dihydrofolate antimetabolite

8.2.4 Two Types of Drug Resistances: Genetic Resistance and Phenotypic Resistance or Drug Tolerance

It must be pointed that that there are two types of drug resistance, genetic drug resistance and phenotypic drug resistance or tolerance, which hinder the effectiveness of treatment [32]. Genetic drug resistance refers to resistance in growing bacteria that grow in the presence of the antibiotic, and is caused by mutations or mobile genetic elements and is thus heritable. However, phenotypic drug resistance or drug tolerance refers to resistance in non-growing bacteria (persisters), which do not grow in the presence of the antibiotic but the persisters are not killed, and is caused by physiological changes in quiescent or dormant bacteria or persisters, but the bacteria remain susceptible to the same antibiotic when bacteria start growing again, and is thus non-heritable or phenotypic. There is increasing recent interest in drug tolerance in persister bacteria. This is because the pesister bacteria are underlying persistent and recurrent infections that can frequently lead to genetic resistance.

Although it is widely accepted that persistence or drug tolerance is a major problem for TB treatment causing prolonged therapy, the mechanism of drug tolerance in *M. tuberculosis* is not well understood. There is significant interest in understanding the biology of persisters and developing new drugs targeting persisters [33]. Various genes and pathways were found to be involved in persistence in *M. tuberculosis* [33]. In addition, lack of drug uptake, *i.e.*, rifampin, linezolid, and quinolone, in starvation-induced dormant TB bacteria, could underlie drug tolerance in such organisms [34]. Due to space limitation, the possible mechanisms of drug tolerance will not be covered here. For more information on this topic, please refer to recent reviews on this topic [33].

Understanding the mechanisms of drug resistance to antituberculosis drugs not only enables the development of rapid molecular diagnostic tests and helps to prevent further spread of drug resistance but also has implications for designing new TB drugs [35]. In this chapter, I will provide an update on the mechanisms of action and resistance on the most important first-line and second-line agents in *M. tuberculosis*, discuss the clinical relevance of drug resistance in TB, and provide a brief overview of current new TB drugs in clinical development.

8.2.4.1 Isoniazid

INH has been the cornerstone of all effective regimens for the treatment of TB disease and latent TB infection. *M. tuberculosis* is highly susceptible to INH (MICs = 0.02–0.2 μg/ml). INH is only active against growing *M. tuberculosis* but has no activity against non-growing stationary phase bacilli. INH is a prodrug that is activated by KatG encoded by the *katG* gene [18] to generate a range of highly reactive species which then attack multiple targets in *M. tuberculosis* [36]. The primary target of INH is the InhA enzyme (enoyl acyl carrier protein reductase), involved in elongation of fatty acids in mycolic acid synthesis [37]. The active species (isonicotinic acyl radical or anion) derived from KatG-mediated INH activation reacts with NAD(H) forming an INH-NAD adduct which then inhibits InhA, leading to inhibition of cell wall mycolic acid synthesis [38, 39]. It was shown recently that this INH tolerance in non-growing mycobacteria is caused by mycobacterial DNA-binding protein 1 (MDP1), a histone-like protein, which negatively regulates *katG* transcription and leads to tolerance to INH [40].

Resistance to INH occurs frequently at 1 in 10^{5-6} bacilli *in vitro* [41]. INH-resistant *M. tuberculosis* often lose catalase and peroxidase enzyme activity [42] encoded by *katG*, especially in high-level resistant strains (MIC>1–5 µg/ml) [41]. Low-level resistant strains (MICs<1 µg/ml) may still retain catalase activity [41] (see later). Mutation in *katG* is the major mechanism of INH resistance (Table 8.1) [18, 43, 44]. KatG S315 mutation is the most common mutation in INH-resistant strains, accounting for 50–95% of INH-resistant clinical isolates [36, 43, 44]. Mutations in the *katG* promoter region *furA-katG* intergenic region that affect KatG expression were occasionally found in INH-resistant strains [45]. Resistance to INH can also occur by mutations in the promoter region of *mabA/inhA* operon causing over-expression of InhA, or by mutations at the InhA active site lowering the InhA affinity to INH-NAD adduct [37, 39]. Mutations in *inhA* or its promoter region are usually associated with low-level resistance (MICs = 0.2–1 µg/ml) and less frequent than *katG* mutations (Table 8.1) [36, 43, 44]. Mutations in *inhA* not only cause INH resistance but also confer cross-resistance to the structurally related drug, ETH [37]. In KatG-negative INH-resistant strains, mutations in the promoter region of *ahpC*, encoding an alkylhydroperoxide reductase, leading to increased expression of AhpC, serve as a compensation for the lack of KatG in such strains [46, 47]. About small percentage of low level INH-resistant strains do not have mutations in *katG* or *inhA* [43] and may be due to new mechanism(s) of resistance or due to heteroresistance. Mutations in *mshA* encoding glycosyltransferase and *mshC* encoding cysteine ligase involved in mycothiol biosynthesis confer low level INH resistance and high level ETH resistance in *M. tuberculosis* strains *in vitro* [48, 49]; however, their role in clinical INH resistance has not been found so far. Recently, sigma factor I (SigI) was found to regulate expression of *katG* and *sigI* deletion caused a low level of resistance to INH (MIC = 0.18 µg/ml) compared with the wild-type (MIC = 0.04 µg/ml) [50]. It is unknown if *sigI* is involved in INH resistance in clinical isolates.

8.2.4.2 Rifampin

RIF is bactericidal for *M. tuberculosis* with a very low MIC (0.05–1 µg/ml), but the MIC is much higher in egg-based media (MIC = 2.5–10 µg/ml). Strains with MICs <1 µg/ml in liquid or agar medium, or < 40 µg/ml in Lowenstein–Jensen medium are considered RIF susceptible. RIF is active against growing populations as well as stationary phase bacilli with low metabolic activity. The latter activity is related to its sterilizing activity and its ability to shorten TB treatment from 12 to 18 months to 9 months [51].

RIF interferes with RNA synthesis by binding to the β-subunit of the RNA polymerase. The RNA polymerase is an oligomer consisting of a core enzyme formed by four chains ($\alpha 2\beta\beta'$) in association with an σ-subunit to specifically initiate transcription from promoters. The RIF binding site is located upstream of the catalytic center and physically blocks the elongation of the RNA chain. In *M. tuberculosis*, resistance to RIF occurs at a frequency of 10^{-7} to 10^{-8}. Mutations in a defined 81 bp region of the *rpoB* are found in about 96% of RIF-resistant *M. tuberculosis* isolates [52]. Mutations at position 531, 526, and 516 are among the most frequent mutations in RIF-resistant strains. Mutation in *rpoB* generally results cross-resistance to all rifamycins.

The report of RIF-dependent or enhanced strains of *M. tuberculosis* in some MDR-TB strains is potentially worrying [53, 54]. These strains have poor growth on culture media without RIF but grew better in the presence of RIF. RIF-dependent strains may be more prevalent than currently realized since the current diagnostic practice uses only drug-free media. The circumstances whereby the RIF-dependent strains arise remain unclear but they often occur as MDR-TB and seem to develop upon repeated treatment with rifamycins in re-treatment patients. Continued use of rifamycins for the treatment of patients harboring RIF-dependent strains can worsen the disease [55]. It would be of interest to determine the mechanism of RIF dependence/enhancement and assess the contribution of such strains to treatment failure. RIF resistance due to *rpoB* mutations could alter the fitness of *M. tuberculosis* and cause compensatory mutations in *rpoC* or *rpoA* [56].

8.2.4.3 Pyrazinamide

PZA has played a unique role in shortening TB therapy from 9 to12 months to 6 months because it kills populations of persisters in acidic environments that are not killed by other drugs [51]. PZA is an unconventional and paradoxical persister drug that has high sterilizing activity *in vivo* [57] but no activity against TB bacilli under normal culture conditions near neutral pH [58]. PZA is only active against *M. tuberculosis* at acid pH (*e.g.*, 5.5) [59] and even at acid pH, its anti-TB activity is poor with high MIC (6.25–50 µg/ml) [57]. In addition to acid pH, PZA activity can be enhanced by various factors such as low oxygen or anaerobic conditions [60], weak acids [61], and energy inhibitors such as DCCD, azide, and rotenone [62].

PZA is a prodrug that has to be converted to its active form, pyrazinoic acid (POA), by the pyrazinamidase/nicotinamidase enzyme encoded by the *pncA* gene of *M. tuberculosis* [19]. POA produced intracellularly gets to the cell surface through passive diffusion and a defective efflux [63], and if extracellular pH is acidic, it facilitates the formation of uncharged protonated POA, which then permeates through the membrane and causes accumulation of POA inside the cell. The protonated POA brings protons into the cell and could eventually cause cytoplasmic acidification and de-energize the membrane by collapsing the proton motive force, which affects membrane transport [62]. Although fatty acid synthase I (Fas-I) was initially proposed as a target for PZA [64] it was subsequently shown to be a target of 5-Cl-PZA but not PZA [65]. A new target of PZA, RpsA (ribosomal protein S1) involved in trans-translation, was recently identified [66]. Overexpression of RpsA caused PZA resistance in *M. tuberculosis*. In addition, a low level PZA-resistant clinical strain DHM444 without *pncA* mutations [67] contained a deletion of alanine at the 438th residue (438 ΔA) due to a 3-bp GCC in the C-terminus of the RpsA [66]. POA bound to the wild-type RpsA but not the mutant RspAΔA438 from the PZA-resistant strain and specifically inhibited the trans-translation of *M. tuberculosis* but not the canonical translation. Trans-translation is a process that removes toxic protein products formed under stress conditions by adding a tmRNA tag, which is the protease recognition sequence, and then sending them for degradation by proteases [68]. Trans-translation is dispensable during active growth but becomes important for managing stalled ribosomes or damaged mRNA and proteins under stress conditions [69, 70]. More recently, a new gene, *panD* encoding aspartate decarboxylase was found to be associated with PZA resistance [71]. PanD is involved in the synthesis of β-alanine that is a precursor for pantothenate and co-enzyme A biosynthesis. PanD may serve as a target of PZA, while POA inhibition of the synthesis of pantothenate and co-enzyme A may be critical for eliminating persister TB bacteria. The observations that POA inhibits multiple novel targets such as membrane potential [62], trans-translation [66], and possibly pantothenate and CoA synthesis [71] in *M. tuberculosis* help to explain the unique sterilizing activity of PZA [62, 72].

Mutation in the *pncA* gene encoding pyrazinamidase/nicotinamidase is the major mechanism for PZA resistance in *M. tuberculosis* [19, 67, 73] [74–82]. The natural PZA resistance in *Mycobacterium bovis* is due to a characteristic mutation of "C" to "G" at nucleotide position 169 of the *pncA* gene, causing amino acid substitution at position 57 of the PncA sequence [19]. The *pncA* mutations are highly diverse and scattered along the gene, which is unique to PZA resistance. Despite the highly diverse and scattered distribution of *pncA* mutations, there is some degree of clustering of mutations at three regions of the PncA, 3–17, 61–85, and 132–142 [67, 76]. Most PZA-resistant *M. tuberculosis* strains (72–99%) have mutations in *pncA* [67, 73, 74, 76–78, 80–82]; however, some resistant strains do not have *pncA* mutations. Recently, it was shown that some PZA-resistant clinical isolates such as DHM444 without *pncA* mutations [67] had mutations in the drug target RpsA [66, 83]. RpsA target mutations are usually associated with low-level PZA resistance (MIC = 200–300 µg/ml PZA). Most recently, a new gene *panD* encoding aspartate decarboxylase, was found to be involved in PZA resistance [71]. *panD* mutations were identified in naturally PZA-resistant *Mycobacterium canettii* strains and a PZA-resistant MDR-TB strain. Future studies are needed to address the role of *panD* mutations in PZA resistance.

8.2.4.4 Ethambutol

EMB ([S,S']-2,2′[ethylenediimino]di-1-butanol) is mainly used in combination with INH, RIF, and PZA to prevent emergence of drug resistance. The MICs of EMB for *M. tuberculosis* are in the range 0.5–2 µg/ml. Strains with MICs >7.5 µg/ml are considered EMB resistant [84]. EMB is a bacteriostatic agent that is active against growing mycobacteria. EMB interferes with the biosynthesis of cell wall arabinogalactan [85]. It inhibits the polymerization of cell wall arabinan of arabinogalactan and of lipoarabinomannan and induces accumulation of D-arabinofuranosyl-P-decaprenol, an intermediate in arabinan biosynthesis [86, 87]. Arabinosyl transferase, encoded by *embB*, an enzyme involved in synthesis of arabinogalactan, is the target of EMB in *M. tuberculosis* [88]. In *M. tuberculosis*, *embB* is organized into an operon with *embC* and *embA* in the order *embCAB*. *embC*, *embB*, and *embA* share over 65% amino acid identity with each other and are predicted to encode transmembrane proteins [88].

Mutation to EMB resistance occurs at a frequency of 10^{-5}. Mutations in the *embCAB* operon, in particular *embB*, and occasionally *embC*, are responsible for resistance to EMB [88]. *embB* codon 306 mutation is most frequent in clinical isolates resistant to EMB, accounting for as much as 68% of resistant strains [89, 90]. While mutations at EmbB 306 leading to certain amino acid changes caused EMB resistance, other amino acid substitutions had little effect on EMB resistance [91]; however, about 35% of EMB-resistant strains (MIC<10 µg/ml) do not have *embB* mutations [92], suggesting other mechanisms of EMB resistance. Further studies are needed to identify potential new mechanisms of EMB resistance. Like quinolone resistance (see later), multiple mutations in functionally related genes, *embB* and *ubiA*, involved in arabinogalactan synthesis, can cause a higher level of EMB resistance [30].

8.2.4.5 Aminoglycosides (Streptomycin, Kanamycin/Amikacin) and Capreomycin

SM kills actively growing *M. tuberculosis* with MICs of 2–4 µg/ml [84], but is inactive against nongrowing or intracellular bacilli [51]. SM inhibits protein synthesis by binding to the 30S subunit of bacterial ribosome causing misreading of the mRNA message [93]. The site of action of SM is the 30S subunit of the ribosome at the ribosomal protein S12 and the 16S rRNA. Resistance to SM is caused by mutations in the S12 protein encoded by the *rpsL* gene and 16S rRNA encoded by the *rrs* gene [94]. Mutations in *rpsL* and *rrs* are the major mechanism of SM resistance, [94–96] accounting for about 50% and 20% of SM-resistant strains, respectively [94–96]. The most common mutation in *rpsL* is a substitution in codon 43 from lysine to arginine, [94–96] causing high-level resistance to SM. Mutation in codon 88 is also common [94–96]. Mutations of the *rrs* gene occur in the loops of the 16S rRNA and are clustered in two regions around nucleotides 530 and 915 [94–96]. SM-dependent SM-resistant *M. tuberculosis* strains seem to be caused by a "C" insertion in the 530 loop [97]. However, about 20–30% of SM-resistant strains with low-level resistance (MICs<32 μg/ml) do not have mutations in *rpsL* or *rrs* [98], which indicates other mechanism of resistance. Recently, mutation in *gidB* encoding a conserved 7-methylguanosine (m[7]G) methyltransferase specific for 16S rRNA, has been found to cause low-level SM resistance in 33% of resistant *M. tuberculosis* isolates [99]. In addition, some low-level SM resistance seems to be caused by increased efflux, because efflux pump inhibitors caused increased sensitivity to SM, though the exact mechanism remains to be identified [100]. Recently, it has been shown that mutations in the promoter region of *whiB7* contribute to cross-resistance to SM and kanamycin due to increased expression of the *tap* efflux gene, which is controlled by *whiB7* [101].

Kanamycin and its derivative amikacin also inhibit protein synthesis through modification of ribosomal structures at 16S rRNA [102, 103]. Mutations at 16S rRNA (*rrs*) position 1400 are associated with high-level resistance to kanamycin and amikacin [102, 103]. SM-resistant strains are usually still susceptible to kanamycin and amikacin. Mutations in the promoter region of the *eis* gene, encoding aminoglycoside acetyltransferase, caused low-level resistance to kanamycin but not to amikacin [104].

Capreomycin is a polypeptide antibiotic. A gene called *tlyA* encoding rRNA methyltransferase was shown to be involved in resistance to capreomycin [105]. The rRNA methyltransferase modifies nucleotide C1409 in helix 44 of 16S rRNA and nucleotide C1920 in helix 69 of 23S rRNA [106]. Variable cross-resistance may be observed between kanamycin, amikacin, capreomycin, or viomycin [41]. Mutants resistant to capreomycin and viomycin could have *tlyA*, C1402T, or G1484T *rrs* mutations, while mutants resistant to capreomycin but not viomycin could have an A1401G *rrs* mutation [107]. Mutants with an A1401G mutation could cause resistance to kanamycin and capreomycin but not viomycin [107]. Mutants resistant to capreomycin, kanamycin, and viomycin, could have either a C1402T or a G1484T mutation in the *rrs* gene [107]. Multiple mutations may occur in the *rrs* gene in one strain, conferring cross-resistance among these agents [107].

8.2.4.6 Fluoroquinolones

Topoisomerases regulate DNA supercoiling and unlink tangled nucleic acid strands to meet replicative and transcriptional needs [108]. Fluoroquinolones inhibit DNA gyrase (topoisomerase II) and topoisomerase IV resulting in bacterial death. DNA gyrase is a tetrameric A2B2 protein. The A subunit carries the breakage–reunion active site, whereas the B subunit promotes ATP hydrolysis. *M. tuberculosis gyrA* and *gyrB* encode the A and B subunits, respectively [109]. A conserved region, quinolone-resistance-determining region (QRDR) of *gyrA* (320 bp) and *gyrB* (375 bp), is involved in fluoroquinolone resistance in *M. tuberculosis* [109]. Mutations within the QRDR of *gyrA*, largely clustered at codons 90, 91, 94, [109–113] with Asp94 being rather frequent, has been identified in *M. tuberculosis* [111, 114]. Others involved in resistance include codons 74, 83, and 87 [110, 114, 115]. Mutation at codon 95 is a polymorphism not involved in quinolone resistance [116]. *gyrB* mutations are much less frequent in clinical isolates [112, 113, 117]. Two mutations in *gyrA* or concomitant mutations in *gyrA* plus *gyrB* are required for the development of higher levels of resistance [109, 118].

A novel mechanism of quinolone resistance mediated by MfpA was recently identified [119]. MfpA is a member of the pentapeptide repeat protein from *M. tuberculosis*, whose expression causes resistance to fluoroquinolones. MfpA binds to DNA gyrase and inhibits its activity by DNA mimicry, which explains its inhibitory effect on DNA gyrase and quinolone resistance [119]; however, clinical strains that elaborate MfpA to develop resistance to quinolones remain to be identified. Other mechanisms responsible for mycobacterial resistance to fluoroquinolones, such as decreased cell wall permeability to drug, efflux pump, drug sequestration, or perhaps even drug inactivation are possible [108].

8.2.4.7 Ethionamide (2-Ethylisonicotinamide, ETH)/Prothionamide (2-Ethyl-4-pyridinecarbothioamide, PTH) and Thioamides (Thiacetazone)

ETH is a derivative of isonicotinic acid and is a bactericidal agent only against *M. tuberculosis*. The MICs of ETH for *M. tuberculosis* are 0.5–2 µg/ml in liquid medium and 2.5–10 µg/ml in 7H11 agar, and 5–20 µg/ml in Lowenstein–Jensen medium. Like INH, ETH is also a prodrug that is activated by EtaA/EthA (a mono-oxygenase) [120, 121] and inhibits the same target as INH, *i.e.*, InhA of the mycolic acid synthesis pathway [37]. PTH shares almost the identical structure and activity as ETH. EthA is an FAD-containing enzyme that oxidizes ETH to the corresponding *S*-oxide, which is further oxidized to 2-ethyl-4-amidopyridine, presumably *via* the unstable oxidized sulfinic acid intermediate [122]. EthA also activates thiacetazone, thiocarlide, thiobenzamide, and perhaps other thioamide drugs [122], which explains the cross-resistance between ETH and thiacetazone, thiocarlide, and other thioamides and thioureas (*e.g.*, isoxyl) [123]. Thiacetazone is one of the many thiosemicarbazones discovered by Gerhard Domagk. Thiacetazone is a bacteriostatic drug that is active against *M. tuberculosis* with MIC of 1 µg/ml [124]. Mutations in the drug activating enzyme EtaA/EthA [120, 121] cause resistance to ETH and other thioamides. In addition, mutations in the target InhA confer resistance to both ETH and INH.

8.2.4.8 Cycloserine (CS) and Terizidone

CS is a second-line agent used in combination with other TB drugs for the treatment of MDR-TB. Terizidone is a combination of two molecules of CS. CS is mainly a bacteriostatic drug and can be used to prevent resistance to other second-line drugs. CS is active against a broad range of bacterial species (*e.g.*, *Escherichia coli* and *Staphylococcus aureus*) and is more active against mycobacteria than other bacteria. The MIC of CS for *M. tuberculosis* ranges widely from 1.5 to 30 µg/ml depending on the culture medium. The cutoff for resistance in 7H11 agar is 60 µg/ml [84]. CS inhibits the synthesis of cell wall peptidoglycan by blocking the action of D-alanine racemase (Alr) and D-alanine: D -alanine ligase (Ddl) [125, 126]. Alr is involved in conversion of L-alanine to D-alanine, which then serves as a substrate for Ddl. The D-alanine racemase encoded by *alrA* from *Mycobacterium smegmatis* was cloned and its over-expression in *M. smegmatis* and *M. bovis* BCG caused resistance to CS [127]. In *M. smegmatis* inactivation of *alrA* [128] or *ddl* [129] resulted in increased sensitivity to CS. Over-expression of Alr conferred higher resistance to CS than Ddl overexpression in *M. smegmatis*, suggesting Alr might be the primary target of CS [130]. Recently, *cycA* encoding D-serine, L- and D-alanine, and glycine transporter involved in the uptake of D-cycloserine, was found to be defective in *M. bovis* BCG which could be related to its natural resistance to CS [131], though the mechanism of CS resistance in *M. tuberculosis* remains to be determined.

8.2.4.9 Para-aminosalicylic Acid (PAS)

PAS (Figure 8.2) was discovered as an effective TB drug in 1946 [132] based on a curious observation made by Bernheim that salicylic acid stimulated the oxygen consumption of tubercle bacillus [133]. PAS is a bacteriostatic agent with MIC of 0.5–2 µg/ml for *M. tuberculosis* [124]. PAS has some activity against other mycobacteria but no activity against other bacterial species. Like sulfa drugs, the antituberculosis activity of PAS is antagonized by structurally related para-aminobenzoic acid (PABA). Interference with folic acid biosynthesis [134, 135] and inhibition of iron uptake [136] have been proposed as two possible mechanisms of action [41]. Recently, PAS is shown to be a prodrug that inhibits folic acid metabolism by incorporation into the folate pathway by dihydropteroate synthase and dihydrofolate synthase to generate a toxic hydroxydihydrofolate antimetabolite that inhibits dihydrofolate reductase and then targets dihydrofolate reductase [DHFR, encoded by *dfrA* (Rv2763c)] [137]. Mutations causing PAS resistance occur at a frequency of 10^{-5-9} [135]. The mechanism of PAS resistance is not well understood. Mutations in *thyA* encoding thymidylate synthase that reduce the utilization of tetrahydrofolate, were found to be responsible for about 37% clinical isolates of *M. tuberculosis* resistant to PAS [134, 138]. Mutations in dihydrofolate synthase, an enzyme thought to activate PAS to hydroxydihydrofolate, were found to cause PAS resistance in *M. tuberculosis* [139]. More studies are needed to validate the identified PAS resistance genes in clinical isolates.

8.3 Clinical Impact of Drug Resistance and Management Strategies

8.3.1 Clinical Impact of Drug Resistance

The most important treatment strategy for TB has been drug combination as it not only prevents drug resistance but also improves efficacy of the therapy. Despite drug combination, drug resistance can still occur in clinical settings. MDR/XDR-TB, which could occur due to poor patient compliance or inappropriate treatment, pose significant challenges for treatment.

INH resistance is the most common drug resistance encountered [3]. INH-monoresistant TB is relatively easy to treat. Standard short-course chemotherapy can achieve a good success when all four drugs – INH, RIF, PZA, EMB – are used throughout the six months of treatment [140]. When the four drugs are reduced to RIF and INH after two months, the relapse rate after six-month therapy rises to 10% [141].

RIF-resistant TB has a more serious prognosis because the outcome of standard chemotherapy for this disease is poor in terms of disease status after six-month treatment and relapse [142]. RIF monoresistance in *M. tuberculosis* is relatively rare, except perhaps in HIV-infected patients [143, 144] and thus RIF resistance could serve as a surrogate marker for MDR-TB (resistance to both RIF and INH) [145]. Second-line drugs plus PZA and EMB are required for the treatment of MDR-TB, which takes a lengthy 20–24 months with significant side effects and a less than ideal average cure rate

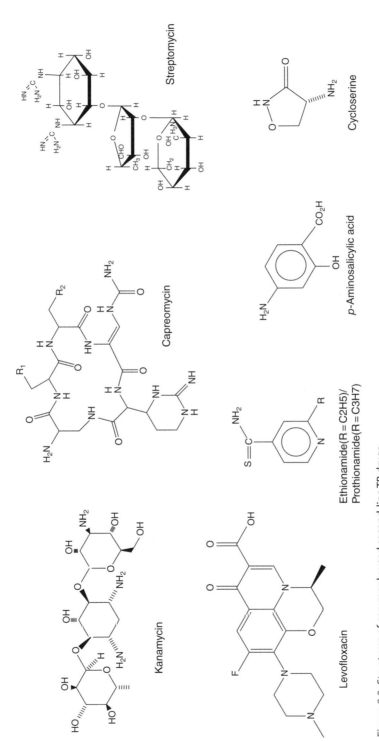

Figure 8.2 Structures of commonly used second-line TB drugs.

of about 62% [146]. Although *rpoB* is an essential gene and some mutations may affect the fitness of RIF-resistant strains *in vitro*, some RIF-resistant strains seem to become more virulent and acquire some peculiar properties such as RIF-dependence or enhancement where continued RIF treatment may cause worsening of the disease [55].

Resistance to PZA is not commonly determined due to difficulty in its drug suscepti-bility testing, but detection of *pncA* mutations by DNA sequencing is a good surrogate test for PZA resistance [147]. Resistance to PZA seems to have poorer treatment success rate than PZA susceptible MDR-TB patients [148]. It has been recently proposed to divide MDR-TB into PZA-susceptible and PZA-resistant MDR-TB for improved treat-ment of MDR-TB [149]. The diagnosis of resistance to PZA and/or EMB has prognostic value in MDR-TB, as such resistance in addition to the dual resistance to INH and RIF generally have even worse prognosis [150], especially when patients receive only "standard" second-line drug regimens [151]. A high prevalence of resistance to PZA would also hamper the efficacy of standard six-month short-course chemotherapy, as PZA has a unique role in sterilizing TB lesions and in preventing disease relapse [57].

Fluoroquinolone-resistant MDR-TB likely results from the use of suboptimal second-line drug regimens that have an inadequate number and/or dosage of accompanying agents to fluoroquinolone [151]. Alternatively, development of fluoroquinolone-resistant TB might be related to frequent use of this class of antibiotics in the treatment of lower respiratory tract infections [152, 153] or other community-acquired infections. Resistance to fluoroquinolones can predict a poor outcome in the treatment of MDR-TB [154–156].

Since aminoglycosides or capreomycin have potent antituberculosis activity, losing these second-line injectables through suboptimal use in the management of MDR-TB would result in XDR-TB with a much worse prognosis than MDR-TB [157]. Currently, the cure rate of XDR-TB is generally less than 50% [158].

Resistance development to some drugs such as INH for KatG mutations and RIF may affect the fitness of some strains *in vitro*, but the resistant strains can acquire compensa-tory mutations such that the resistant strains with drug resistance mutations may still be virulent and cause active disease. Although MDR-TB and XDR-TB transmit readily in immunocompromised patients, such as HIV-positive individuals, they can also be actively transmitted to seemingly healthy people without immunocompromised con-ditions. This suggests that these highly drug-resistant strains are still fully virulent and highly transmissible and deadly due to the selection of more fit strains *in vivo* as a result of plasticity and compensatory changes that occur in such strains.

8.3.2 Management Strategies

8.3.2.1 Rapid Molecular Detection of MDR/XDR-TB

Understanding of the molecular mechanisms of drug resistance in *M. tuberculosis* has provided the basis for rapid detection of drug-resistant strains. Although there are differ-ent molecular methods available to detect drug resistance, currently two commercial tests are available and endorsed by WHO for rapid diagnosis of drug-resistant TB. One is GenoType® MTBDRplus (Hain Life Science, Germany) and the other Xpert® MTB/RIF assay (Cepheid, CA, USA) for rapid detection of RIF resistance as a surrogate marker for MDR-TB. A positive test in Xpert MTB/RIF would diagnose MDR but some strains that are RIF mono-resistant but INH susceptible would thus miss the benefit of using INH for the treatment of such cases. The advantage of the Xpert test is its rapidity (completed in two hours) and convenience, *i.e.*, sputum processing, DNA extraction and PCR, and detection (molecular beacon format) all performed in a single cartridge. But the disadvantage is its cost and detection of only RIF (*rpoB* mutations) resistance and not other drug resistances. MTBDRplus detects both INH resistance (KatG315 and *inhA* promoter mutations) and RIF (*rpoB* mutations) resistance. In addition, an expanded version of the Hain test is GenoType MTBDR*sl* for detection of XDR-TB, which includes mutations causing resistance to second-line agents fluoroquinolone and injectables (kanamycin, amikacin, and capreomycin) besides the mutations for INH and RIF resistance as in MTBDRplus as above. Although the above molecular tests are useful for guiding the treatment of drug-resistant TB, they do not cover the drug susceptibility profile for all available TB drugs, which limits the choice of drugs that the bacteria are still sensitive to for improved personalized treatment. The recent advances in DNA sequencing technology may allow all drug resistance mutations to be detected to better guide the treatment of the disease in the near future.

8.3.2.2 New TB Drug Development
The increasing emergence of drug-resistant TB has highlighted the need for new and more effective TB drugs. There is considerable interest in developing new TB drugs that are not only active against drug-resistant TB but more importantly shorten TB therapy [35, 159–161]. Various drug candidates and active compounds have been identified. For a more detailed review, please refer to the recent reviews [35, 159–161] and the websites (http://www.newtbdrugs.org/project.php?id=186; http://www.tballiance.org/downloads/Pipeline/TBA-Pipeline-Q3-2013.pdf).

Here only the most promising drug candidates that are in clinical trials will be briefly mentioned (Table 8.2, Figure 8.3). In addition, some drugs used for other indications

Table 8.2 Promising new TB drug candidates in clinical trials.

Name	Developer	Properties/Mode of action	Stage of development
Moxifloxacin, Gatifloxacin	Bayer Pharmaceuticals, Bristol-Myers Squibb	Inhibition of DNA synthesis	Phase 3
Rifapentine	Sanofi-Aventis	Inhibition of RNA synthesis	Phase 3
Delamanid (OPC-67683)	Otsuka Pharmaceutical	Inhibition of mycolate synthesis; NO donor	Phase 3
PA-824	TB Alliance	Inhibition of mycolate synthesis; NO donor	Phase 2
Bedaquiline (bedaquiline)	Tibotec, Johnson & Johnson	Inhibition of ATP synthase	Phase 3
SQ109	Sequella, Inc.	Inhibition of trehalose monomycolate transporter MmpL3	Phase 2
Oxazolidinones Sutezolid (PNU-100480), AZD5847	Pfizer, AstraZeneca	Inhibition of 50S ribosome, 23S RNA	Phase 2

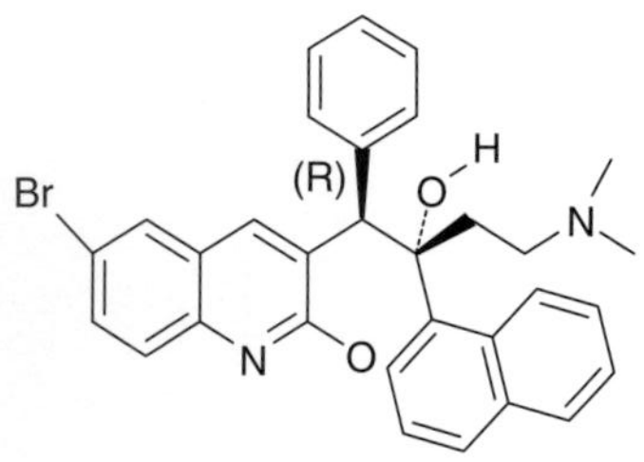

TMC207 (Bedaquiline)

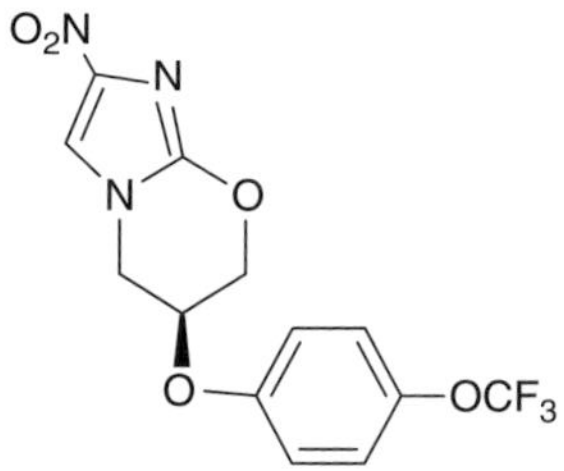

PA-824

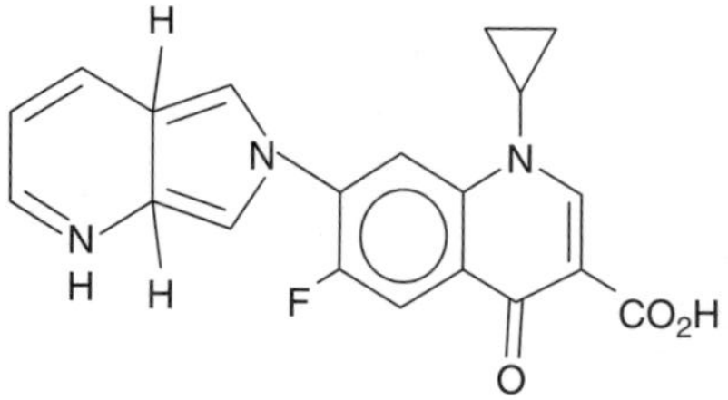

Moxifloxacin

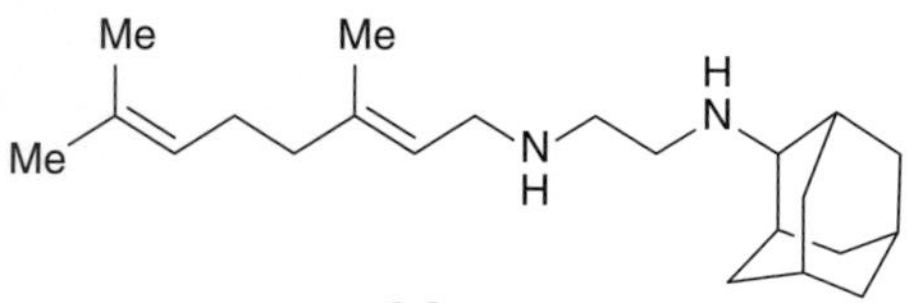

SQ-109

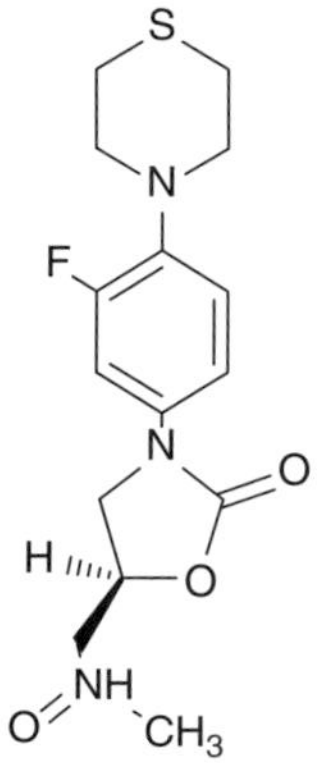

Oxazolidinone U-100480

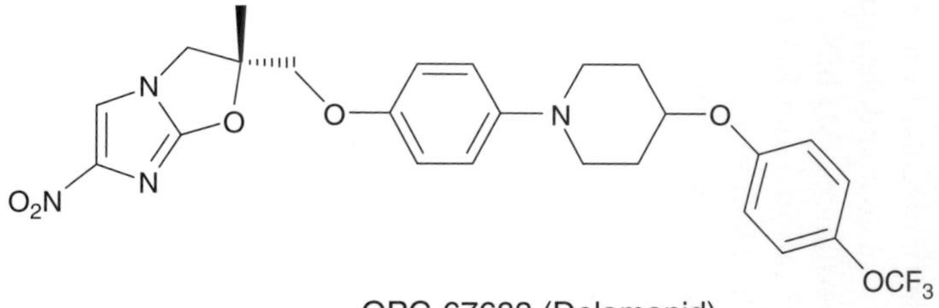

OPC-67683 (Delamanid)

Figure 8.3 Structures of new drug candidates in clinical trials.

such as clofazimine used for leprosy treatment and linezolid for Gram-positive bacteria have been found to have good therapeutic effects for MDR/XDR-TB. Besides adding new drug candidates to existing regimens, novel drug combinations/regimens, such as PA-824/Moxifloxacin/Pyrazinamide (PaMZ), REMox TB [Isoniazid/ Rifampin/Pyrazinamide/Moxifloxacin (HRZM); Ethambutol/Rifampin/Pyrazinamide/ Moxifloxacin (ERZM)], are also being evaluated in Phase 2 and Phase 3 studies, respectively. These efforts offer promise for more effective treatment of both drug susceptible and drug resistant TB.

8.4 Conclusions

Drug-resistant TB presents a serious challenge to the effective treatment and control of the disease. Although significant progress has been made in our understanding of the mechanisms of drug action and resistance in *M. tuberculosis*, some aspects of drug action and drug resistance remain to be identified. It is increasingly recognized that drug resistance development in the clinical setting is more complex, dynamic, and heterogeneous than previously thought, where different mutations of varying hierarchy could occur that may all contribute to the resistance phenotype and compensate for fitness loss during resistance development. The impact of drug resistance on the fitness and transmission of the resistant strains requires further study. While the two commercial tests MTBDRplus and Xpert for rapid detection of MDR-TB are important developments in diagnosis of MDR-TB, more high-throughput, point-of-care, and affordable molecular tests that detect all drug resistance mutations (not just INH and RIF resistance) and that can be applied in the field are urgently needed to tackle and contain the enormous and increasing problem of drug-resistant TB. High-throughput DNA sequencing technology has been applied to whole genome sequencing of drug-resistant strains and provided important new insights into unknown mechanisms of drug resistance, and identified mutations that are associated with drug resistance, which will help to detect drug resistance more efficiently. Besides improved understanding of mechanisms of drug action and drug resistance and rapid detection of drug resistance, new drug development and reformulation of optimal regimens that improve the treatment of MDR/XDR-TB and shorten therapy are vitally important. Several new drug candidates are in clinical trials and show significant promise; however, these drug candidates have limited activity against the persister problem, and an improved understanding of the biology of persistence that provides crucial targets for the design of new drugs targeting persisters and of combined chemical and immune attack are required for developing more effective treatment. A sustained funding environment and collaborations between academia, pharma, and funding agencies are critical for maintaining the current progress and for translating the basic science into much needed drugs for better treatment of both drug-resistant and drug-susceptible TB in the future.

Acknowledgements

This work was supported in part by NIH grants AI99512 and AI108535.

References

1 Dye, C. and Williams, B.G. (2010). The population dynamics and control of tuberculosis. *Science* 328: 856–861.

2 Zignol, M., Hosseini, M.S., Wright, A. et al. (2006). Global incidence of multidrug-resistant tuberculosis. *J. Infect. Dis.* 194: 479–485.

3 WHO. 2008. Anti-tuberculosis Drug Resistance in the World, Report No. 4 [Online]. Available: http://www.who.int/tb/publications/2008/drs_report4_26feb08.pdf.

4 Schatz, A., Bugie, E., and Waksman, S. (1944). Streptomycin, a substance exhibiting antibiotic activity against gram-positive and gram-negative bacteria. *Proc. Soc. Exp. Biol. Med.* 55: 66–69.

5 Chorine, V. (1945). Action de l'amide nicotinique sur les bacilles du genre Mycobacterium. *CR Acad. Sci. Paris* 220: 150–151.

6 Liebermann, D., Moyeux, M., Rist, N., and Grumbach, F. (1956). Sur la preparation de nouveaux thioamides pyridineques acitifs dans la tuberculose experimentale. *C. R. Acad. Sci.* 242: 2409–2412.

7 Thomas, J., Baughn, C., Wilkinson, R., and Shepherd, R. (1961). A new synthetic compound with antituberculous activity in mice: ethambutol (dextro-2,2′-(ethylenediimino)-di-butanol). *Am. Rev. Respir. Dis.* 83: 891–893.

8 Kurosawa, H. (1952). Studies on the antibiotic substances from actinomyces. XXIII. The isolation of an antibiotic produced by a strain of streptomyces "K 30". *J. Antibiot. Ser. B* 5: 682–688.

9 Umezawa, H., Ueda, M., Maeda, K. et al. (1957). Production and isolation of a new antibiotic, kanamycin. *J. Antibiot.* 10: 181–188.

10 Bartz, Q., Ehrlich, J., Mold, J. et al. (1951). Viomycin, a new tuberculostatic antibiotic. *Am. Rev. Tuberc.* 63: 4–6.

11 Herr, E., Haney, M., Pittenger, G., and Higgens, C. (1960). Isolation and characterization of a new peptide antibiotic. *Proc. Indiana Acad. Sci.* 69: 134.

12 Sensi, P., Margalith, P., and Timbal, M. (1959). Rifomycin, a new antibiotic. Preliminary report. *Farmaco Sci.* 14: 146–147.

13 Sensi, P., Maggi, N., Furesz, S., and Maffii, G. (1966). Chemical modifications and biological properties of rifamycins. *Antimicrob. Agents Chemother. (Bethesda)* 6: 699–714.

14 Tsunekawa, H., Miyachi, T., Nakamura, E. et al. (1987). Therapeutic effect of ofloxacin on 'treatment-failure' pulmonary tuberculosis. *Kekkaku* 62: 435–439.

15 Yew, W.W., Kwan, S.Y., Ma, W.K. et al. (1990). In-vitro activity of ofloxacin against *Mycobacterium tuberculosis* and its clinical efficacy in multiply resistant pulmonary tuberculosis. *J. Antimicrob. Chemother.* 26: 227–236.

16 Fox, W., Ellard, G.A., and Mitchison, D.A. (1999). Studies on the treatment of tuberculosis undertaken by the British Medical Research Council tuberculosis units, 1946–1986, with relevant subsequent publications. *Int. J. Tuberc. Lung Dis.* 3: S231–S279.

17 Wade, M.M. and Zhang, Y. (2004b). Mechanisms of drug resistance in *Mycobacterium tuberculosis*. *Front. Biosci.* 9: 975–994.

18 Zhang, Y., Heym, B., Allen, B. et al. (1992). The catalase-peroxidase gene and isoniazid resistance of *Mycobacterium tuberculosis*. *Nature* 358: 591–593.

19 Scorpio, A. and Zhang, Y. (1996). Mutations in pncA, a gene encoding pyrazinamidase/nicotinamidase, cause resistance to the antituberculous drug pyrazinamide in tubercle bacillus. *Nat. Med.* 2: 662–667.

20 Goodwin, A., Kersulyte, D., Sisson, G. et al. (1998). Metronidazole resistance in *Helicobacter pylori* is due to null mutations in a gene (rdxA) that encodes an oxygen-insensitive NADPH nitroreductase. *Mol. Microbiol.* 28: 383–393.

21 Kwon, D.H., Kato, M., El-Zaatari, F.A. et al. (2000). Frame-shift mutations in NAD(P)H flavin oxidoreductase encoding gene (frxA) from metronidazole resistant *Helicobacter pylori* ATCC43504 and its involvement in metronidazole resistance. *FEMS Microbiol. Lett.* 188: 197–202.

22 Brennan, P.J. and Nikaido, H. (1995). The envelope of mycobacteria. *Annu. Rev. Biochem.* 64: 29–63.

23 Segura, C., Salvado, M., Collado, I. et al. (1998). Contribution of beta-lactamases to beta-lactam susceptibilities of susceptible and multidrug-resistant *Mycobacterium tuberculosis* clinical isolates. *Antimicrob. Agents Chemother.* 42: 1524–1526.

24 Madsen, C.T., Jakobsen, L., Buriankova, K. et al. (2005). Methyltransferase Erm(37) slips on rRNA to confer atypical resistance in *Mycobacterium tuberculosis. J. Biol. Chem.* 280: 38942–38947.

25 Buriankova, K., Doucet-Populaire, F., Dorson, O. et al. (2004). Molecular basis of intrinsic macrolide resistance in the *Mycobacterium tuberculosis* complex. *Antimicrob. Agents Chemother.* 48: 143–150.

26 Morris, R.P., Nguyen, L., Gatfield, J. et al. (2005). Ancestral antibiotic resistance in *Mycobacterium tuberculosis. Proc. Natl. Acad. Sci. U. S. A.* 102: 12200–12205.

27 Geiman, D.E., Raghunand, T.R., Agarwal, N., and Bishai, W.R. (2006). Differential gene expression in response to exposure to antimycobacterial agents and other stress conditions among seven *Mycobacterium tuberculosis* whiB-like genes. *Antimicrob. Agents Chemother.* 50: 2836–2841.

28 Zhang, Y. and Yew, W. (2009a). Mechanisms of drug resistance in *Mycobacterium tuberculosis. Int. J. Tuberc. Lung Dis.* 13 (11): 1320–1330.

29 Zhang, H., Li, D., Zhao, L. et al. (2013a). Genome sequencing of 161 *Mycobacterium tuberculosis* isolates from China identifies genes and intergenic regions associated with drug resistance. *Nat. Genet.* 45: 1255–1260.

30 Safi, H., Lingaraju, S., Amin, A. et al. (2013). Evolution of high-level ethambutol-resistant tuberculosis through interacting mutations in decaprenylphosphoryl-beta-D-arabinose biosynthetic and utilization pathway genes. *Nat. Genet.* 45: 1190–1197.

31 Sun, G., Luo, T., Yang, C. et al. (2012). Dynamic population changes in *Mycobacterium tuberculosis* during acquisition and fixation of drug resistance in patients. *J. Infect. Dis.* 206: 1724–1733.

32 Zhang, Y. (2014). Persisters, persistent infections and the Yin-Yang model. *Emerg. Microbes. Infect.* 3: e3. https://doi.org/10.1038/emi.2014.3.

33 Zhang, Y., Yew, W.W., and Barer, M.R. (2012b). Targeting persisters for tuberculosis control. *Antimicrob. Agents Chemother.* 56: 2223–2230.

34 Sarathy, J., Dartois, V, Dick, T., and Gengenbacher, M. (2013). Reduced drug uptake in phenotypically resistant nutrient-starved nonreplicating *Mycobacterium tuberculosis. Antimicrob. Agents Chemother.* 57: 1648–1653.

35 Zhang, Y. (2005). The magic bullets and tuberculosis drug targets. *Annu. Rev. Pharmacol. Toxicol.* 45: 529–564.

36 Zhang, Y. and Telenti, A. (2000). Genetics of drug resistance in *Mycobacterium tuberculosis*. In: *Molecular Genetics of Mycobacteria* (ed. G. Hatfull and W.R. Jacobs), 235–254. Washington, DC: ASM Press.

37 Banerjee, A., Dubnau, E., Quemard, A. et al. (1994). inhA, a gene encoding a target for isoniazid and ethionamide in *Mycobacterium tuberculosis*. *Science* 263: 227–230.

38 Rawat, R., Whitty, A., and Tonge, P.J. (2003). The isoniazid-NAD adduct is a slow, tight-binding inhibitor of InhA, the *Mycobacterium tuberculosis* enoyl reductase: adduct affinity and drug resistance. *Proc. Natl. Acad. Sci. U. S. A.* 100: 13881–13886.

39 Rozwarski, D.A., Grant, G.A., Barton, D.H. et al. (1998). Modification of the NADH of the isoniazid target (InhA) from *Mycobacterium tuberculosis*. *Science* 279: 98–102.

40 Niki, M., Tateishi, Y., Ozeki, Y. et al. (2012). A novel mechanism of growth phase-dependent tolerance to isoniazid in mycobacteria. *J. Biol. Chem.* 287: 27743–27752.

41 Winder, F. (1982). Mode of action of the antimycobacterial agents and associated aspects of the molecular biology of mycobacteria. In: *The Biology of Mycobacteria*, vol. 1 (ed. C. Ratledge and J. Stanford), 353–438. New York: Academic Press.

42 Middlebrook, G. (1954). Isoniazid resistance and catalase activity of tubercle bacilli. *Am. Rev. Tuberc.* 69: 471–472.

43 Hazbon, M.H., Brimacombe, M., Bobadilla Del Valle, M. et al. (2006). Population genetics study of isoniazid resistance mutations and evolution of multidrug-resistant *Mycobacterium tuberculosis*. *Antimicrob. Agents Chemother.* 50: 2640–2649.

44 Zhang, Y. and Yew, W.W. (2009b). Mechanisms of drug resistance in Mycobacterium tuberculosis. *Int. J. Tuberc. Lung Dis.* 13: 1320–1330.

45 Ando, H., Kitao, T., Miyoshi-Akiyama, T. et al. (2011). Downregulation of katG expression is associated with isoniazid resistance in *Mycobacterium tuberculosis*. *Mol. Microbiol.* 79: 1615–1628.

46 Deretic, V., Philipp, W., Dhandayuthapani, S. et al. (1995). Mycobacterium tuberculosis is a natural mutant with an inactivated oxidative-stress regulatory gene: implications for sensitivity to isoniazid. *Mol. Microbiol.* 17: 889–900.

47 Sherman, D.R., Sabo, P.J., Hickey, M.J. et al. (1995). Disparate responses to oxidative stress in saprophytic and pathogenic mycobacteria. *Proc. Natl. Acad. Sci. U. S. A.* 92: 6625–6629.

48 Vilcheze, C., Av-Gay, Y., Attarian, R. et al. (2008). Mycothiol biosynthesis is essential for ethionamide susceptibility in *Mycobacterium tuberculosis*. *Mol. Microbiol.* 69: 1316–1329.

49 Vilcheze, C., Av-Gay, Y., Barnes, S.W. et al. (2011). Coresistance to isoniazid and ethionamide maps to mycothiol biosynthetic genes in *Mycobacterium bovis*. *Antimicrob. Agents Chemother.* 55: 4422–4423.

50 Lee, J.H., Ammerman, N.C., Nolan, S. et al. (2012). Isoniazid resistance without a loss of fitness in *Mycobacterium tuberculosis*. *Nat. Commun.* 3: 753.

51 Mitchison, D.A. (1985). The action of antituberculosis drugs in short course chemotherapy. *Tubercle* 66: 219–225.

52 Telenti, A., Imboden, P., Marchesi, F. et al. (1993). Detection of rifampicin-resistance mutations in *Mycobacterium tuberculosis*. *Lancet* 341: 647–650.

53 Nakamura, M., Harano, Y., and Koga, T. (1990). Isolation of a strain of M. tuberculosis which is considered to be rifampicin-dependent, from a patient with long-lasted smear positive and culture difficult (SPCD) mycobacteria. *Kekkaku* 65: 569–574.

54 Zhong, M., Wang, Y., Sun, C. et al. (2002). Growth of rifampin-dependent *Mycobacterium tuberculosis* in conditions without rifampin. *Zhonghua Jie He He Hu Xi Za Zhi* 25: 588–590.

55 Zhong, M., Zhang, X., Wang, Y. et al. (2010). An interesting case of rifampicin-dependent/-enhanced multidrug-resistant tuberculosis. *Int. J. Tuberc. Lung Dis.* 14: 40–44.

56 De Vos, M., Muller, B., Borrell, S. et al. (2013). Putative compensatory mutations in the rpoC gene of rifampin-resistant *Mycobacterium tuberculosis* are associated with ongoing transmission. *Antimicrob. Agents Chemother.* 57: 827–832.

57 Zhang, Y. and Mitchison, D. (2003). The curious characteristics of pyrazinamide: a review. *Int. J. Tuberc. Lung Dis.* 7: 6–21.

58 Tarshis, M.S. and Weed, W.A. Jr. (1953). Lack of significant *in vitro* sensitivity of *Mycobacterium tuberculosis* to pyrazinamide on three different solid media. *Am. Rev. Tuberc.* 67: 391–395.

59 Mcdermott, W. and Tompsett, R. (1954). Activation of pyrazinamide and nicotinamide in acidic environment *in vitro*. *Am. Rev. Tuberc.* 70: 748–754.

60 Wade, M.M. and Zhang, Y. (2004a). Anaerobic incubation conditions enhance pyrazinamide activity against *Mycobacterium tuberculosis*. *J. Med. Microbiol.* 53: 769–773.

61 Wade, M.M. and Zhang, Y. (2006). Effects of weak acids, UV and proton motive force inhibitors on pyrazinamide activity against *Mycobacterium tuberculosis in vitro*. *J. Antimicrob. Chemother.* 58: 936–941.

62 Zhang, Y., Wade, M.M., Scorpio, A. et al. (2003). Mode of action of pyrazinamide: disruption of *Mycobacterium tuberculosis* membrane transport and energetics by pyrazinoic acid. *J. Antimicrob. Chemother.* 52: 790–795.

63 Zhang, Y., Scorpio, A., Nikaido, H., and Sun, Z. (1999). Role of acid pH and deficient efflux of pyrazinoic acid in unique susceptibility of *Mycobacterium tuberculosis* to pyrazinamide. *J. Bacteriol.* 181: 2044–2049.

64 Zimhony, O., Cox, J.S., Welch, J.T. et al. (2000). Pyrazinamide inhibits the eukaryotic-like fatty acid synthetase I (FASI) of *Mycobacterium tuberculosis*. *Nat. Med.* 6: 1043–1047.

65 Boshoff, H.I., Mizrahi, V., and Barry, C.E. 3rd (2002). Effects of pyrazinamide on fatty acid synthesis by whole mycobacterial cells and purified fatty acid synthase I. *J. Bacteriol.* 184: 2167–2172.

66 Shi, W., Zhang, X., Jiang, X. et al. (2011). Pyrazinamide inhibits trans-translation in *Mycobacterium tuberculosis*. *Science* 333: 1630–1632.

67 Scorpio, A., Lindholm-Levy, P., Heifets, L. et al. (1997). Characterization of pncA mutations in pyrazinamide-resistant *Mycobacterium tuberculosis*. *Antimicrob. Agents Chemother.* 41: 540–543.

68 Keiler, K.C. (2008). Biology of trans-translation. *Annu. Rev. Microbiol.* 62: 133–151.

69 Muto, A., Fujihara, A., Ito, K.I. et al. (2000). Requirement of transfer-messenger RNA for the growth of *Bacillus subtilis* under stresses. *Genes Cells* 5: 627–635.

70 Thibonnier, M., Thiberge, J.M., and De Reuse, H. (2008). Trans-translation in *Helicobacter pylori*: essentiality of ribosome rescue and requirement of protein tagging for stress resistance and competence. *PLoS One* 3: e3810.

71 Zhang, S., Chen, J., Shi, W. et al. (2013b). Mutations in panD encoding aspartate decarboxylase are associated with pyrazinamide resistance in *Mycobacterium tuberculosis*. *Emerg. Microbes Infect.* 2: e34. https://doi.org/10.1038/emi.2013.38.

72 Zhang, Y., Permar, S., and Sun, Z. (2002). Conditions that may affect the results of susceptibility testing of *Mycobacterium tuberculosis* to pyrazinamide. *J. Med. Microbiol.* 51: 42–49.

73 Cheng, S.J., Thibert, L., Sanchez, T. et al. (2000). pncA mutations as a major mechanism of pyrazinamide resistance in *Mycobacterium tuberculosis*: spread of a monoresistant strain in Quebec, Canada. *Antimicrob. Agents Chemother.* 44: 528–532.

74 Hirano, K., Takahashi, M., Kazumi, Y. et al. (1997). Mutation in pncA is a major mechanism of pyrazinamide resistance in *Mycobacterium tuberculosis*. *Tuber. Lung Dis.* 78: 117–122.

75 Jureen, P., Werngren, J., Toro, J.C., and Hoffner, S. (2008). Pyrazinamide resistance and pncA gene mutations in *Mycobacterium tuberculosis*. *Antimicrob. Agents Chemother.* 52: 1852–1854.

76 Lemaitre, N., Sougakoff, W., Truffot-Pernot, C., and Jarlier, V. (1999). Characterization of new mutations in pyrazinamide-resistant strains of *Mycobacterium tuberculosis* and identification of conserved regions important for the catalytic activity of the pyrazinamidase PncA. *Antimicrob. Agents Chemother.* 43: 1761–1763.

77 Louw, G.E., Warren, R.M., Donald, P.R. et al. (2006). Frequency and implications of pyrazinamide resistance in managing previously treated tuberculosis patients. *Int. J. Tuberc. Lung Dis.* 10: 802–807.

78 Marttila, H.J., Marjamaki, M., Vyshnevskaya, E. et al. (1999). pncA mutations in pyrazinamide-resistant *Mycobacterium tuberculosis* isolates from northwestern Russia. *Antimicrob. Agents Chemother.* 43: 1764–1766.

79 Mestdagh, M., Fonteyne, P.A., Realini, L. et al. (1999). Relationship between pyrazinamide resistance, loss of pyrazinamidase activity, and mutations in the pncA locus in multidrug-resistant clinical isolates of *Mycobacterium tuberculosis*. *Antimicrob. Agents Chemother.* 43: 2317–2319.

80 Morlock, G.P., Crawford, J.T., Butler, W.R. et al. (2000). Phenotypic characterization of pncA mutants of *Mycobacterium tuberculosis*. *Antimicrob. Agents Chemother.* 44: 2291–2295.

81 Portugal, I., Barreiro, L., Moniz-Pereira, J., and Brum, L. (2004). pncA mutations in pyrazinamide-resistant *Mycobacterium tuberculosis* isolates in Portugal. *Antimicrob. Agents Chemother.* 48: 2736–2738.

82 Sreevatsan, S., Pan, X., Zhang, Y. et al. (1997b). Mutations associated with pyrazinamide resistance in pncA of *Mycobacterium tuberculosis* complex organisms. *Antimicrob. Agents Chemother.* 41: 636–640.

83 Feuerriegel, S., Koser, C.U., Richter, E., and Niemann, S. (2013). *Mycobacterium canettii* is intrinsically resistant to both pyrazinamide and pyrazinoic acid. *J. Antimicrob. Chemother.* 68 (6): 1439–1440.

84 Heifets, L. and Desmond, E. (2005). Clinical mycobacteriology laboratory. In: *Tuberculosis and the Tubercle Bacillus* (ed. S. Cole, K. Eisenach, D. Mcmurray and W. Jacobs Jr.), 49–70. Washington, DC: ASM Press.

85 Takayama, K. and Kilburn, J. (1989). Inhibition of synthesis of arabinogalactan by ethambutol in *Mycobacterium smegmatis*. *Antimicrob. Agents Chemother.* 33: 1493–1499.

86 Mikusov, K., Slayden, R., Besra, G., and Brennan, P. (1995). Biogenesis of the mycobacterial cell wall and the site of action of ethambutol. *Antimicrob. Agents Chemother.* 39: 2484–2489.

87 Wolucka, B., Mcneil, M., De Hoffmann, E. et al. (1994). Recognition of the lipid intermediate for arabinogalactan/arabinomannan biosynthesis and its relation to the mode of action of ethambutol on mycobacteria. *J. Biol. Chem.* 269: 23328–23335.

88 Telenti, A., Philipp, W., Sreevatsan, S. et al. (1997). The emb operon, a unique gene cluster of *Mycobacterium tuberculosis* involved in resistance to ethambutol. *Nat. Med.* 3: 567–570.

89 Ramaswamy, S.V., Amin, A.G., Goksel, S. et al. (2000). Molecular genetic analysis of nucleotide polymorphisms associated with ethambutol resistance in human isolates of *Mycobacterium tuberculosis*. *Antimicrob. Agents Chemother.* 44: 326–336.

90 Sreevatsan, S., Stockbauer, K., Pan, X. et al. (1997c). Ethambutol resistance in *Mycobacterium tuberculosis*: critical role of embB mutations. *Antimicrob. Agents Chemother.* 41: 1677–1681.

91 Safi, H., Sayers, B., Hazbon, M.H., and Alland, D. (2008). Transfer of embB codon 306 mutations into clinical *Mycobacterium tuberculosis* strains alters susceptibility to ethambutol, isoniazid, and rifampin. *Antimicrob. Agents Chemother.* 52: 2027–2034.

92 Alcaide, F., Pfyffer, G.E., and Telenti, A. (1997). Role of embB in natural and acquired resistance to ethambutol in mycobacteria. *Antimicrob. Agents Chemother.* 41: 2270–2273.

93 Davies, J., Gorini, L., and Davis, B. (1965). Misreading of RNA codewords induced by aminoglycoside antibiotics. *Mol. Pharmacol.* 1: 93–106.

94 Finken, M., Kirschner, P., Meier, A. et al. (1993). Molecular basis of streptomycin resistance in *Mycobacterium tuberculosis*: alterations of the ribosomal protein S12 gene and point mutations within a functional 16S ribosomal RNA pseudoknot. *Mol. Microbiol.* 9: 1239–1246.

95 Honore, N. and Cole, S.T. (1994). Streptomycin resistance in mycobacteria. *Antimicrob. Agents Chemother.* 38: 238–242.

96 Nair, J., Rouse, D.A., Bai, G.H., and Morris, S.L. (1993). The rpsL gene and streptomycin resistance in single and multiple drug-resistant strains of *Mycobacterium tuberculosis*. *Mol. Microbiol.* 10: 521–527.

97 Honore, N., Marchal, G., and Cole, S.T. (1995). Novel mutation in 16S rRNA associated with streptomycin dependence in *Mycobacterium tuberculosis*. *Antimicrob. Agents Chemother.* 39: 769–770.

98 Cooksey, R.C., Morlock, G.P., Mcqueen, A. et al. (1996). Characterization of streptomycin resistance mechanisms among *Mycobacterium tuberculosis* isolates from patients in New York City. *Antimicrob. Agents Chemother.* 40: 1186–1188.

99 Okamoto, S., Tamaru, A., Nakajima, C. et al. (2007). Loss of a conserved 7-methylguanosine modification in 16S rRNA confers low-level streptomycin resistance in bacteria. *Mol. Microbiol.* 63: 1096–1106.

100 Spies, F.S., Da Silva, P.E., Ribeiro, M.O. et al. (2008). Identification of mutations related to streptomycin resistance in clinical isolates of *Mycobacterium tuberculosis* and possible involvement of efflux mechanism. *Antimicrob. Agents Chemother.* 52: 2947–2949.

101 Reeves, A.Z., Campbell, P.J., Sultana, R. et al. (2013). Aminoglycoside cross-resistance in *Mycobacterium tuberculosis* due to mutations in the 5′ untranslated region of whiB7. *Antimicrob. Agents Chemother.* 57: 1857–1865.

102 Alangaden, G., Kreiswirth, B., Aouad, A. et al. (1998). Mechanism of resistance to amikacin and kanamycin in *Mycobacterium tuberculosis*. *Antimicrob. Agents Chemother.* 42: 1295–1297.

103 Suzuki, Y., Katsukawa, C., Tamaru, A. et al. (1998). Detection of kanamycin-resistant *Mycobacterium tuberculosis* by identifying mutations in the 16S rRNA gene. *J. Clin. Microbiol.* 36: 1220–1225.

104 Zaunbrecher, M.A., Sikes, R.D. Jr., Metchock, B. et al. (2009). Overexpression of the chromosomally encoded aminoglycoside acetyltransferase eis confers kanamycin resistance in *Mycobacterium tuberculosis. Proc. Natl. Acad. Sci. U. S. A.* 106: 20004–20009.

105 Maus, C.E., Plikaytis, B.B., and Shinnick, T.M. (2005b). Mutation of tlyA confers capreomycin resistance in *Mycobacterium tuberculosis. Antimicrob. Agents Chemother.* 49: 571–577.

106 Johansen, S., Maus, C., Plikaytis, B., and Douthwaite, S. (2006). Capreomycin binds across the ribosomal subunit interface using tlyA-encoded 2′-O-methylations in 16S and 23S rRNAs. *Mol. Cell* 23: 173–182.

107 Maus, C.E., Plikaytis, B.B., and Shinnick, T.M. (2005a). Molecular analysis of cross-resistance to capreomycin, kanamycin, amikacin, and viomycin in *Mycobacterium tuberculosis. Antimicrob. Agents Chemother.* 49: 3192–3197.

108 Drlica, K. and Malik, M. (2003). Fluoroquinolones: action and resistance. *Curr. Top. Med. Chem.* 3: 249–282.

109 Takiff, H., Salazar, L., Guerrero, C. et al. (1994). Cloning and nucleotide sequence of *Mycobacterium tuberculosis gyrA* and *gyrB* genes and detection of quinolone resistance mutations. *Antimicrob. Agents Chemother.* 38: 773–780.

110 Alangaden, G.J., Manavathu, E.K., Vakulenko, S.B. et al. (1995). Characterization of fluoroquinolone-resistant mutant strains of *Mycobacterium tuberculosis* selected in the laboratory and isolated from patients. *Antimicrob. Agents Chemother.* 39: 1700–1703.

111 Cheng, A.F., Yew, W.W., Chan, E.W. et al. (2004). Multiplex PCR amplimer conformation analysis for rapid detection of gyrA mutations in fluoroquinolone-resistant *Mycobacterium tuberculosis* clinical isolates. *Antimicrob. Agents Chemother.* 48: 596–601.

112 Lee, A.S., Tang, L.L., Lim, I.H., and Wong, S.Y. (2002). Characterization of pyrazinamide and ofloxacin resistance among drug resistant *Mycobacterium tuberculosis* isolates from Singapore. *Int. J. Infect. Dis.* 6: 48–51.

113 Pitaksajjakul, P., Wongwit, W., Punprasit, W. et al. (2005). Mutations in the gyrA and gyrB genes of fluoroquinolone-resistant *Mycobacterium tuberculosis* from TB patients in Thailand. *Southeast Asian J. Trop. Med. Public Health* 36 (Suppl 4): 228–237.

114 Sun, Z., Zhang, J., Zhang, X. et al. (2008). Comparison of gyrA gene mutations between laboratory-selected ofloxacin-resistant *Mycobacterium tuberculosis* strains and clinical isolates. *Int. J. Antimicrob. Agents* 31: 115–121.

115 Sulochana, S., Narayanan, S., Paramasivan, C.N. et al. (2007). Analysis of fluoroquinolone resistance in clinical isolates of *Mycobacterium tuberculosis* from India. *J. Chemother.* 19: 166–171.

116 Sreevatsan, S., Pan, X., Stockbauer, K.E. et al. (1997a). Restricted structural gene polymorphism in the *Mycobacterium tuberculosis* complex indicates evolutionarily recent global dissemination. *Proc. Natl. Acad. Sci. U. S. A.* 94: 9869–9874.

117 Wang, J.Y., Lee, L.N., Lai, H.C. et al. (2007). Fluoroquinolone resistance in *Mycobacterium tuberculosis* isolates: associated genetic mutations and relationship to antimicrobial exposure. *J. Antimicrob. Chemother.* 59: 860–865.

118 Kocagoz, T., Hackbarth, C., Unsal, I. et al. (1996). Gyrase mutations in laboratory selected, fluoroquinolone-resistant mutants of *Mycobacterium tuberculosis* H37Ra. *Antimicrob. Agents Chemother.* 40: 1768–1774.

119 Hegde, S.S., Vetting, M.W., Roderick, S.L. et al. (2005). A fluoroquinolone resistance protein from *Mycobacterium tuberculosis* that mimics DNA. *Science* 308: 1480–1483.

120 Baulard, A., Betts, J., Engohang-Ndong, J. et al. (2000). Activation of the pro-drug ethionamide is regulated in mycobacteria. *J. Biol. Chem.* 275: 28326–28331.

121 Debarber, A., Mdluli, K., Bosman, M. et al. (2000). Ethionamide activation and sensitivity in multidrug-resistant *Mycobacterium tuberculosis*. *Proc. Natl. Acad. Sci. U. S. A.* 97: 9677–9682.

122 Vannelli, T., Dykman, A., and Ortiz De Montellano, P. (2002). The antituberculosis drug ethionamide is activated by a flavoprotein monooxygenase. *J. Biol. Chem.* 277: 12824–12829.

123 Trnka, L. (1988). Thiosemicarbazones. In: *Antituberculosis Drugs* (ed. K. Bartmann), 92–103. Berlin: Spring-Verlag.

124 Kucers, A., Crowe, S., Grayson, M., and Hoy, J. (1997). *The Use of Antibiotics*. Oxford: Butterworth Heinemann.

125 David, H.L., Takayama, K., and Goldman, D. (1969). Susceptibility of mycobacterial D-alanyl-D-alanine synthetase to D-cycloserine. *Am. Rev. Respir. Dis.* 100: 579–581.

126 Strych, U., Penland, R., Jimenez, M. et al. (2001). Characterization of the alanine racemases from two Mycobacteria. *FEMS Microbiol. Lett.* 196: 93–98.

127 Caceres, N., Harris, N., Wellehan, J. et al. (1997). Overexpression of the D-alanine racemase gene confers resistance to D-cycloserine in *Mycobacterium smegmatis*. *J. Bacteriol.* 179: 5046–5055.

128 Chacon, O., Feng, Z., Harris, N. et al. (2002). *Mycobacterium smegmatis* D-alanine racemase mutants are not dependent on D-alanine for growth. *Antimicrob. Agents Chemother.* 46: 47–54.

129 Belanger, A., Porter, J., and Hatfull, G. (2000). Genetic analysis of peptidoglycan biosynthesis in mycobacteria: characterization of a ddlA mutant of *Mycobacterium smegmatis*. *J. Bacteriol.* 182: 6854–6856.

130 Feng, Z. and Barletta, R. (2003). Roles of *Mycobacterium smegmatis* D-alanine:D-alanine ligase and D-alanine racemase in the mechanisms of action of and resistance to the peptidoglycan inhibitor D-cycloserine. *Antimicrob. Agents Chemother.* 47: 283–291.

131 Chen, J.M., Uplekar, S., Gordon, S.V., and Cole, S.T. (2012). A point mutation in cycA partially contributes to the D-cycloserine resistance trait of *Mycobacterium bovis* BCG vaccine strains. *PLoS One* 7: e43467.

132 Lehmann, J. (1946). Determination of pathogenicity of tubercle bacilli by their intermediate metabolism. *Lancet* 250: 14–15.

133 Bernheim, F. (1940). The effect of salicylate on the oxygen uptake of the tubercle bacillus. *Science* 92: 204.

134 Rengarajan, J., Sassetti, C.M., Naroditskaya, V. et al. (2004). The folate pathway is a target for resistance to the drug para-aminosalicylic acid (PAS) in mycobacteria. *Mol. Microbiol.* 53: 275–282.

135 Trnka, L. and Mison, P. (1988). P-Aminosalicylic acid (PAS). In: *Antituberculosis Drugs* (ed. K. Bartmann), 51–68. Berlin: Springer-Verlag.

136 Ratledge, C. (2004). Iron, mycobacteria and tuberculosis. *Tuberculosis (Edinb)* 84: 110–130.

137 Zheng, J., Rubin, E.J., Bifani, P. et al. (2013). Para-Aminosalicylic acid is a prodrug targeting dihydrofolate reductase in *Mycobacterium tuberculosis*. *J. Biol. Chem.* 288: 23447–23456.

138 Mathys, V., Wintjens, R., Lefevre, P. et al. (2009). Molecular genetics of para-aminosalicylic acid resistance in clinical isolates and spontaneous mutants of *Mycobacterium tuberculosis*. *Antimicrob. Agents Chemother.* 53: 2100–2109.

139 Zhao, F., Wang, X.D., Erber, L.N. et al. (2014). Binding pocket alterations in dihydrofolate synthase confer resistance to para-aminosalicylic acid in clinical isolates of *Mycobacterium tuberculosis*. *Antimicrob. Agents Chemother.* 58: 1479–1487.

140 (1987). Five-year follow-up of a controlled trial of five 6-month regimens of chemotherapy for pulmonary tuberculosis. Hong Kong Chest Service/British Medical Research Council. *Am. Rev. Respir. Dis.* 136: 1339–1342.

141 Hong Kong Chest Service/British Medical Research Council (1991). Controlled trial of 2, 4, and 6 months of pyrazinamide in 6 month, three-times-weekly regimens for smear-positive pulmonary tuberculosis, including an assessment of a combined preparation of isoniazid, rifampin and pyrazinamide. *Am. Rev. Respir. Dis.* 143: 700–706.

142 Mitchison, D.A. and Nunn, A.J. (1986). Influence of initial drug resistance on the response to short-course chemotherapy of pulmonary tuberculosis. *Am. Rev. Respir. Dis.* 133: 423–430.

143 Barnes, P.F., Bloch, A.B., Davidson, P.T., and Snider, D.E. Jr. (1991). Tuberculosis in patients with human immunodeficiency virus infection. *N. Engl. J. Med.* 324: 1644–1650.

144 Vernon, A., Burman, W., Benator, D. et al. (1999). Acquired rifamycin monoresistance in patients with HIV-related tuberculosis treated with once-weekly rifapentine and isoniazid. Tuberculosis Trials Consortium. *Lancet* 353: 1843–1847.

145 Traore, H., Fissette, K., Bastian, I. et al. (2000). Detection of rifampicin resistance in *Mycobacterium tuberculosis* isolates from diverse countries by a commercial line probe assay as an initial indicator of multidrug resistance. *Int. J. Tuberc. Lung Dis.* 4: 481–484.

146 World Health Organization (2011). *Guidelines for the Programmatic Management of Drug-Resistant Tuberculosis*. Geneva, Switzerland: World Health Organization.

147 Chang, K.C., Yew, W.W., and Zhang, Y. (2011). Pyrazinamide susceptibility testing in *Mycobacterium tuberculosis*: a systematic review with meta-analyses. *Antimicrob. Agents Chemother.* 55: 4499–4505.

148 Chang, K.C., Leung, C.C., Yew, W.W. et al. (2012). Pyrazinamide may improve fluoroquinolone-based treatment of multidrug-resistant tuberculosis. *Antimicrob. Agents Chemother.* 56: 5465–5475.

149 Zhang, Y., Chang, K., Leung, C. et al. (2012a). "ZS-MDR-TB" versus "ZR-MDR-TB": improving treatment of MDR-TB by identifying pyrazinamide susceptibility. *Emerg. Microbes Infect.* 1: e5. https://doi.org/10.1038/emi.2012.18.

150 Migliori, G.B., Besozzi, G., Girardi, E. et al. (2007). Clinical and operational value of the extensively drug-resistant tuberculosis definition. *Eur. Respir. J.* 30: 623–626.

151 Caminero, J.A. (2008). Likelihood of generating MDR-TB and XDR-TB under adequate National Tuberculosis Control Programme implementation. *Int. J. Tuberc. Lung Dis.* 12: 869–877.

152 Ginsburg, A.S., Hooper, N., Parrish, N. et al. (2003). Fluoroquinolone resistance in patients with newly diagnosed tuberculosis. *Clin. Infect. Dis.* 37: 1448–1452.

153 Wang, J.Y., Hsueh, P.R., Jan, I.S. et al. (2006). Empirical treatment with a fluoroquinolone delays the treatment for tuberculosis and is associated with a poor prognosis in endemic areas. *Thorax* 61: 903–908.

154 Leimane, V., Riekstina, V., Holtz, T.H. et al. (2005). Clinical outcome of individualised treatment of multidrug-resistant tuberculosis in Latvia: a retrospective cohort study. *Lancet* 365: 318–326.

155 Park, S.K., Lee, W.C., Lee, D.H. et al. (2004). Self-administered, standardized regimens for multidrug-resistant tuberculosis in South Korea. *Int. J. Tuberc. Lung Dis.* 8: 361–368.

156 Yew, W.W., Chan, C.K., Chau, C.H. et al. (2000). Outcomes of patients with multidrug-resistant pulmonary tuberculosis treated with ofloxacin/levofloxacin-containing regimens. *Chest* 117: 744–751.

157 Migliori, G.B., Lange, C., Centis, R. et al. (2008). Resistance to second-line injectables and treatment outcomes in multidrug-resistant and extensively drug-resistant tuberculosis cases. *Eur. Respir. J.* 31: 1155–1159.

158 Kim, D.H., Kim, H.J., Park, S.K. et al. (2008). Treatment outcomes and long-term survival in patients with extensively drug-resistant tuberculosis. *Am. J. Respir. Crit. Care Med.* 178: 1075–1082.

159 Duncan, K. and Barry, C.E. 3rd (2004). Prospects for new antitubercular drugs. *Curr. Opin. Microbiol.* 7: 460–465.

160 GATB (2001). Tuberculosis. Scientific blueprint for tuberculosis drug development. *Tuberculosis (Edinb)* 81 (Suppl 1): 1–52.

161 O'brien, R. and Nunn, P. (2001). The need for new drugs against tuberculosis. Obstacles, opportunities, and next steps. *Am. J. Respir. Crit. Care Med.* 163: 1055–1058.

9

Multidrug Resistance

Robert L. Marshall[1] and Vassiliy N. Bavro[2]

[1] *Institute of Microbiology and Infection, University of Birmingham, Birmingham, UK*
[2] *School of Biological Sciences, University of Essex, Colchester, UK*

9.1 Introduction

As discussed in earlier chapters on molecular mechanisms of antibiotic resistance, bacteria employ several different approaches to evade antibiotic pressure.

Resistance to more than one antibiotic is referred to as multidrug resistance (MDR). While bacterial resistance to a single antibiotic may cause a nuisance to clinicians, a second antibiotic can usually be utilized to circumvent the problem. In the case of MDR, such circumvention is not always possible and all available treatments may fail, thus increasing both morbidity and mortality of infections. Increased prevalence of MDR leads to more cases of failed treatment, and there are now widespread fears that we may return to a situation whereby there are no effective treatments for most infections, the long predicted "post-antibiotic era" [1]. This situation has been recognized by the World Health Organization as one of the major threats to human health in the modern world [2].

MDR can arise through two main routes. The first is sequential accumulation of many genes or mutations that cause resistance to individual antibiotics, which is a very slow process. The second is by reducing the concentration of the antimicrobial compounds in the cell, either by lowering the membrane permeability, or by active export of these xenobiotics out of the cell by specialized transporters.

9.1.1 Defining Drug Resistance

When studying drug resistance, it is important to consider the context in which we approach the subject. Given the significance of bacteria in human health and disease, studies into drug resistance are often of medical interest, for the prevention or treatment of infection by use of administered substances. Antibiotic resistance is therefore defined from clinically relevant inhibitory concentrations, whereby the ability of a bacterial strain to survive a drug at its breakpoint concentration classifies it as resistant. This classification allows for the categorization as "resistant" or "susceptible" for

Bacterial Resistance to Antibiotics – From Molecules to Man, First Edition.
Edited by Boyan B. Bonev and Nicholas M. Brown.
© 2020 John Wiley & Sons Ltd. Published 2020 by John Wiley & Sons Ltd.

isolates based upon whether or not the infection would likely be treatable by a certain antibiotic, and is therefore highly relevant for many clinical applications.

In studies of the evolution or function of drug resistance mechanisms, relative susceptibilities of various strains are usually compared, assessing how tolerant to a given antibiotic each strain is. This allows more gradual classification of how resilient or sensitive to a drug any bacterial strain might be, rather than using a broad "resistant" or "susceptible" label. Investigating drug tolerance rather than resistance is useful for such studies, as smaller differences become apparent and therefore finer detail of the mechanisms of resistance can be determined.

9.1.2 Evolution of Multidrug Resistance

It is easy to think only about how bacteria respond to changes that we as humans impose upon them, be they in a clinical or laboratory setting, and to forget that most interactions involving bacteria occur without any intentional manipulation on our part. While anthropogenic influence cannot be denied, it is the interaction with the natural environments which drove bacterial evolution, forcing bacteria to develop an intrinsic resistance to toxic xenobiotic substances. Perhaps the most notable cell structure hypothesized to have developed as a result of xenobiotic-mediated selection pressure is the defining feature of Gram-negative bacteria – the outer membrane [3]. According to this hypothesis, its development introduced a second permeability barrier against the influx of xenobiotics, thus enabling the survival of early didermic (two-membraned) bacteria in chemically hostile environments.

Owing to its unique composition, which is discussed below, the outer membrane is much less permeable than the inner (cytoplasmic) membrane. Any hydrophilic compounds, as well as some hydrophobic substances, require dedicated pores and channels to cross it in many didermic bacteria. These pore proteins can also be exploited for the entry of potentially toxic substances, thus increasing drug susceptibility [4]. Some channel proteins have evolved to form essential parts of the so-called multidrug efflux pumps aka multi-drug-resistance pumps (MDR-pumps), which actively remove toxic substances from within the cell. In combination with an already efficient influx barrier, the MDR-pumps substantially improve the survival of Gram-negative bacteria in the presence of toxic substances, including antibiotics. Over-expression of, or even a point mutation in these proteins has been shown to cause clinically significant MDR and may change the drug susceptibility profile of the strain [5–7].

9.2 A Permeability Barrier to Decrease Drug Accumulation

9.2.1 Gram-Negative Outer Membrane

Several outer membrane structures have been found to exist in nature. Perhaps the most widely known of these is that of the Gram-negative bacteria. It is an asymmetric bilayer, of which the inner leaflet consists of glycerophospholipids, enriched in phosphatidylethanolamine and saturated fatty acids relative to the inner membrane. The outer leaflet, which is most responsible for the permeability barrier of this membrane, is formed of lipopolysaccharide (LPS). LPS is a complex molecule consisting, as its

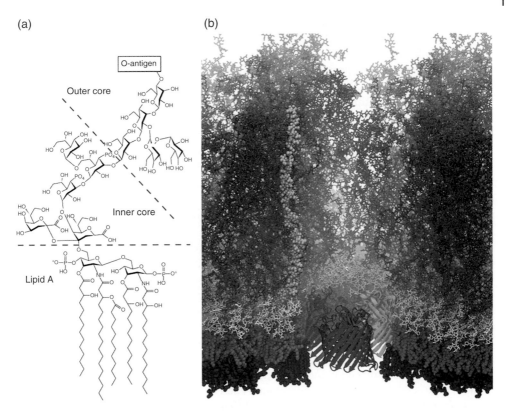

Figure 9.1 (a) A representative skeletal formula of an LPS molecule, excluding detail of the O-antigen. Many modifications are possible to the LPS molecule, including additional acylation, acylation with different chain lengths, incorporation of extra or alternative sugar subunits, and further phosphorylation. (b) A full atomic model of a section of outer membrane from *Escherichia coli*. The full atomic model highlights a single O-antigen chain (pink) and the acyl chains (gray) in a space-filling representation, while the core (yellow) and all other LPS molecules are shown in line format only. A trimer of OmpF is shown in ribbon format with each protomer in a different color. Clearing of O-antigen above the porin can be seen. Please note that the perceived density of packing of LPS molecules visualized here is lower than reality, due to the use of line rather than space-filling representation. *Source:* 9.1a kindly provided by James C. Gumbart (Georgia Tech, USA), personal communication.

name suggests, of lipid and polysaccharide parts (see Figure 9.1). The lipid part of the structure anchors it in the membrane and is referred to as Lipid A, to which a core oligosaccharide is attached, which in turn is conjugated to a variable-length repeating oligosaccharide structure, known as O-antigen, extending outside the cell.

LPS varies between species and strains, affecting a range of biological processes. Variation is seen throughout the LPS molecule, but characteristic variations include the length and composition of the O-antigen and the acylation of Lipid A. The acyl-chains form a hydrophobic interaction interface with the fatty acids of the inner leaflet glycerophospholipids to form the functional bilayer of the outer membrane.

Lipid A consists of a disaccharide of hexosamines, which are covalently bound *via* a 1–6′ ether bond. Each of the hexosamines are acylated by amide and ester bonds at their

2- and 3- positions, respectively, providing the minimum of tetra-acylation seen naturally. A maximum of eight acyl chains have been observed in natural LPS – ester bonds can be made between the 3-hydroxy position of one fatty acid and the carboxylic acid of another, forming an acyloxyacyl moiety that can be attached to the dihexosamine at each of the four acylation positions. The fatty acids found in LPS are often, though not always, saturated, and thus can pack closer together for a greater combined hydrophobicity. This contributes to the decreased fluidity and permeability of the outer membrane compared to cytoplasmic membranes. The LPS acyl chains are usually between 10 and 16 carbon atoms long, but chains as long as 21 carbon atoms have been reported. The hexosamines of Lipid A are also phosphorylated, increasing hydrophilicity of the backbone region at the external face of the membrane, creating negative charge, and decreasing permeability to hydrophobic substances. The negative charge attracts a high density of divalent ions, which tightly bridge the Lipid A head-groups, greatly limiting their mobility.

The core oligosaccharide of LPS is attached to the free 6-hydroxy position of the dihexosamine. The core region consists of a series of 1–5′ linked other sugars, which vary by species. In enteric bacteria, this is usually a series of three heptose sugars followed by up to six other sugars. The core region displays great variability, in terms of length, composition, and branching, and in the interest of space will therefore not be discussed further. However, strains with shorter core regions are more susceptible to lipophilic drugs, indicating that the LPS core region plays a role in preventing access of drugs into the periplasm.

On the outside of the cell is the O-antigen, which can vary from lacking entirely (in "rough-type" cells) to as many as 50 repeats each consisting of up to eight sugar units [8]. The number of different sugars, modifications, and interlinking compounds (such as organic acids) that can make up each subunit, combined with the possible variations in number of sugars per subunit, provides huge scope for variation in O-antigen structure between different bacterial serotypes. This variation determines the immunogenicity of the outer membrane but is not generally regarded as significant in decreasing permeability of the membrane to drugs, despite evidence that permeation of the outer membrane by macrolides is dependent upon the length of the O-antigen [9].

LPS creates a very effective permeability barrier that prevents both toxic substances and beneficial nutrients and water from crossing the outer membrane. To circumvent this barrier, Gram-negative bacteria have developed specialized outer membrane pore proteins to allow the uptake of nutrients. Most outer membrane pores are channel-forming proteins known as porins, which fold into β-barrels embedded into the outer membrane. While the outside of the porin barrel facing the membrane is highly hydrophobic, the inner surface is predominantly hydrophilic, allowing the introduction of a water-filled pore through the otherwise highly hydrophobic membrane. The variable loops connecting the strands that form the barrel create a selectivity plug in the middle of the barrel, which acts as a filter to provide selectivity of solute transport. Porins differ in both diameter and electrostatic nature of their pores, thus accounting for high specificity of transport. Some of the porins are therefore either cation- or anion-specific, while others are sugar-specific (*e.g.*, OmpB in *Escherichia coli*), and some even transport hydrophobic (*e.g.*, FadL) or non-polar (*e.g.*, CymA) substrates [10, 11], while others do not show solute specificity and are regarded as general porins (*e.g.*, OmpF and OmpC in

E. coli). General porins tend to have a wider accessible opening than specific porins, and are a route of entry for many drugs. Loss or down-regulation of porins is associated with drug resistance, and the low expression levels of general porins in *Pseudomonas spp.* and *Acinetobacter spp.* is attributable for their increased drug tolerance compared to *E. coli* [12]. As a result, the permeability barrier of the outer membrane in its native state very much depends on the limits imposed by the porin channels, and that generally limits the passive diffusion to molecules of up to approximately 600 Da – what has become known as the "OM size-exclusion limit," although there are notable exceptions to this rule [13]. For molecules larger than this molecular cut-off, active importers are required.

Size, number, and specificity of the porins contribute greatly in determining how tolerant a cell will be to antibiotics, but changes in the porin profile are relatively slow in the context of responding to an environmental change. For a quicker response, porins can be closed by ligand binding under certain conditions and cannot be thought of as constitutively open holes in the membrane [11].

Not all drugs can fit through porin channels, due to either their size or charge. Macrolides, for example, access the cell by permeating the lipid bilayer itself rather than utilizing a protein as route of entry [14]. Vancomycin is a large drug which cannot easily pass through either the porins or the outer membrane itself and as a result is rather ineffective against Gram-negative infections. However, it may gain entry to the Gram-negative bacterium through a mutated or gated drug efflux channel protein, which is normally sealed except for during efflux events [15], as discussed in Section 9.4.1.1.

In summary, the cell envelope of Gram-negative bacteria forms an effective permeability barrier, without which the cells would be sensitive to a range of external factors, including antibiotics. The barrier effect of the outer membrane is caused by both the highly dense oligosaccharide layer which plays the role of a molecular sieve that all but shuts down the free diffusion of metabolites and xenobiotics, as well as by the decreased fluidity of the outer membrane, which in turn is caused by the tighter packing of fatty acids in LPS compared to conventional phospholipid bilayers. Low membrane permeability introduces the requirement for porins, both for uptake of nutrients and to allow exit of toxic metabolic waste products. Porins are used as a point of entry by most drug molecules, and porins therefore represent a factor in antibiotic resistance.

9.2.2 Other Outer Membranes

Members of the *Corynebacterineae* (*Mycobacteria, Corynebacteria, Nocardia*) have an outer membrane containing mycolic acid derivatives, which are covalently bonded to the peptidoglycan cell wall *via* arabinogalactan. This outer membrane lacks phospholipids [16]. The exact composition of the mycolic membrane is strain-dependent and is affected by a variety of growth conditions [17, 18], but always contains mycolic acids in both leaflets. In the inner leaflet, the mycolic acids are attached to the arabinogalactan, while in the outer leaflet they are esterified with trehalose or glycerol. On the outermost surface ("top layer") of these membranes is a matrix consisting mainly of polysaccharides, although loose lipids and proteins are also present [19]. The rigid structure of the outer membrane, combined with the composition of and possible protein function within the extracellular "top layer," make the entire cell envelope of *Corynebacterineae* an effective barrier against permeation by potentially toxic substances. However, it also forms an effective barrier against permeation by required nutrients, and, similarly to the Gram-negative bacteria, *Corynebacterineae* also heavily rely on porins (known as Msp). Deletional mutagenesis of the genes encoding these pore proteins has shown that they are used as a route of entry by antibiotic substances [20, 21], further indicating the effectiveness of the outer membrane as a permeability barrier. The high intrinsic drug resistance of *Corynebacterineae*, and in particular the *Mycobacteria*, is usually attributed to their cell envelope.

Other outer "membrane" structures, such as the proteinaceous toga structure seen in certain thermophiles, are known to exist but have yet to be characterized [22]. Their possible roles in resistance to toxic substances cannot be determined until their composition and general structure are known.

9.3 Drug Efflux Pumps

Transmembrane (TM) pump proteins are present and common throughout all kingdoms of life and can be responsible for either active uptake of substances such as nutrients and vitamins, or for active efflux of unwanted (toxic) substances [23]. Of particular interest in the context of MDR are the drug efflux-pumps, which are grouped into six main families, based upon their phylogenetic connection and correspondingly shared architectures (Figure 9.2).

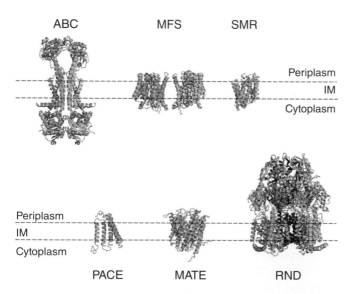

Figure 9.2 General architecture and relative sizes of the principal classes of MDR-transporters (drawn at the same scale). Where oligomeric state is known, separate protomers have been colored distinctly. ABC-, MFS-, and SMR-families appear to function as dimers, while the RND-transporter family members form trimeric assemblies. As stoichiometry of PACE- and MATE-transporters hasn't been conclusively established, they are shown as monomers. The position of the inner membrane (IM) is relative.

9.3.1 ABC Transporters

Of these protein families, just one generates the energy required for substrate trans-location directly – namely by hydrolysis of ATP. The members possess a signature ATP-recognition sequence motif and are thus nominated **ATP-b**inding **c**assette (ABC) transporters. ABC transporters are a varied group, and only a few have been directly implicated in multidrug efflux, *e.g.*, the ABC transporters of the MacB-family, which form part of a macrolide efflux pump. In common with the majority of ABC transporters, the MacB-like transporters are dimeric [24], with each protomer providing half of the ATP-binding site. However, and in contrast to the canonical ABC transporters, which have extensive TM domains of 5 to 10 helices per protomer, each monomer of MacB has just four TM helices [25]. Also in contrast to the rest of the ABC transporters, its cytoplasmic nucleotide binding domain is located N-terminally [26], and the protein has a prominent mixed α/β periplasmic domain, which shares structural similarity with the RND-transporter family [27]. The periplasmic domain interacts with partner proteins, while the nucleotide binding domain binds and hydrolyses nucleotides, primarily ATP [24, 28]. While the direct data on cargo binding of MacB is missing, it is likely that, similar to other better-studied members of the ABC-family such as MsbA and the P-glycoprotein, the TM helices of both subunits form a clam-like arrangement to bind the substrate in the inner leaflet of the membrane and, following ATP-hydrolysis, undergo a conformational change to release it on the outer side [29]. ABC transporters are a varied family of transporters; some members function by themselves, while others form part of complex assemblies.

9.3.2 MFS Transporters

All the other five known families of MDR-pumps are secondary transporters that utilize pre-created ion gradients for symport or antiport. The largest of these families is the *major facilitator superfamily* (MFS). The wider superfamily contains examples of all three possible transport types (symport, antiport, and facilitated diffusion), indicating that although a single superfamily has common architecture and a common mechanism for the transport action, both the directionality of and energy source for the transport event may vary [23].

The crystal structures of members of this superfamily show that the proteins are formed almost entirely of a bundle of TM helices [30]. They have virtually no periplasmic or cytoplasmic domains except for minor protrusions of the TM helices and their interconnecting loops. Most proteins in this family contain either 12 or 14 TM helices. In the *E. coli* MFS transporter EmrD these helices account for 240 out of 394 residues, though most pumps in this family are slightly larger, up to 600 amino acids long. The EmrD crystal structure indicates a dimeric assembly [30], although its physiological relevance is not fully proven.

Within each MFS protomer there is a narrow pore, which has been identified as the route through which the substrate translocates. This pore is open to only one side of the protein at a time, for binding or release of the substrate respectively. A rocker-switch motion rearranges the helices around the active site, temporarily occluding the protein lumen during the translocation process, until the pore re-opens at the other side of the protein. The use of a rocker-switch type movement of a single binding site, which alternates the side of the membrane to which the binding pocket is open [31] is a unifying feature of the MFS transporters and for antiport-driven pumps, this process will then reverse with the binding of the antiporting substrate. Like ABC transporters, the MFS transporters may function alone or as part of tripartite assemblies, *e.g.*, in *E. coli* the EmrB pump associates with the adaptor protein EmrA and outer membrane protein TolC.

9.3.3 SMR Transporters

The smallest proteins currently identified as multidrug transporters are the so-called *s*mall *m*ultidrug *r*esistance (SMR) family [32], which is part of the drug/metabolite transporter (DMT) superfamily. SMR pumps have a more limited range of substrates than the other families of transporters and are only known to transport lipophilic compounds, *e.g.*, quaternary ammonium compounds and cationic dyes. Despite this, single members within this family are capable of transporting several structurally distinct substrates and are therefore bona fide multidrug transporters. SMR transporters are encoded both chromosomally and on mobile genetic elements (plasmids and integrons), creating potential for transfer of resistance. All members of this family for which the transport type is known, function by drug/proton antiport.

Like MFS transporters, SMR proteins contain virtually no cytoplasmic or periplasmic regions, but are dramatically smaller, with only four TM helices per monomer. The functional oligomeric state of SMR proteins is unclear, as the multimerization state determined experimentally is dependent upon the conditions used in the investigation [33]. The best-studied example of the group – the *E. coli* EmrE – is also special as it seems to insert itself into the membrane in two different ways – a dual-topology arrangement [34].

There have been several mechanisms of pump activity proposed over the years, but more recent studies are converging toward rocker-switch type mechanism like that in MFS transporters, but with a binding pocket and pore formed between dimers [35, 36].

9.3.4 MATE Transporters

*M*ultidrug *a*nd *t*oxic compound *e*xtrusion (MATE) efflux pumps consist of three subtypes: cluster 2 proteins are eukaryotic, and clusters 1 and 3 include bacterial and archaeal proteins. Cluster 1 proteins (of the NorM-type) follow the same phylogenetic patterns as the organisms from which they originate, while cluster 3 proteins (of the DinF-type) follow no such pattern. MATE transporters utilize electrochemical potential, antiporting protons, Na^+, or K^+. One example protein (KetM from *Klebsiella pneumoniae*) has been found to be energized by any one of protons, Na^+, K^+, or Li^+, and to pump several structurally related compounds [37]. While other pumps within the MATE-family have been shown to pump a variety of structurally unrelated substrates [38] and are therefore recognized as multidrug transporters, MATE-pumps seemingly contribute little to the overall intrinsic resistance of bacteria to toxic compounds.

Crystal structures of MATE-transporters from different organisms are available, displaying different substrate binding states [39–42]. These reveal the protein to consist of 12 TM helices, which undergo major conformational changes upon substrate binding. Uniquely among antiporters, MATE proteins have been shown bound to both substrate and cation simultaneously, although the binding of the cation causes such a significant conformational change as to force release of the substrate and enable continuation of the cycle. MATE proteins of the NorM family have also shown for the first time that $Na^+–\pi$ bonding can occur in membrane proteins, thus indicating that selectivity for Na^+ as an energy source is dependent on the presence of an aromatic residue at the cation-binding position. The drug and cation binding sites in the NorM family of MATE transporters are separated by a helix – so that the two moieties interact neither with each other nor with the same binding pocket of the protein. The DinF family utilizes a very different mechanism: competition between protonation and substrate binding causes release of the separate moieties, from the same binding site [40]. The NorM-type and DinF-type subfamilies of MATE transporters are therefore clearly distinct, despite the structural similarity.

9.3.5 PACE Transporters

The most recently discovered family of multidrug efflux pumps is the *proteobacterial antimicrobial compound efflux* (PACE) family. The nomenclature may require redefining, however, as this family is not limited to the proteobacteria, nor do all proteobacteria encode these pumps [43]. The natural substrates for the PACE pumps have not yet been determined, and currently have only been found to pump synthetic biocides [44]. The only factor apparently common to the identified substrates is the presence of at least one aromatic ring. As of early 2016, no structural or mechanistic information is available for this family of pumps; however, they seem to share remote sequence similarity to the MATE family, and homology modeling (Marshall and Bavro, *unpublished*) suggests a likely identical topological organization to the same.

9.3.6 RND Transporters

Of all the MDR proteins present in Gram-negative bacteria, the *resistance-nodulation-cell division* (RND) transporters provide the greatest contribution to MDR. The extent of this is such that any effect of all the other previously discussed efflux-pumps is usually masked in wild-type cells, and their studies often require use of strains in which the major RND genes have been deleted. RND pumps are easily up-regulated in response to different stimuli, providing wide non-specific resistance [45]. RND transporters involved in MDR are often capable of transporting a wide variety of unrelated compounds, using proton gradients as the energy source. While in most secondary transporters the directionality of the translocation is provided by the cargo-binding affinity at the respective membrane leaflet and is thought to be reversible, the RND transporters seem to be a notable exception from the rule as they are unidirectional and they only pump substrates across the inner membrane in outer direction.

Transporters of this superfamily are found across all kingdoms of life and are involved in a highly diverse range of functions – which often are combined in even a single protein. These functions include export of multiple, unrelated toxic compounds (Figure 9.3); transport of unfolded proteins; transport of outer membrane lipid- and lipooligosaccharide-components; export of bacterial quorum sensing molecules and export of eukaryotic lipopeptides and steroids. Their role in the transport of outer membrane lipids in the *Corynebacterineae* (and *Mycobacteriaceae* in particular) is an example of how an RND pump can contribute to MDR without pumping the drugs [46, 47].

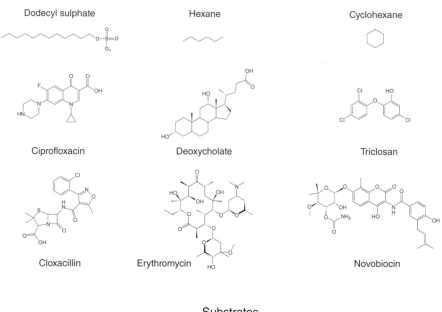

Figure 9.3 Substrates, inhibitors, and a non-substrate of common RND multi-drug transporters. Great variation in size, shape, and hydrophobicity can be seen between different substrates. Phenylalanine-arginine beta-naphthylamide (PAβN) and pyranopyridines are competitive inhibitors of the pumps, although PAβN is also a poor substrate. Vancomycin is not an efflux substrate, but utilizes open complexes required for RND efflux pump activity to gain entry to the cell.

Given the significant contribution of RND transporters to MDR, they have been extensively studied, and their mechanism of cargo translocation is relatively well under-stood. RND transporters function as trimeric assemblies, each protomer of which has three main domains [48].Two of those reside in the periplasm and form the presumed assembly point for auxiliary periplasmic and outer membrane proteins and also form the main drug binding and extrusion conduits, while the third domain is fully embedded in the inner membrane and composed of 12 TM helices per protomer. The sequence and topology of the RND transporters reveals an early gene duplication event in formation of the individual protomers. Communication and cooperative activity of all three protomers is essential for effective function as demonstrated by functional concatenation and cross-linking experiments [49].

The membrane domain couples proton translocation to drug extrusion. The periplasmic domains are formed of two large loops located between TM-helices TM 1 and TM 2, and TM 7 and TM 8 respectively. These loops account for around 60% of the amino acids in each protomer – a unique and defining feature among the efflux-transporters.

The upper part of the periplasmic structure has historically been known as "TolC-docking domain" [48], and is thought to form the platform for association with the *outer membrane factor* (OMF) family of channel proteins of which TolC is a member; as well as for the periplasmic adaptor proteins or PAPs [50]. As discussed later, this interpretation has been challenged by some of the more recent models of pump complex assembly.

The lower part of the periplasmic domain forms the principal part of drug binding and extrusion conduits, and has been shown to undergo a coordinated peristaltic pump-like structural transition during the efflux cycle [51, 52]. The mechanism is also often described as an "alternating-access cycle" due to the conformation of the periplasmic domain cycling between three forms in a sequence that results in an antibiotic molecule being captured on the periplasmic side and subsequently squeezed through the protein toward the center-top of the RND-trimer and eventually released into the exit conduit formed by the outer membrane proteins of OMF-family.

Coupling of the motion of the TM domain to that of the periplasmic domains reveals that the proton transport also proceeds according to the "alternating access mechanism" [53]. Cycle coupling appears to be mediated by a piston-like motion of TM 2 and TM 8, which is generated by the movement of protons through a conserved "proton relay" network formed of several charged residues. The two cycles are tightly interlinked, despite being separated by about 50 Å. In the archetypal *E. coli* RND-transporter AcrB, these are formed by D407, D408, K940, E346, and D924. This charge network is very well conserved across the RND family – in particular, two aspartic acid residues essential for AcrB activity (D407, D408) seem to be absolutely conserved in all RND multidrug efflux pumps from alpha-, beta-, gamma-, delta-, and epsilon-proteobacteria. The protonation states of these residues are tightly connected to conformational transitions in the TM domains and are communicated to the drug-binding pockets in the periplasmic domains above, effectively coupling the proton motion to efflux. While the exact energetics of RND dependent efflux are still unclear, the available structural data are consistent with a two-proton conformational switch mechanism [53].

In most studied MDR RND pumps, including the prototypical AcrB multidrug transporter, the cargoes are loaded from the periplasmic side of the inner membrane. However, at least some of the members of the RND family, such as the metal efflux pump CusA from *E. coli*, are also able to load a substrate on the cytoplasmic side of the inner membrane [54]. The structures of the pumps in apo, substrate- and inhibitor-bound state have revealed that there are several access channels through which the drug could be taken in and expelled through the protomer [48, 55–57].

RND transporters involved in MDR appear to have much lower substrate specificity than either the ABC or MFS transporters, which underpins their function as main *multidrug* efflux pumps. The principal substrate translocation channel forms a couple of distinct binding pockets within each protomer designated as "proximal" and "distal" pockets, which are separated by a so called "switch-loop." Although the cargo must move through both pockets on its way out of the cell, the pockets seem to have clear preference for particular subtypes of cargo: higher molecular mass drugs tend to bind predominantly to the proximal pocket, and smaller, more lipophilic compounds bind transiently to the distal pocket [55, 57]. The binding pockets are only accessible from the periplasm during the initial phase of the extrusion cycle when the protomer is in its "drug-binding capable" state. As the cycle progresses, the conformational change within the protomer leads to a rearrangement of the channels and the periplasmic access tunnel is sealed, while an outward facing one toward the auxiliary OMF is open for the drugs to move through. As mentioned earlier, the RND transporters display intrinsic low substrate-specificity and affinity. Any tight binding of a substrate in the respective binding pockets may increase its retention time and lower the efficiency of efflux, thus potentially turning it into an inhibitory compound, *e.g.*, by exploiting the so-called "hydrophobic trap" [56]. Conversely, any changes within these drug-binding pockets could impact how effectively the pump will transport any given substance, which can in turn have a significant impact on clinical outcome [5].

9.4 Moving a Substrate beyond the Cell Envelope

All the families of pumps indicated earlier have a TM domain to facilitate transport across the cytoplasmic membrane. However, in the case of Gram-negative cells, they do not span the entire cell envelope on their own and would, therefore, not remove the drug past the main permeability barrier of the outer membrane. For many toxic substances, their removal from the cytoplasm would be sufficient to prevent their action, as any drug which acts at the level of DNA, RNA, protein synthesis, or on any other cytoplasmic function would be removed from its location of activity. Such transporter action would, however, be completely ineffective against antibiotics with a periplasmic target. Even for drugs with a cytoplasmic target, this movement overcomes only a lesser permeability barrier, and the drug may be able to re-enter the cytoplasm relatively quickly.

As an extra level of protection in Gram-negative bacteria, the ABC, RND, and some MFS transporters have evolved to form part of a tripartite complex capable of spanning the entire cell envelope, and thus be able to transport their substrates to the external environment. Tripartite assemblies consist of the inner membrane transporter protein belonging to one of the families described earlier; an outer membrane channel protein of the OMF-family, such as TolC [58], and a *p*eriplasmic *a*daptor *p*rotein (PAP; also known for historical reasons as a membrane fusion protein [MFP]), which links the pump and channel proteins to ensure a continuous conduit from the inner membrane to the extracellular environment (Figure 9.4) [59]. Some RND-transporter based complexes may have an additional small modulatory protein bound to the TM part of the transporter; however, these have been shown to be non-essential for efflux activity. These small proteins are currently poorly understood but effect the efflux profile of the pump [60]. For drug efflux to be detectable *in vivo*, all the transporter, PAP, and channel proteins must be expressed, and an energy source must be present. Such efflux activity can be derived through several approaches, including assays to determine minimum inhibitory concentration (MIC), fluorescent substrate accumulation rate, and removal of fluorescent substrate from preloaded cells.

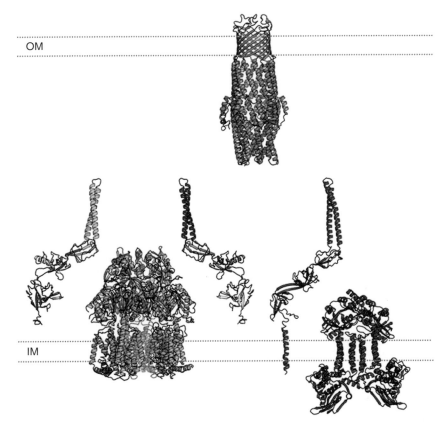

Figure 9.4 Exploded view of tripartite complexes. The distance between inner (IM) and outer (OM) membranes is not to scale, and the distances between proteins should not therefore be taken as representative of a model of interaction. Two transporters are shown: the RND transporter AcrB (teal) and ABC transporter MacB (purple), as the physiologically relevant trimer and dimer, respectively. The outer membrane channel protein TolC (green) is shown in the closed state of its physiological trimer. The PAPs MacA (pink) and two copies of AcrA (both multi-colored) are shown as monomers. The N-terminal lipidation of AcrA is not shown in either monomer. The topology of AcrA can be seen from the rainbow coloring, indicating how each domain is formed of both N- and C-terminal halves of the protein. The four domains of AcrA are shown on the four-colored monomer of AcrA: hairpin (red), lipoyl (orange), β-barrel (brown) and membrane-proximal (yellow).

9.4.1 Additional Components of Tripartite Complexes

9.4.1.1 Outer Membrane Factors (OMFs)

As stated earlier, tripartite complexes consist of the transporter, an outer membrane channel protein of the TolC-family also known as OMF, and an PAP.

Although the OMF is sometimes encoded on the same operon along with a dedicated efflux system, more often it is a promiscuous component, which can pair with various different transporter–adaptor pairs, thus serving as a universal exit duct out of the cell. For example, in *Salmonella enterica*, at least seven transporter/PAP pairs utilize one channel protein, TolC [61]. This may be a contributing factor toward the OMF channel itself having no substrate specificity and that its only limitations on the cargo are those imposed by the steric restrictions of passage of the pumped xenobiotics through its channel.

OMFs of the TolC family have a distinctive structural architecture, which separates them from other outer membrane proteins. Like the RND transporters, the OMFs are functional trimers and each protomer displays internal pseudo-symmetry, resulting in overall pseudo-sixfold symmetry of the assembled trimer. These trimers fold into extremely elongated tubular structures, which, while embedded in the outer membrane using a porin-fold-like β-barrel, have a unique α-barrel protrusion that extends more than 100 Å into the periplasm [62]. The OMF β-barrel is unique among the outer membrane proteins as, despite presenting a typical porin fold, each protomer contributes 4 β-strands to build a 12-stranded antiparallel barrel. The α-domain is also unique to this family of outer membrane proteins and is made up of 12 helices with each protomer contributing two sets of double helical hairpins. Each hairpin pair is formed of one continuous helix spanning the full length of the periplasmic domain, while the second helix is interrupted, and formed of two shorter helices stacked end-to-end to match the length of the first one. Spliced into this interruption is a small α/β extension, which along with a part of the C-terminal tail of the protein forms the "equatorial domain."

The mixed α-helical/β-stranded "equatorial domain" forms a pseudo-continuous ring structure around the α-barrel tube, approximately halfway down its periplasmic portion. While parts of the equatorial domain are essential for TolC function [63], the exact role of this domain remains unclear.

To prevent the channel from simply acting as a large porin, the periplasmic end of the channel in the resting state is tapered and occluded [62]. This is primarily achieved through the natural twist of the α-helical hairpins, which form the lower portion of the periplasmic domain. This curvature is stabilized in the resting state of the channel, *via* an extended network of charged and polar residues, which interact to hold the channel in a closed conformation, and which must be disrupted before any efflux activity is possible [64]. Mutational disruption of this network can result in a channel that is constitutively open [64, 65]. A few helical turns up the inside of the coiled coils are a pair of rings of aspartate residues, or one acidic and one hydrophobic ring [66, 67], which also require disruption to facilitate channel opening. Opening of the wild-type channel requires interaction with an adaptor protein [15, 68], with a possible contribution from the transporter, at least for RND transporters [50].

9.4.1.2 Periplasmic Adaptor Proteins

PAPs have a highly modular architecture owning to the fact that they must interact specifically with both their cognate transporter and OMF channel proteins. They are elongated proteins, which can be composed of a number of domains, and while some core domains are shared by all members of the family, the others are not always present and their presence appear to be correlated with the type of transporter with which they function [59].

PAPs have a very distinctive topological organization. The polypeptide chain is folded uniquely in a hairpin-fashion, approximately in half. Each domain is therefore formed of contributions from both the N- and C-terminal parts of the protein. Most PAPs are anchored in the inner membrane by either a TM helix or *via* a lipid moiety covalently attached to an N-terminal cysteine. The protein resides in the periplasm and is composed of three principal domains. In order of distance from the inner membrane, these domains are: first, as its name suggests, the "membrane proximal domain" [69]; followed by the "β-barrel" and the "lipoyl" domains [70, 71]. All three domains are predominantly β-stranded structures. In addition, most PAPs possess an α-helical hairpin domain of varying length that is spliced into the lipoyl domain and contacts the channel protein. The length of this hairpin, determined in helical turns, appears to correlate with the size of the periplasmic domain of the cognate transporter and, correspondingly, the distance which is required to be bridged and sealed to form a continuous efflux conduit from the transporter to the OMF [71]. It thus follows that PAPs that function with RND transporters, which have large periplasmic domains, require relatively short hairpins, while those that function with an MFS transporter require much longer hairpins as they must bridge a substantial gap between the transporter and the channel proteins.

The requirement for the hairpin domain is not absolute. In some pumps, such as the *E. coli* metal-ion transporter CusBA [54], the hairpin of the PAP CusB is very much shorter than that of the closely related AcrA that functions in conjunction with the AcrB pump discussed previously [72]. In some systems, the PAPs have lost the hairpin domain entirely, instead having a short flexible loop, as reported for the spirochete *Borrelia burgdorferi* [73]. The existence of such PAPs in what appears to be a functional efflux system raises questions regarding the assembly mechanisms involved. Homology modeling suggests that the cognate channel protein, BesC, lacks some of the internal interactions that maintain the closed state in other TolC-family members [74]. Implications of the lack of the hairpin domain on the stability and recognition of the OMF-PAP and transporter-PAP interactions remain to be fully elucidated. Furthermore, the role of the hairpin domain may not be restricted to providing a linkage to the OMF, as PAPs containing hairpin domains are also found in some Gram-positive organisms [75].

All PAPs characterized to date contain both lipoyl and β-barrel domains. Each of these domains readily cross-link to the cognate transporter protein, suggesting that one of the roles of these domains is to interact with the transporter component of the complex [69]. Additional direct evidence for this role within the β-barrel domain is provided by adaptive mutagenesis data, whereby the PAP is paired with a non-cognate transporter and mutations are tested for gain-of-function to determine which positions are required to determine specificity between two proteins [76].

Recent bioinformatic analysis indicates that the membrane-proximal domain (MPD) is only present in PAPs working in conjunction with transporters that have large periplasmic domains (such as RND transporters and the MacB-family of ABC transporters) and take their cargoes from the periplasmic side of the membrane [59]. The role of these domains may then be connected not only with the necessity to span the diameter and accommodate the larger transporter, but to also present the cargo and/or activate the pump itself. It has been found that the MPD of MacA is crucial in stimulating the ATPase activity of the MacB transporter [77], while active presentation of the substrate by MPD has been demonstrated for several PAPs associated with metal-ion pumps, for example ZneB [78]. In addition, there is evidence that MPDs are crucial at the level of assembly of the tripartite complex [79]. MPDs therefore provide direct interaction and coupling with the transporter. At the same time, PAPs associated with transporters lacking large periplasmic domains contain additional N-terminal TM helices, the function of which are not fully established, though plausibly provide better anchorage and a means of communication with their cognate transporter within the membrane leaflet.

Thus, the roles of the PAP appear to be diverse and complex. At the simplest level, these proteins interact with the other components of tripartite complexes, stabilizing the assembly [80]. *In vitro* studies have shown that the PAP is required for activation of the transporter protein [24, 77], but only when a substrate is also present [81]. Activation of the transporter is independent of the outer membrane channel protein, as activity can be seen *in vitro* in its absence. At the same time, the PAP is required for removal of the constriction of the channel protein thus priming the channel open and ready for an efflux event [68].This opening of the channel protein appears to be independent of the presence of either a substrate or the transporter protein, as shown by determination of susceptibility to vancomycin (a drug that is too large to efficiently traverse the outer membrane unless a wide channel such as TolC is available) for combinations of complex components and mutants thereof [15].

9.4.2 Models of Complete Assembly

Broadly, there are two main models of how the complete tripartite complex assembles (Figure 9.5). While both place the transporters within the inner membrane and the channel protein with its β-barrel in the outer membrane, they differ greatly in the level of interaction between the PAPs and the outer membrane channel. The original model proposed predicted an extensive interaction between the surface of the OMF channels

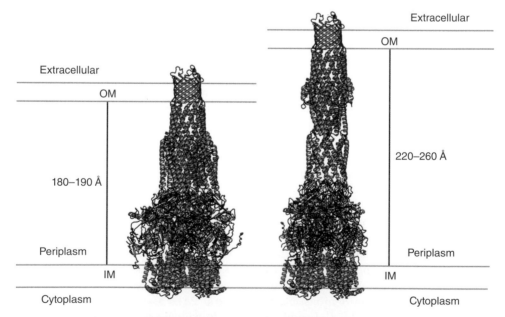

Figure 9.5 Representative depictions of the bundling (left) and tip-to-tip (right) models of assembly for the AcrAB-TolC efflux complex. The inter-membrane distance is indicated for each model in Ångstroms. AcrB (cyan) is shown spanning the inner membrane (IM), TolC (green) is shown as a pore in the outer membrane (OM), and AcrA (brown) is tethered to the inner membrane at its N-terminal.

and the helical hairpin domains of the PAPs, resulting in formation of robust helical bundles [69, 82]. This "deep interpenetration" model of interaction brings the OMFs closer to the inner membrane and even allows for direct interaction between the channel and the large transporters, *e.g.*, members of the RND family. The alternative, later model, was first proposed based upon crystal packing of PAPs observed in structural studies [83], and speculates that the interaction between the PAP and the channel takes place only at the tips of their respective coiled-coil structures, thus spacing the transporter and OMF-channel proteins apart by at least the length of the PAP hairpin domain [84]. Henceforth, these models of assembly will be referred to as the "bundling" and "tip-to-tip" models, respectively. As we will see later, there is evidence in support of each model; however, at present there is no conclusive proof for either model and the topic remains a hotly debatable one as some evidence for each model seems to contradict the other. While clearly only future research would conclusively settle this debate, the two models perhaps reflect different stages of the dynamic functional cycling and rearrangements associated with the pump-complex assembly process.

9.4.2.1 Stoichiometry of Complex Assembly

The stoichiometry of the complex is yet to be definitively determined, though the general consensus is that 3:6:3 RND transporter:PAP:OMF is most likely in either model. Such consensus derives from multiple lines of evidence: crystallographic studies show that PAPs often form dimers and some can form hexamers when crystallized [70, 72, 83], cross-linking, mass-spectrometry, and atomic-force microscopy also identified hexameric PAP assemblies [24, 85], while surface plasmon resonance (SPR)and isothermal calorimetry revealed two-step association processes consistent with dimeric and hexameric PAP species [86]. Furthermore, genetic and functional analysis utilizing AcrA–AcrA fusions shows that a forced dimer is functional, a complex which requires two unique PAPs has been identified [87], and electron microscopy images attempting to show the fully assembled complex directly show a 3:6:3 stoichiometry [88–90]. Despite this, it has also been shown that the AcrAB-TolC complex in *E. coli* is capable of functioning with 3:3:3 stoichiometry, as AcrA–AcrB fusions remain functional without any other copy of *acrA* [91]. The relevance of each of these investigations to the natural physiological state is therefore still not fully settled.

9.4.2.2 Specificity of Interactions

While OMF channel proteins are capable of interacting with a variety of transporter-PAP pairs, they maintain a level of selectivity for these interactions. This requirement for specificity imposes some limitations on the nature of the putative interaction interface, which may appear to be self-contradictory. First, the interface on different PAPs that share a common OMF channel must be very similar, or at least share similar crucial recognition determinants; second, the interfaces on PAPs that are not capable of functioning with the same channels must be quite different. For example, AcrA and MacA from *E. coli* both function with TolC [92, 93], but do not function with OprM from *Pseudomonas aeruginosa* [94]; MexA from *P. aeruginosa* functions with OprM but not TolC [95]. It would therefore be expected that the interfaces presented by AcrA and MacA would be very similar while that presented by MexA would be quite different. In *P. aeruginosa*, the requirement for specificity is exaggerated by the presence of multiple channel proteins and RND systems [96] that are not cross-reactive despite their structural similarity [97–99]. This necessity for expanded combinatorial recognition, as well as the considerations of purely energetic nature – as the PAP-OMF interaction has the principal contribution for the complex stability in majority of the models – suggest that such an interface must be fairly extensive.

Projecting these considerations into the two main models of assembly suggests that the interpenetrative bundling model may seem more likely than the tip-to-tip, given the relative sizes of their interfaces. As early evidence in support of the tip-to-tip model identified a conserved motif at the tip of the hairpin of several PAPs [100], it was proposed that this motif would be conserved in all PAPs and that it represented the key recognition determinant on the PAP side of the interface. However, such a highly conserved interface would seem to contradict the apparent specificity of the interaction, as all PAPs possessing this conserved motif should function with all of the channels that have a cognate PAP that contains it [59]. Supporting such interpretation, later developments showed that this motif is neither always present, nor always located at the tip of the PAP hairpin, even when comparing PAPs that pair with a common channel [101].

When examining the inter-component specificity within the pumps, it is important to separate the matters of interaction and functionality. A lack of activity for a set of proteins observed in a specific assay is not necessarily an indication of absence of complex formation. Indeed, several non-functional complexes do form from non-cognate components [67, 94, 95]. This indicates that while interaction interfaces are often sufficiently preserved to maintain the complex stability, a level of fine tuning is involved that separates the cognate from non-cognate partners and therefore non-functional from functional complexes.

9.4.2.3 Evidence in Support of Either Model

Historically, several approaches have been used to validate the proposed models. However, the recent advances in cryo-electron microscopy have provided perhaps the most compelling and visually appealing evidence to date and we will focus our discussion on these results.

The breakthrough work by Du et al. [89] revealed for the first time what appears to be a complete assembly composed of an OMF-trimer (TolC), a hexameric ring of PAPs (AcrA), and a trimeric RND transporter (AcrB) as well as an auxiliary membrane modulator protein AcrZ. The complex has been resolved to approximately 16 Å and its stoichiometric arrangement seems to be a standard 3 : 6 : 3 assembly. However, the structure also revealed a peculiar feature – the lipoyl and β-barrel domains of the PAPs formed a tight hexameric ring on top of the RND transporter, effectively isolating it from the outer membrane OMF-channel, and thus excluding a direct interaction between the two. A similar arrangement of PAPs on top of the RND-transporter was previously observed in the crystal structure of the CusBA metal-pump [54], but there, due to the unusually short hairpin domains of the PAP CusB and the lack of the OMF in the crystallized complex, it was difficult to deduce whether the OMF may still contact the RND transporter upon docking. In the Du et al. structure, however, the separation of the OMF from the top of the RND transporter is nearly 70 Å, which rules out any direct interaction between the two unless a dramatic rearrangement of the structure takes place. Importantly, the hairpin domains of the PAPs associate into a continuous tube, of similar diameter to that of the OMF, and although there is a limited interpenetration observed with the helical hairpins of the OMF, the cryo-EM structure appears to strongly support a tip-to-tip model of assembly with only a limited interaction interface between the two proteins.

Despite its visual appeal, electron microscopy faces some intrinsic limitations, pertaining to the requirements of sample monodispersity and stability, especially as very dilute samples are used for single particle reconstructions. The attempts to circumvent these bottlenecks have sometimes prompted the use of highly engineered proteins [89, 90, 102], raising questions regarding the physiological relevance of the complexes captured on the grid. Indeed, to date, only a single electron-microscopy study was able to reconstitute the entire complex assembly using wild-type proteins (Figure 9.6) [88]. While all these data seem to point toward a tip-to-tip mode of assembly, these studies

Figure 9.6 Model of the tripartite assembly fitted into experimental cryo-EM electron density map, suggesting a tip-to-tip like association of the PAP-OMF component in complex. *Source:* Docking based on data from [88] with permission.

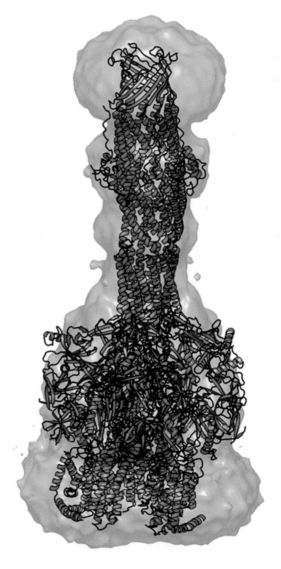

are yet to reach a level of resolution that could allow identification of specific side-chain interactions that might solve the mystery of reconciliation of the limited interaction interface with apparent high affinity and high specificity.

Additionally, as we will see later, despite their increasing popularity, the tip-to-tip interaction models seem to contradict a wealth of data generated using functional, biochemical, and mutagenesis approaches. One explanation for this conflict could be that the cryo-EM structures, which are captured in the absence of an energy source, represent a certain low-energy state of the complex, and alternative configurations of the complex are possible during its functional cycle.

Indeed, evidence from cross-linking, pull-down, and SPR experiments, as well as isothermal titration calorimetry show that the transporter and channel proteins are in close proximity and interact directly [103–105]. Perhaps most strikingly, the tip-to-tip models seem to directly contradict the apparent difference in affinities between PAPs with different hairpin lengths toward the same OMF as detected by SPR [86, 104].

Furthermore, cross-linking experiments have decisively shown that positions along the entire length of the PAP hairpin can be cross-linked to the OMF-channel, while conversely, positions on the OMF-barrel, as high as the equatorial domain, can be cross-linked to the PAP [85, 106]. Mutagenesis studies both with cognate partners and generating gain-of-function mutants to function with non-cognate components, show that positions along the entire PAP hairpin and throughout both the coiled-coil and equatorial domains of channel proteins are required for functionality [63, 67, 95, 107]. Taken together, these results seem to be incompatible with the tip-to-tip model. Such evidence for the interpenetrative model might indicate that *in vivo*, at least during a part of the cycle, the complex exists in a form consistent with the bundling model, while the conditions used *in vitro* are less favorable. Alternatively, these data perhaps hint at the possibility that the complex is highly dynamic during efflux, allowing structural switches between the two models.

9.4.3 Energy Requirements Within the Complex

To transport a substance against its concentration gradient, as is the case during antibiotic efflux, energy is required by the pump protein. As mentioned earlier, this energy may be provided either by ATP hydrolysis, the proton motive force, or another electrochemical gradient, dependent upon the pump type. The requirements for assembly of a PAP and OMF channel were first elucidated for a Type 1 secretion system, which is closely related to the tripartite efflux pump assembly, where the substrate was found to be required for assembly but ATPase activity was not [108]. If the same were true in multidrug transporters, then the transporter and the PAP would be expected to assemble as an inner membrane complex, which would then engage the channel protein upon binding of the substrate, regardless of whether or not an energy source was present. Investigations on the energy requirements of the assembly and efflux, however, revealed a rather complex pattern of dependencies.

OMF–transporter association: Reports of direct interaction of the OMFs and transporters are limited to the RND family and although most of them are based on circumstantial evidence [50, 91], some report short-range *in vivo* cross-linking indicative of tight association [103].

PAP–transporter association: PAPs clearly play an important role in activation of both the ABC and RND transporters [24, 77, 109, 110]. Specifically, the MPD domain of the PAPs has been shown to be responsible for stimulating the ATPase activity of the ABC transporters [77]. However, while the affinity of association between the PAP MacA and the ABC transporter MacB was found to increase upon ATP binding, it didn't appear to change during ATP hydrolysis [111]. Therefore, there is no obvious requirement for the ATP hydrolysis (and hence energy) for assembly of the ABC transporter-PAP pair.

PAP–OMF interactions: PAPs and RND transporters seem to associate even in the absence of the OMF and furthermore, the basal activity of the transporter is activated by the PAPs in the absence of OMFs [110]. However, *in vitro* studies using isolated components have demonstrated that the OMF enhances the activity of the RND transporter as well as the time that it pumps, but only if the PAP is also present ([109]). Logic dictates that the OMF channel must not be dilated prior to complex formation as it will compromise the permeability barrier of the outer membrane. In agreement with this, an artificially stabilized open-state channel was found unable to efficiently bind either the PAP or transporter protein [80]. Which component drives the opening of the channel and what is the energy requirement for that remained enigmatic until very recently.

One reasonable expectation for channel opening would be that the PAPs are transducing energy from the energized transporter to the OMF channel protein *via* conformational transition to induce opening of the channel. However, it has been conclusively demonstrated that opening of the channel requires no such transduction of conformational information. Indeed OMFs, including both TolC from *E. coli* and MtrE from *N. gonorrhoeae* have been shown to open in the absence of a functional cognate transporter protein, indicating that transporter activity and channel opening are not energetically coupled [15, 80]. The same studies also show that PAPs must be present to induce channel opening, in an energy-independent manner.

In RND complexes, each component was able to interact with each of the others, even in the absence of either the third component or a substrate [104]. The *in vitro* affinity of PAP–OMF interaction is dramatically enhanced at mildly acidic pH, which may suggest that either a protonation-dependent conformational switch is involved or that charged residues are involved in stabilization of the complex and are being titrated by the low pH [86]. However, the experiments described earlier, using isolated components, indicate that energy is not required to promote association. Consistent with that, the assembled structure did not require energy for its maintenance, either *in vivo*, where once assembled it can be extracted intact [80], or *in vitro* after reconstitution [88, 112]

So, what is the energy being used for? Perhaps somewhat counter-intuitively, it is required for re-setting of the complex and its dissociation after a successful efflux event. This has been revealed by using an MtrE-channel mutant that becomes open when its cognate PAP, MtrC, is also present, as witnessed by its hypersensitivity to vancomycin [15]. Cells expressing this MtrE mutant, MtrC, and the RND transporter MtrD appear to be vancomycin-sensitive (suggesting the MtrE channel is open), while those expressing an inactivated mutant of MtrD, in which the proton relay network is disrupted, are resistant. This is explained by the transporter occluding the end of the channel thus preventing vancomycin influx – the inactive mutant always "plugs" the leaky MtrE channel because the lack of energy transduction prevents complex dissociation, while the active transporter dissociates leaving the channel leaky, which allows the vancomycin to enter the cell. These experiments are explained visually in Figure 9.7.

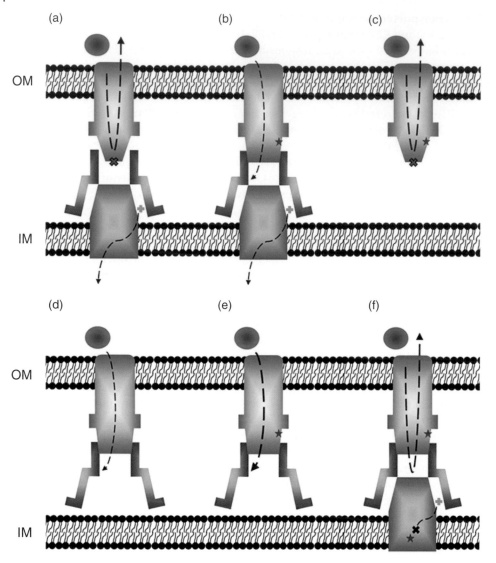

Figure 9.7 Cartoon demonstrating the results from Janganan et al. [15]. Light blue – MtrD RND-pump; brown – MtrC PAP; green – MtrE OMF; dark blue – the large efflux-restricted drug vancomycin; green cross – protons; red asterisks indicate the approximate positions of the described mutations; purple cross indicates that the channel is occluded and arrow weight indicates relative vancomycin diffusion rate through the channel. The wild-type MtrCDE complex (a) does not allow penetration of vancomycin. Introduction of the E434K mutation in MtrE to the entire complex (b) causes a low increase in susceptibility to vancomycin, although the mutation does not cause the channel to become constitutively open (c). MtrCE only (d), lacking the pump component, shows low-level susceptibility, indicating that the pump component is not required for channel opening. MtrC-MtrE(E434K) causes much higher susceptibility to vancomycin (e), suggesting that the mutation prevents channel closing. An inactive mutant of MtrD blocks the periplasmic end of the channel to prevent vancomycin entry to the cell (f) even when the channel cannot close – combined with the information from panels (b) and (e), this indicates that energy is required for disassembly of the complex and normal closure of the channel protein.

This interpretation of the energy requirement for dissociation of the complex has recently been confirmed directly using pull-down assays with both wild-type and proton-relay-deficient RND transporters in an *in vitro* system utilizing OMFs and transporter-PAP pairs incorporated into separate liposomes [112]. There, the trans-liposome tripartite OMF–PAP–transporter complex assembled in the absence of a pH-pulse, allowing for the channel to be pulled down. Addition of a pH-pulse creating a proton gradient across the liposome membrane and activating the transporter resulted in the loss of OMF association.

This dissociation of the complex is not an effect of the pH per se, as might be suspected, but rather a result of the proton-cycle-dependent conformational transition, as it is obliterated when RND pumps with an inactivated proton relay were used, and they remained associated even in the presence of a pH-pulse [112]. While this aspect of the cycle has been clarified, there are still questions remaining about how the actual substrate efflux is synchronized with this complex recycling event. While such a "reset system" may appear wasteful as clearly a longer pump association would increase processivity, there might be a functional reason for that. One plausible explanation might be that complex recycling is suppressed in the presence of substrates, and the assembled tripartite pump keeps on pumping until there are no more substrates present, which leads to an abortive cycle resulting in a complex reset. The fact that some OMFs, such as TolC, are used by numerous different efflux systems [61] might suggest that such a reset may allow the OMF to switch between different transporters depending on the prevalent substrate present at any time.

Although the use of an outer membrane channel protein allows extrusion of the drug directly into the extracellular environment, the channel protein is not essential for transporter activity, as seen *in vitro* [81]. A transporter, PAP, proton gradient, and a substrate must all be present, however, for this activity to be observable.

In summary, while all three pump components are required for a successful efflux event to take place, the specificity of cargo selection is predominantly determined by the PAP–transporter pair, which is also responsible for the energy utilization. This energy, at least within RND transporter-based efflux systems, appears to be required for both substrate translocation and for complex dissociation. Energy, however, is *not* required for complex assembly or channel opening.

9.5 Potential Drug Design

9.5.1 Efflux Pump Inhibitors (EPIs)

The central role that efflux pumps play in antimicrobial resistance, and RND transporter-based pumps in particular, makes them attractive targets for drug discovery. However, efforts to identify inhibitors of pump activity have yielded mixed results. While compounds have been identified that achieve this function, few are inhibitors of the pumps in the real sense, their effect instead coming by disruption of the energy source. An inhibitor commonly used in research studies, carbonyl cyanide m-chlorophenyl hydrazone (CCCP), is a protonophore acting by decoupling the proton gradient. While this makes it an effective inhibitor for *in vitro* studies, it is not a specific inhibitor of efflux pumps and hence cannot be considered for combinatorial therapy in clinical practice [113]. Similarly, non-hydrolysable ATP analogues could be used against ABC-based transporters; however, they also perturb nearly all cellular processes relying on ATP for energy.

There are, however, some substances that have been shown to both competitively and non-competitively bind to RND transporters, such as another inhibitor commonly used in research studies, the peptidomimetic phenylalanine-arginine-β-naphthylamide (PAβN). PAβN has been shown to competitively inhibit RND transporters such as AcrB [114, 115], thus potentiating the effect of antibiotics, especially fluoroquinolones. However, given that it is a substrate of AcrB [116], it is only effective at relatively high concentrations. Some researchers also question whether the apparent decrease in efflux activity caused by PAβN is purely due to its pump inhibition, as PAβN may permeabilize the outer membrane [117], thus increasing influx. Objective assessment is further complicated, as most commonly used experimental set-ups indirectly measure the net movement of molecules, without separation of influx or efflux, and hence the increased rate of influx may appear the same as a decrease in efflux activity. In addition, high levels of side effects and toxicity associated with the PAβN and related arylamines have precluded their development for medical practice [118].

Despite these setbacks, the EPI field, given its high promise, is rapidly developing and new classes of compounds are entering test stages. Along with less toxic arylamine derivatives, e.g., MC-04124 [119], other classes of EPIs have been actively screened, including arylpiperidines, pyranopyridine compounds (e.g., MBX2319), and pyridopyrimidine derivatives (e.g., D13-9001) [56, 120–123].

As discussed earlier, a key element of multidrug pump success is their low level of specific interaction with cargoes. Therefore, increasing the specificity or affinity of the substrate for the drug binding pockets of the transporter effectively turns this substrate into a competitive inhibitor. Crystallography and molecular dynamics studies have highlighted the importance of the distal binding pocket, especially its bottom part, the so-called "hydrophobic trap" in inhibitor binding [56, 115, 124]. Drug discovery has been further accelerated by the recent advances in our understanding of substrate binding and recognition by the efflux pumps, combined with computational studies, have allowed the predicting and further refining of EPI properties.

For example, this approach has been successfully used as a departure point for the design of a new generation of pyranopyridine EPIs utilizing crystallographic input and in silico prediction of their binding within a phenylalanine-rich cage adjacent to the distal binding pocket. The resulting new compounds show higher potency and fill the binding pocket to lock the conformational state of the protein, thereby preventing the peristaltic action of the RND-pump [125]. The hydrophobic trap in RND transporters may prove to be the most promising target for competitive inhibition, as it has recently been shown that PAβN also binds to the same region, although it also appears to interfere with the binding of other substrates to the upper part of the binding pocket [116].

Despite these advances, the highly adaptable nature of the drug binding pocket complicates the design of an effective competitive inhibitor. It would seemingly require only a relatively minor mutation to arise for the inhibitor to be rendered ineffective by altering efflux profiles; there have been precedents in clinical practice for such mutations arising [5]. This poses a threat to the potential clinical use of competitive EPIs. It also remains to be seen whether mutations in, or over-expression of, alternative efflux pumps may occur to ultimately mask the effect of the inhibitor in a clinical setting, as such an inhibitor may prove to be highly selective for only certain transporters.

9.5.2 Inhibiting Complex Assembly

As an alternative to inhibiting the pump activity directly, it may be possible to target assembly of the complex instead, at the level of the protein–protein interactions. Such an approach targets larger interfaces and is less susceptible to point-mutant adaptations of the bacteria, making re-emergence of resistance less likely.

While there are several theoretical interfaces (transporter–PAP, transporter–channel, PAP–channel, PAP–PAP), their experimental description remains a challenge. There are additional difficulties for the development of effective pump-assembly inhibitors – from the actual design of such a molecule, to the delivery of this potentially large molecule to the periplasm.

The efflux system is clearly dynamic, undergoing numerous conformational changes during the cycle and a potential peptidomimetic or interfering peptide may have to target a conformation that is not currently represented in the available structures, complicating design. Although molecular dynamics simulations are becoming more powerful and are useful at a single protein level, their predictive power diminishes with the increase of the complexity of the simulated system.

The PAP–transporter interface may be the most promising interface to be targeted as it is the best characterized and by far the largest; however, its disruption may not be trivial, precisely because of the extensive surface involved. On the other hand, the channel–PAP interface may present a good target, as irrespective of the association model, the helical hairpins of the channel protein need to undergo a significant rearrangement to dilate the channel, and helical bundles could plausibly be formed with the help of small helical peptides or peptidomimetics to stabilize its open form, thus creating leaky channels, and allowing access to the periplasm for large drugs such as vancomycin. Furthermore, such peptidomimetics might interfere directly with the PAP binding, effectively decoupling the pump assembly.

Regardless of the interaction that is being disrupted, the administration of such a protein–protein interaction inhibitor would not be trivial – the whole point of targeting a tripartite efflux pump is that it removes toxic substances to the outside of a permeability barrier; that same permeability barrier will also prevent access of the inhibitor to its site of action. Biotinylation of peptides has proven to enable peptides of up to 31 amino acids long to enter the Gram-negative cell, making use of the biotin uptake system, though this has not yet been shown to localize the peptide to the periplasm [126]. While large peptides generally have difficulty penetrating the Gram-negative membranes, some specialized import systems may be hijacked for the purpose, including colicins and siderophores, which provide a proven strategy for periplasmic cargo delivery with at least one compound – monosulfactam BAL30072 – in clinical trials [127]. In addition, sideromycins are widely spread natural conjugates of siderophores to antibiotics and are effective against Gram-negative bacteria [128]. Importantly, some naturally occurring microcins are delivered as siderophore-peptide conjugates and exceed in size even the largest of PAP hairpins, which they may mimic and competitively inhibit – *e.g.*, *Klebsiella*-secreted microcin E492m – an 84-residue toxin [129]. While functional protein–protein pump inhibitors are still some time away, the feasibility of the approach has been demonstrated by the success of specific targeting of the periplasmic part of the LptD protein in Gram-negative bacteria using non-membranolytic hairpin peptidomimetics [130]. These hairpins also showed potent antimicrobial activity in a mouse septicemia infection model, with median effective ED_{50} in the range of 0.25–0.5 mg kg^{-1}.

So, while at present both competitive and non-competitive efflux-pump inhibitors are still awaiting their entry into clinical practice, there is every reason to be optimistic that they will soon take their position among the growing arsenal of other antibiotic-potentiating treatments [131].

9.6   Impact of Multidrug Resistance

9.6.1   Clinical Impact

Our reliance on antibiotics has become so absolute that we no longer appreciate their value; once major killers such as typhoid fever and bacillary dysentery are now considered barely more than an inconvenience – if somebody contracts the illness, then antibiotics can treat it.

However, the continuing advance of Gram-negative MDR infections, such as *Pseudomonas, Acinetobacter*, and different *Enterobacteriaceae* species such as carbapenem-resistant *Klebsiella* causes justified alarm [132]. Such infections cause protracted hospitalization times, significantly increasing the cost of treatment and hence the economic burden of infection. MDR infections are also associated with much higher morbidity and mortality, with some reports citing mortality rates as high as 50% [133]. If unchecked, the spread of MDR may overwhelm health systems globally, killing up to 10 million people annually, and is projected to cost the global economy up to US$100 billion a year, or 3.5% of the global GDP, by 2050 [134].

Combating this requires a coordinated global effort to stem the rise of antimicrobial resistance. Antibiotics are a cornerstone of modern medicine, as even routine surgical operations are reliant upon antibiotics to prohibit post-surgical infection. At the same time, many health conditions cause either chronic or acute immunodeficiency, leaving patients susceptible to infections that would normally be overcome by the body's own defenses – for these patients too, antibiotics are essential survival tools. While resistance to one drug may be relatively trivial to overcome, MRD poses a serious threat to all disciplines of physical medicine.

9.6.2   Animal Impact

Intensive farming and meat production relies heavily on maintaining animals in high-density settings, which leads to an increased chance of infection spread. To control infection, antibiotics have been routinely applied for prophylaxis [135], as their use was also found to promote faster weight gain. Thus, farmed livestock is becoming one of the most significant focal points in studies into antibiotic resistance.

Although transfer of resistance is usually attributable to mobile genetic elements, continued exposure of bacteria to a mild selection pressure, as seen with prophylactic use of antibiotics, could overcome the general fitness cost associated with increased efflux pump expression. Bacteria with chromosomal mutations that increase efflux pump expression are therefore more likely to develop, survive, and spread with the continued prophylactic use of antibiotics. Although at a high metabolic cost, the expression of the MDR pumps allows enough time for additional, specific resistance mechanisms to develop. Such developments may ultimately be catastrophic for the meat industry. This impact of MDR infections is not limited only to the farmed species, as many bacterial pathogens can jump trans-species borders affecting wildlife, as well as humans, and reside in the environment [136], threatening entire species.

9.6.3 A Human Benefit from Multidrug Resistance Pumps

Given all the above, it may appear strange that MDR pumps may have benefits for humans in certain applications. Biotechnology frequently makes use of bacteria for a variety of tasks, from bioremediation to production of recombinant protein or biofuel. Utilization of MDR mechanisms can, in the correct conditions, enable new products to be made by bacteria or improve the efficiency of existing processes. The AcrAB–TolC MDR system has been mutated to improve secretion of butanol for production of biofuels [137], for example, and the future manipulation of MDR-pumps for human benefit is an intriguing possibility.

9.6.4 Why Is it So Difficult to Stop the Emergence and Spread of Multidrug Resistance?

The co-evolution of bacterial and human populations is a complex process in which the advent of antibiotics has temporarily distorted the balance in our favor. But while we have become dependent upon antibiotics to support our population growth, the bacteria have become dependent upon their resistance mechanisms to survive our use of antibiotics. It is a variation of the *Red Queen model of evolution*, in which a species must continue to evolve purely to survive the evolution of its ecological neighbors, except that in this case, the "evolution" of human use of antibiotics is technological rather than genetic.

However, after the triumphs of twentieth century medicine, the situation now appears to be reversing, with the bacteria winning this arms race, as their natural evolution of resistance has outpaced our ability to find new and effective drugs. The broad specificity of antibiotics in targeting different bacterial species also does not aid our cause. Our bodies naturally contain more bacterial than human cells, of many different species and performing a variety of commensal tasks. Antibacterial agents used to treat an infection do not only affect the pathogen in question, but most of the other bacteria present too, whether they are causing harm at the time or not. This causes directionality in the evolution of the microbiota (toward antibiotic resistant) and provides clearances and new bacterially naïve niches to open for colonization by bacteria that were not previously a part of the native microbiota, increasing the threat for subsequent infection.

Multidrug efflux pumps are a major contributor to MDR, but they also have other, possibly more ancient, functions that were discovered subsequent to their role in drug resistance [138]. These functions are likely linked to the primary roles of these pumps in bacterial physiology. Roles in quorum sensing [139], biofilm formation [140], and host colonization [141], all mean that multidrug efflux pumps play an essential role outside the laboratory environment. The capacity of a number of multidrug efflux pumps, including MFS-based and RND-based transporters to pump non-polar and detergent-like molecules likely reflects their biological role in surviving the host's digestive tract conditions, at least in *Enterobacteriaceae*, which may have evolved efficient pumping systems to withstand the detergent-like bile salts [142].

9.7   Summary

Bacteria, especially Gram-negative bacteria, have evolved their cellular envelopes into formidable permeability barriers. They possess multilayered defense mechanisms, which combine basic permeability barriers, limiting the entry of toxic compounds into the cell, with active transport of noxious compounds against their respective concentration gradients out of the cell. This active efflux is performed by so-called multidrug efflux pumps, which are energized multi-protein molecular machines with broad substrate specificity. These pumps contribute significantly to MDR, which has become a major threat to human and animal health globally.

Advances in the structural biology of the pumps have elucidated all the pump components in isolation and have allowed the proposition of several models of the full assembly, greatly clarifying the energy requirements and mechanisms of efflux; however, future research will have to deliver a conclusive model of these dynamic complexes. A better understanding of the mechanisms of their function and assembly paves the way for targeting efflux pumps in a combinatorial therapy approach and specifically tailored efflux pump inhibitors promise to be an important tool in the race to overcome the rise of MDR.

References

1 Fox, R. (1996). The post-antibiotic era beckons. *Journal of the Royal Society of Medicine* 89: 602–603.

2 World Health Organization (2014). *Antimicrobial Resistance: Global Report on Surveillance*. Geneva: WHO Press.

3 Gupta, R.S. (2011). Origin of diderm (Gram-negative) bacteria: antibiotic selection pressure rather than endosymbiosis likely led to the evolution of bacterial cells with two membranes. *Antonie van Leeuwenhoek International Journal of General and Molecular Microbiology* 100: 171–182.

4 Danilchanka, O., Pires, D., Anes, E., and Niederweis, M. (2015). The *Mycobacterium tuberculosis* outer membrane channel protein CpnT confers susceptibility to toxic molecules. *Antimicrobial Agents and Chemotherapy* 59: 2328–2336.

5 Blair, J.M.A., Bavro, V.N., Ricci, V. et al. (2015a). AcrB drug-binding pocket substitution confers clinically relevant resistance and altered substrate specificity. *Proceedings of the National Academy of Sciences of the United States of America* 112: 3511–3516.

6 Blair, J.M.A., Smith, H.E., Ricci, V. et al. (2015b). Expression of homologous RND efflux pump genes is dependent upon AcrB expression: implications for efflux and virulence inhibitor design. *Journal of Antimicrobial Chemotherapy* 70: 424–431.

7 Piddock, L. (2006). Clinically relevant chromosomally encoded multidrug resistance efflux pumps in bacteria. *Clinical Microbiology Reviews* 19: 382–402.

8 Caroff, M. and Karibian, D. (2003). Structure of bacterial lipopolysaccharides. *Carbohydrate Research* 338: 2431–2447.

9 Tsujimoto, H., Gotoh, N., and Nishino, T. (1999). Diffusion of macrolide antibiotics through the outer membrane of *Moraxella catarrhalis*. *Journal of Infection and Chemotherapy* 5: 196–200.

10 Bhamidimarri, S.P., Prajapati, J.D., van den Berg, B. et al. (2016). Role of Electroosmosis in the permeation of neutral molecules: CymA and Cyclodextrin as an example. *Biophysical Journal* 110: 600–611.

11 Lepore, B.W., Indic, M., Pham, H. et al. (2011). Ligand-gated diffusion across the bacterial outer membrane. *Proceedings of the National Academy of Sciences of the United States of America* 108: 10121–10126.

12 Delcour, A.H. (2009). Outer membrane permeability and antibiotic resistance. *Biochimica Et Biophysica Acta-Proteins and Proteomics* 1794: 808–816.

13 van den Berg, B., Bhamidimarri, S.P., Prajapati, J.D. et al. (2015). Outer-membrane translocation of bulky small molecules by passive diffusion. *Proceedings of the National Academy of Sciences of the United States of America* 112: E2991–E2999.

14 Nikaido, H. (2003). Molecular basis of bacterial outer membrane permeability revisited. *Microbiology and Molecular Biology Reviews* 67: 593–656.

15 Janganan, T.K., Bavro, V.N., Zhang, L. et al. (2013). Tripartite efflux pumps: energy is required for dissociation, but not assembly or opening of the outer membrane channel of the pump. *Molecular Microbiology* 88: 590–602.

16 Marchand, C.H., Salmeron, C., Raad, R.B. et al. (2012). Biochemical disclosure of the mycolate outer membrane of *Corynebacterium glutamicum*. *Journal of Bacteriology* 194: 587–597.

17 Laneelle, M., Tropis, M., and Daffe, M. (2013). Current knowledge on mycolic acids in *Corynebacterium glutamicum* and their relevance for biotechnological processes. *Applied Microbiology and Biotechnology* 97: 9923–9930.

18 Yang, Y., Shi, F., Tao, G., and Wang, X. (2012). Purification and structure analysis of mycolic acids in *Corynebacterium glutamicum*. *Journal of Microbiology* 50: 235–240.

19 Burkovski, A. (2013). Cell envelope of corynebacteria: structure and influence on pathogenicity. *ISRN Microbiology* 2013: 935736.

20 Costa-Riu, N., Burkovski, A., Kramer, R., and Benz, R. (2003). PorA represents the major cell wall channel of the gram-positive bacterium *Corynebacterium glutamicum*. *Journal of Bacteriology* 185: 4779–4786.

21 Frenzel, E., Schmidt, S., Niederweis, M., and Steinhauer, K. (2011). Importance of Porins for biocide efficacy against *Mycobacterium smegmatis*. *Applied and Environmental Microbiology* 77: 3068–3073.

22 Petrus, A.K., Swithers, K.S., Ranjit, C. et al. (2012). Genes for the major structural components of *Thermotogales* species' togas revealed by proteomic and evolutionary analyses of OmpA and OmpB homologs. *PLoS One* 7: e40236.

23 Saier, M.H. Jr., Reddy, V.S., Tamang, D.G., and Vaestermark, A. (2014). The transporter classification database. *Nucleic Acids Research* 42: D251–D258.

24 Lin, H.T., Bavro, V.N., Barrera, N.P. et al. (2009). MacB ABC transporter is a dimer whose ATPase activity and macrolide-binding capacity are regulated by the membrane fusion protein MacA. *Journal of Biological Chemistry* 284: 1145–1154.

25 Okada, U., Yamashita, E., Neuberger, A. et al. (2017). Crystal structure of tripartite-type ABC transporter MacB from *Acinetobacter baumannii*. *Nature Communications* 8: 1336.

26 Kobayashi, N., Nishino, K., Hirata, T., and Yamaguchi, A. (2003). Membrane topology of ABC-type macrolide antibiotic exporter MacB in *Escherichia coli*. *FEBS Letters* 546: 241–246.

27 Xu, Y., Sim, S., Nam, K.H. et al. (2009). Crystal structure of the periplasmic region of MacB, a noncanonic ABC transporter. *Biochemistry* 48: 5218–5225.

28 Lu, S. and Zgurskaya, H.I. (2013). MacA, a periplasmic membrane fusion protein of the macrolide transporter MacAB-TolC, binds lipopolysaccharide core specifically and with high affinity. *Journal of Bacteriology* 195: 4865–4872.

29 Seeger, M.A. and van Veen, H.W. (2009). Molecular basis of multidrug transport by ABC transporters. *Biochimica Et Biophysica Acta-Proteins and Proteomics* 1794: 725–737.

30 Yin, Y., He, X., Szewczyk, P. et al. (2006). Structure of the multidrug transporter EmrD from *Escherichia coli*. *Science* 312: 741–744.

31 Law, C.J., Maloney, P.C., and Wang, D. (2008). Ins and outs of major facilitator superfamily, antiporters. *Annual Review of Microbiology* 62: 289–305.

32 Bay, D.C., Rommens, K.L., and Turner, R.J. (2008). Small multidrug resistance proteins: a multidrug transporter family that continues to grow. *Biochimica Et Biophysica Acta-Biomembranes* 1778: 1814–1838.

33 Bay, D.C., Budiman, R.A., Nieh, M., and Turner, R.J. (2010). Multimeric forms of the small multidrug resistance protein EmrE in anionic detergent. *Biochimica Et Biophysica Acta-Biomembranes* 1798: 526–535.

34 Woodall, N.B., Yin, Y., and Bowie, J.U. (2015). Dual-topology insertion of a dual-topology membrane protein. *Nature Communications* 6: 8099.

35 Dastvan, R., Fischer, A.W., Mishra, S. et al. (2016). Protonation-dependent conformational dynamics of the multidrug transporter EmrE. *Proceedings of the National Academy of Sciences of the United States of America* 113: 1220–1225.

36 Gayen, A., Leninger, M., and Traasethl, N.J. (2016). Protonation of a glutamate residue modulates the dynamics of the drug transporter EmrE. *Nature Chemical Biology* 12: 141–145.

37 Ogawa, W., Minato, Y., Dodan, H. et al. (2015). Characterization of MATE-type multidrug efflux pumps from *Klebsiella pneumoniae* MGH78578. *PLoS One* 10: e0121619.

38 Kuroda, T. and Tsuchiya, T. (2009). Multidrug efflux transporters in the MATE family. *Biochimica Et Biophysica Acta-Proteins and Proteomics* 1794: 763–768.

39 He, X., Szewczyk, P., Karyakin, A. et al. (2010). Structure of a cation-bound multidrug and toxic compound extrusion transporter. *Nature* 467: 991–994.

40 Lu, M. (2016). Structures of multidrug and toxic compound extrusion transporters and their mechanistic implications. *Channels* 10: 88–100.

41 Lu, M., Radchenko, M., Symersky, J. et al. (2013a). Structural insights into H$^+$-coupled multidrug extrusion by a MATE transporter. *Nature Structural & Molecular Biology* 20: 1310–1317.

42 Lu, M., Symersky, J., Radchenko, M. et al. (2013b). Structures of a Na$^+$-coupled, substrate-bound MATE multidrug transporter. *Proceedings of the National Academy of Sciences of the United States of America* 110: 2099–2104.

43 Hassan, K.A., Elbourne, L.D.H., Li, L. et al. (2015a). An ace up their sleeve: a transcriptomic approach exposes the AceI efflux protein of *Acinetobacter baumannii* and reveals the drug efflux potential hidden in many microbial pathogens. *Frontiers in Microbiology* 6: 333.

44 Hassan, K.A., Liu, Q., Henderson, P.J.F., and Paulsen, I.T. (2015b). Homologs of the *Acinetobacter baumannii* AceI transporter represent a new family of bacterial multidrug efflux systems. *mBio* 6: e01982–e01914.

45 Blair, J.M.A., Webber, M.A., Baylay, A.J. et al. (2015c). Molecular mechanisms of antibiotic resistance. *Nature Reviews Microbiology* 13: 42–51.

46 Varela, C., Rittmann, D., Singh, A. et al. (2012). MmpL genes are associated with mycolic acid metabolism in *Mycobacteria* and *Corynebacteria*. *Chemistry & Biology* 19: 498–506.

47 Yang, L., Lu, S., Belardinelli, J. et al. (2014). RND transporters protect *Corynebacterium glutamicum* from antibiotics by assembling the outer membrane. *MicrobiologyOpen* 3: 484–496.

48 Murakami, S., Nakashima, R., Yamashita, E., and Yamaguchi, A. (2002). Crystal structure of bacterial multidrug efflux transporter AcrB. *Nature* 419: 587–593.

49 Takatsuka, Y. and Nikaido, H. (2009). Covalently linked trimer of the AcrB multidrug efflux pump provides support for the functional rotating mechanism. *Journal of Bacteriology* 191: 1729–1737.

50 Weeks, J.W., Bavro, V.N., and Misra, R. (2014). Genetic assessment of the role of AcrB beta hairpins in the assembly of the TolC-AcrAB multidrug efflux pump of *Escherichia coli*. *Molecular Microbiology* 91: 965–975.

51 Murakami, S., Nakashima, R., Yamashita, E. et al. (2006). Crystal structures of a multidrug transporter reveal a functionally rotating mechanism. *Nature* 443: 173–179.

52 Seeger, M.A., Schiefner, A., Eicher, T. et al. (2006). Structural asymmetry of AcrB trimer suggests a peristaltic pump mechanism. *Science* 313: 1295–1298.

53 Eicher, T., Seeger, M.A., Anselmi, C. et al. (2014). Coupling of remote alternating-access transport mechanisms for protons and substrates in the multidrug efflux pump AcrB. *eLife* 3: https://doi.org/10.7554/eLife.03145.

54 Su, C., Long, F., Zimmermann, M.T. et al. (2011). Crystal structure of the CusBA heavy-metal efflux complex of *Escherichia coli*. *Nature* 470: 558–562.

55 Eicher, T., Cha, H., Seeger, M.A. et al. (2012). Transport of drugs by the multidrug transporter AcrB involves an access and a deep binding pocket that are separated by a switch-loop. *Proceedings of the National Academy of Sciences of the United States of America* 109: 5687–5692.

56 Nakashima, R., Sakurai, K., Yamasaki, S. et al. (2013). Structural basis for the inhibition of bacterial multidrug exporters. *Nature* 500: 102–106.

57 Nakashima, R., Sakurai, K., Yamasaki, S. et al. (2011). Structures of the multidrug exporter AcrB reveal a proximal multisite drug-binding pocket. *Nature* 480: 565–569.

58 Zgurskaya, H.I., Krishnamoorthy, G., Ntreh, A., and Lu, S. (2011). Mechanism and function of the outer membrane channel TolC in multidrug resistance and physiology of Enterobacteria. *Frontiers in Microbiology* 2: 189–189.

59 Symmons, M.F., Marshall, R.L., and Bavro, V.N. (2015). Architecture and roles of periplasmic adaptor proteins in tripartite efflux assemblies. *Frontiers in Microbiology* 6: 513.

60 Hobbs, E.C., Yin, X., Paul, B.J. et al. (2012). Conserved small protein associates with the multidrug efflux pump AcrB and differentially affects antibiotic resistance. *Proceedings of the National Academy of Sciences of the United States of America* 109: 16696–16701.

61 Horiyama, T., Yamaguchi, A., and Nishino, K. (2010). TolC dependency of multidrug efflux systems in *Salmonella enterica* serovar Typhimurium. *Journal of Antimicrobial Chemotherapy* 65: 1372–1376.

62 Koronakis, V., Sharff, A., Koronakis, E. et al. (2000). Crystal structure of the bacterial membrane protein TolC central to multidrug efflux and protein export. *Nature* 405: 914–919.

63 Yamanaka, H., Morisada, N., Miyano, M. et al. (2004). Amino-acid residues involved in the expression of the activity of *Escherichia coli* TolC. *Microbiology and Immunology* 48: 713–722.

64 Andersen, C., Koronakis, E., Bokma, E. et al. (2002a). Transition to the open state of the TolC periplasmic tunnel entrance. *Proceedings of the National Academy of Sciences of the United States of America* 99: 11103–11108.

65 Bavro, V.N., Pietras, Z., Furnham, N. et al. (2008). Assembly and channel opening in a bacterial drug efflux machine. *Molecular Cell* 30: 114–121.

66 Andersen, C., Koronakis, E., Hughes, C., and Koronakis, V. (2002b). An aspartate ring at the TolC tunnel entrance determines ion selectivity and presents a target for blocking by large cations. *Molecular Microbiology* 44: 1131–1139.

67 Vediyappan, G., Borisova, T., and Fralick, J. (2006). Isolation and characterization of VceC gain-of-function mutants that can function with the AcrAB multiple-drug-resistant efflux pump of *Escherichia coli*. *Journal of Bacteriology* 188: 3757–3762.

68 Janganan, T.K., Zhang, L., Bavro, V.N. et al. (2011b). Opening of the outer membrane protein channel in tripartite efflux pumps is induced by interaction with the membrane fusion partner. *Journal of Biological Chemistry* 286: 5484–5493.

69 Symmons, M.F., Bokma, E., Koronakis, E. et al. (2009). The assembled structure of a complete tripartite bacterial multidrug efflux pump. *Proceedings of the National Academy of Sciences of the United States of America* 106: 7173–7178.

70 Akama, H., Matsuura, T., Kashiwagi, S. et al. (2004). Crystal structure of the membrane fusion protein, MexA, of the multidrug transporter in *Pseudomonas aeruginosa*. *Journal of Biological Chemistry* 279: 25939–25942.

71 Higgins, M., Bokma, E., Koronakis, E. et al. (2004). Structure of the periplasmic component of a bacterial drug efflux pump. *Proceedings of the National Academy of Sciences of the United States of America* 101: 9994–9999.

72 Mikolosko, J., Bobyk, K., Zgurskaya, H.I., and Ghosh, P. (2006). Conformational flexibility in the multidrug efflux system protein AcrA. *Structure* 14: 577–587.

73 Greene, N.P., Hinchliffe, P., Crow, A. et al. (2013). Structure of an atypical periplasmic adaptor from a multidrug efflux pump of the spirochete *Borrelia burgdorferi*. *FEBS Letters* 587: 2984–2988.

74 Bunikis, I., Denker, K., Ostberg, Y. et al. (2008). An RND-type efflux system in *Borrelia burgdorferi* is involved in virulence and resistance to antimicrobial compounds. *PLoS Pathogens* 4: e1000009.

75 Zgurskaya, H.I., Yamada, Y., Tikhonova, E.B. et al. (2009). Structural and functional diversity of bacterial membrane fusion proteins. *Biochimica Et Biophysica Acta-Proteins and Proteomics* 1794: 794–807.

76 Krishnamoorthy, G., Tikhonova, E.B., and Zgurskaya, H.I. (2008). Fitting periplasmic membrane fusion proteins to inner membrane transporters: mutations that enable *Escherichia coli* AcrA to function with *Pseudomonas aeruginosa* MexB. *Journal of Bacteriology* 190: 691–698.

77 Modali, S.D. and Zgurskaya, H.I. (2011). The periplasmic membrane proximal domain of MacA acts as a switch in stimulation of ATP hydrolysis by MacB transporter. *Molecular Microbiology* 81: 937–951.

78 De Angelis, F., Lee, J.K., O'Connell, J.D. III et al. (2010). Metal-induced conformational changes in ZneB suggest an active role of membrane fusion proteins in efflux resistance systems. *Proceedings of the National Academy of Sciences of the United States of America* 107: 11038–11043.

79 Ge, Q., Yamada, Y., and Zgurskaya, H. (2009). The C-terminal domain of AcrA is essential for the assembly and function of the multidrug efflux pump AcrAB-TolC. *Journal of Bacteriology* 191: 4365–4371.

80 Tikhonova, E.B. and Zgurskaya, H.I. (2004). AcrA, AcrB, and TolC of *Escherichia coli* form a stable intermembrane multidrug efflux complex. *Journal of Biological Chemistry* 279: 32116–32124.

81 Verchère, A., Broutin, I., and Picard, M. (2012). Photo-induced proton gradients for the *in vitro* investigation of bacterial efflux pumps. *Scientific Reports* 2: 306.

82 Fernandez-Recio, J., Walas, F., Federici, L. et al. (2004). A model of a transmembrane drug-efflux pump from Gram-negative bacteria. *FEBS Letters* 578: 5–9.

83 Yum, S., Xu, Y., Piao, S. et al. (2009). Crystal structure of the periplasmic component of a tripartite macrolide-specific efflux pump. *Journal of Molecular Biology* 387: 1286–1297.

84 Xu, Y., Sim, S., Song, S. et al. (2010). The tip region of the MacA alpha-hairpin is important for the binding to TolC to the *Escherichia coli* MacAB-TolC pump. *Biochemical and Biophysical Research Communications* 394: 962–965.

85 Janganan, T.K., Bavro, V.N., Zhang, L. et al. (2011a). Evidence for the assembly of a bacterial tripartite multidrug pump with a stoichiometry of 3:6:3. *Journal of Biological Chemistry* 286: 26900–26912.

86 Tikhonova, E.B., Dastidar, V., Rybenkov, V.V., and Zgurskaya, H.I. (2009). Kinetic control of TolC recruitment by multidrug efflux complexes. *Proceedings of the National Academy of Sciences of the United States of America* 106: 16416–16421.

87 Weeks, J.W., Nickels, L.M., Ntreh, A.T., and Zgurskaya, H.I. (2015). Non-equivalent roles of two periplasmic subunits in the function and assembly of triclosan pump TriABC from *Pseudomonas aeruginosa*. *Molecular Microbiology* 98: 343–356.

88 Daury, L., Orange, F., Taveau, J. et al. (2016). Tripartite assembly of RND multidrug efflux pumps. *Nature Communications* 7: 10731.

89 Du, D., Wang, Z., James, N.R. et al. (2014). Structure of the AcrAB-TolC multidrug efflux pump. *Nature* 509: 512–515.

90 Kim, J., Jeong, H., Song, S. et al. (2015). Structure of the tripartite multidrug efflux pump AcrAB-TolC suggests an alternative assembly mode. *Molecules and Cells* 38: 180–186.

91 Hayashi, K., Nakashima, R., Sakurai, K. et al. (2016). AcrB-AcrA fusion proteins that act as multidrug efflux transporters. *Journal of Bacteriology* 198: 332–342.

92 Fralick, J.A. (1996). Evidence that TolC is required for functioning of the Mar/AcrAB efflux pump of *Escherichia coli*. *Journal of Bacteriology* 178: 5803–5805.

93 Kobayashi, N., Nishino, K., and Yamaguchi, A. (2001). Novel macrolide-specific ABC-type efflux transporter in *Escherichia coli*. *Journal of Bacteriology* 183: 5639–5644.

94 Stegmeier, J.F., Polleichtner, G., Brandes, N. et al. (2006). Importance of the adaptor (membrane fusion) protein hairpin domain for the functionality of multidrug efflux pumps. *Biochemistry* 45: 10303–10312.

95 Bokma, E., Koronakis, E., Lobedanz, S. et al. (2006). Directed evolution of a bacterial efflux pump: adaptation of the E-coli TolC exit duct to the *Pseudomonas* MexAB translocase. *FEBS Letters* 580: 5339–5343.

96 Yoshihara, E., Maseda, H., and Saito, K. (2002). The outer membrane component of the multidrug efflux pump from *Pseudomonas aeruginosa* may be a gated channel. *European Journal of Biochemistry* 269: 4738–4745.

97 Maseda, H., Yoneyama, H., and Nakae, T. (2000). Assignment of the substrate-selective subunits of the MexEF-OprN multidrug efflux pump of *Pseudomonas aeruginosa*. *Antimicrobial Agents and Chemotherapy* 44: 658–664.

98 Phan, G., Picard, M., and Broutin, I. (2015). Focus on the outer membrane factor OprM, the forgotten player from efflux pumps assemblies. *Antibiotics-Basel* 4: 544–566.

99 Yonehara, R., Yamashita, E., and Nakagawa, A. (2016). Crystal structures of OprN and OprJ, outer membrane factors of multidrug tripartite efflux pumps of *Pseudomonas aeruginosa*. *Proteins* 84 (6): 759–769.

100 Kim, H., Xu, Y., Lee, M. et al. (2010). Functional relationships between the AcrA hairpin tip region and the TolC aperture tip region for the formation of the bacterial tripartite efflux pump AcrAB-TolC. *Journal of Bacteriology* 192: 4498–4503.

101 Kim, J., Song, S., Lee, M. et al. (2016). Crystal structure of a soluble fragment of the membrane fusion protein HlyD in a type I secretion system of Gram-negative bacteria. *Structure* 24: 477–485.

102 Jeong, H., Kim, J., Song, S. et al. (2016). Pseudoatomic structure of the tripartite multidrug efflux pump AcrAB-TolC reveals the intermeshing cogwheel-like interaction between AcrA and TolC. *Structure* 24: 272–276.

103 Tamura, N., Murakami, S., Oyama, Y. et al. (2005). Direct interaction of multidrug efflux transporter AcrB and outer membrane channel TolC detected *via* site-directed disulfide cross-linking. *Biochemistry* 44: 11115–11121.

104 Tikhonova, E.B., Yamada, Y., and Zgurskaya, H.I. (2011). Sequential mechanism of assembly of multidrug efflux pump AcrAB-ToIC. *Chemistry & Biology* 18: 454–463.

105 Touzé, T., Eswaran, J., Bokma, E. et al. (2004). Interactions underlying assembly of the *Escherichia coli* AcrAB-TolC multidrug efflux system. *Molecular Microbiology* 53: 697–706.

106 Lobedanz, S., Bokma, E., Symmons, M.F. et al. (2007). A periplasmic coiled-coil interface underlying ToIC recruitment and the assembly of bacterial drug efflux pumps. *Proceedings of the National Academy of Sciences of the United States of America* 104: 4612–4617.

107 Lee, M., Jun, S., Yoon, B. et al. (2012). Membrane fusion proteins of type I secretion system and tripartite efflux pumps share a binding motif for TolC in Gram-negative bacteria. *PLoS One* 7: e40460.

108 Thanabalu, T., Koronakis, E., Hughes, C., and Koronakis, V. (1998). Substrate-induced assembly of a contiguous channel for protein export from *E. coli*: reversible bridging of an inner-membrane translocase to an outer membrane exit pore. *EMBO Journal* 17: 6487–6496.

109 Verchère, A., Dezi, M., Adrien, V. et al. (2015). *In vitro* transport activity of the fully assembled MexAB-OprM efflux pump from *Pseudomonas aeruginosa*. *Nature Communications* 6: 6890.

110 Zgurskaya, H. and Nikaido, H. (1999). Bypassing the periplasm: reconstitution of the AcrAB multidrug efflux pump of *Escherichia coli*. *Proceedings of the National Academy of Sciences of the United States of America* 96: 7190–7195.

111 Lu, S. and Zgurskaya, H.I. (2012). Role of ATP binding and hydrolysis in assembly of MacAB-TolC macrolide transporter. *Molecular Microbiology* 86: 1132–1143.

112 Enguéné, V.Y.N., Verchère, A., Phan, G. et al. (2015). Catch me if you can: a biotinylated proteoliposome affinity assay for the investigation of assembly of the MexA-MexB-OprM efflux pump from *Pseudomonas aeruginosa*. *Frontiers in Microbiology* 6: 541.

113 Coldham, N.G., Webber, M., Woodward, M.J., and Piddock, L.J.V. (2010). A 96-well plate fluorescence assay for assessment of cellular permeability and active efflux in *Salmonella enterica* serovar Typhimurium and *Escherichia coli*. *Journal of Antimicrobial Chemotherapy* 65: 1655–1663.

114 Renau, T., Leger, R., Flamme, E. et al. (1999). Inhibitors of efflux pumps in *Pseudomonas aeruginosa* potentiate the activity of the fluoroquinolone antibacterial levofloxacin. *Journal of Medicinal Chemistry* 42: 4928–4931.

115 Vargiu, A.V., Ruggerone, P., Opperman, T.J. et al. (2014). Molecular mechanism of MBX2319 inhibition of *Escherichia coli* AcrB multidrug efflux pump and comparison with other inhibitors. *Antimicrobial Agents and Chemotherapy* 58: 6224–6234.

116 Kinana, A.D., Vargiu, A.V., May, T., and Nikaido, H. (2016). Aminoacyl beta-naphthylamides as substrates and modulators of AcrB multidrug efflux pump. *Proceedings of the National Academy of Sciences of the United States of America* 113: 1405–1410.

117 Lamers, R.P., Cavallari, J.F., and Burrows, L.L. (2013). The efflux inhibitor phenylalanine-arginine beta-naphthylamide (PAβN) permeabilizes the outer membrane of Gram-negative bacteria. *PLoS One* 8: e60666.

118 Lomovskaya, O. and Bostian, K. (2006). Practical applications and feasibility of efflux pump inhibitors in the clinic – a vision for applied use. *Biochemical Pharmacology* 71: 910–918.

119 Watkins, W., Landaverry, Y., Leger, R. et al. (2003). The relationship between physicochemical properties, *in vitro* activity and pharmacokinetic profiles of analogues of diamine-containing efflux pump inhibitors. *Bioorganic & Medicinal Chemistry Letters* 13: 4241–4244.

120 Nakayama, K., Ishida, Y., Ohtsuka, M. et al. (2003). MexAB-OprM-specific efflux pump inhibitors in *Pseudomonas aeruginosa*. Part 1: discovery and early strategies for lead optimization. *Bioorganic & Medicinal Chemistry Letters* 13: 4201–4204.

121 Nguyen, S.T., Kwasny, S.M., Ding, X. et al. (2015). Structure activity relationships of a novel pyranopyridine series of Gram-negative bacterial efflux pump inhibitors. *Bioorganic & Medicinal Chemistry* 23: 2024–2034.

122 Opperman, T.J., Kwasny, S.M., Kim, H. et al. (2014). Characterization of a novel pyranopyridine inhibitor of the AcrAB efflux pump of *Escherichia coli*. *Antimicrobial Agents and Chemotherapy* 58: 722–733.

123 Thorarensen, A., Presley-Bodnar, A., Marotti, K. et al. (2001). 3-Arylpiperidines as potentiators of existing antibacterial agents. *Bioorganic & Medicinal Chemistry Letters* 11: 1903–1906.

124 Vargiu, A.V. and Nikaido, H. (2012). Multidrug binding properties of the AcrB efflux pump characterized by molecular dynamics simulations. *Proceedings of the National Academy of Sciences of the United States of America* 109: 20637–20642.

125 Sjuts, H., Vargiu, A.V., Kwasny, S.M. et al. (2016). Molecular basis for inhibition of AcrB multidrug efflux pump by novel and powerful pyranopyridine derivatives. *Proceedings of the National Academy of Sciences of the United States of America* 113: 3509–3514.

126 Walker, J. and Altman, E. (2005). Biotinylation facilitates the uptake of large peptides by *Escherichia coli* and other Gram-negative bacteria. *Applied and Environmental Microbiology* 71: 1850–1855.

127 Page, M.G. (2013). Siderophore conjugates. *Annals of New York Academy of Sciences* 1277: 115–126.

128 Braun, V., Pramanik, A., Gwinner, T. et al. (2009). Sideromycins: tools and antibiotics. *Biometals* 22: 3–13.

129 Nolan, E.M., Fischbach, M.A., Koglin, A., and Walsh, C.T. (2007). Biosynthetic tailoring of microcin e492m: post-translational modification affords an antibacterial siderophore-peptide conjugate. *Journal of the American Chemical Society* 129: 14336–14347.

130 Srinivas, N., Jetter, P., Ueberbacher, B.J. et al. (2010). Peptidomimetic antibiotics target outer-membrane biogenesis in *Pseudomonas aeruginosa*. *Science* 327: 1010–1013.

131 Zabawa, T.P., Pucci, M.J., Parr, T.R.J., and Lister, T. (2016). Treatment of Gram-negative bacterial infections by potentiation of antibiotics. *Current Opinion in Microbiology* 24: 7–12.

132 Munoz-Price, L.S., Poirel, L., Bonomo, R.A. et al. (2013). Clinical epidemiology of the global expansion of *Klebsiella pneumoniae* carbapenemases. *Lancet Infectious Diseases* 13: 785–796.

133 Thabit, A.K., Crandon, J.L., and Nicolau, D.P. (2015). Antimicrobial resistance: impact on clinical and economic outcomes and the need for new antimicrobials. *Expert Opinion on Pharmacotherapy* 16: 159–177.

134 AMR Review (2014). *Antimicrobial Resistance: Tackling a Crisis for the Health and Wealth of Nations.* UK: AMR-Review.

135 Callens, B., Persoons, D., Maes, D. et al. (2012). Prophylactic and metaphylactic antimicrobial use in Belgian fattening pig herds. *Preventive Veterinary Medicine* 106: 53–62.

136 da Costa, P.M., Loureiro, L., and Matos, A.J.F. (2013). Transfer of multidrug-resistant bacteria between intermingled ecological niches: the Interface between humans, animals and the environment. *International Journal of Environmental Research and Public Health* 10: 278–294.

137 Fisher, M.A., Boyarskiy, S., Yamada, M.R. et al. (2014). Enhancing tolerance to short-chain alcohols by engineering the *Escherichia coli* AcrB efflux pump to secrete the non-native substrate n-butanol. *ACS Synthetic Biology* 3: 30–40.

138 Poole, K. (2008). Bacterial multidrug efflux pumps serve other functions. *Microbe* 3: 179–185.

139 Minagawa, S., Inami, H., Kato, T. et al. (2012). RND type efflux pump system MexAB-OprM of *Pseudomonas aeruginosa* selects bacterial languages, 3-oxo-acyl-homoserine lactones, for cell-to-cell communication. *BMC Microbiology* 12: 70.

140 Baugh, S., Ekanayaka, A.S., Piddock, L.J.V., and Webber, M.A. (2012). Loss of or inhibition of all multidrug resistance efflux pumps of *Salmonella enterica* serovar Typhimurium results in impaired ability to form a biofilm. *Journal of Antimicrobial Chemotherapy* 67: 2409–2417.

141 Lin, J., Sahin, O., Michel, L., and Zhang, O. (2003). Critical role of multidrug efflux pump CmeABC in bile resistance and *in vivo* colonization of *Campylobacter jejuni*. *Infection and Immunity* 71: 4250–4259.

142 Lacroix, F., Cloeckaert, A., Grepinet, O. et al. (1996). *Salmonella* Typhimurium *acrB*-like gene: identification and role in resistance to biliary salts and detergents and in murine infection. *FEMS Microbiology Letters* 135: 161–167.

10

Anti-virulence Therapies Through Potentiating ROS in Bacteria

Kristin J. Adolfsen and Mark P. Brynildsen

Department of Chemical and Biological Engineering, Princeton University, Princeton, NJ, USA

10.1 Introduction

Antibiotic resistance is a growing public health concern, and existing approaches to replenish an ever-weakening arsenal are falling short [1–3]. "Superbugs," which exhibit resistance to numerous antibiotics, are increasingly observed in hospital settings [4], and the spread of resistance genes to novel agents is alarmingly rapid [3]. For example, there are currently strains of *Mycobacterium tuberculosis*, the causative agent of tuberculosis, that are resistant to all known treatments [5]. Strains of methicillin-resistant *Staphylococcus aureus* (MRSA) have emerged that are resistant to additional antibiotics, including ciprofloxacin, tetracycline, gentamicin, amikacin, netilmicin, cotrimoxazole, chloramphenicol, and even vancomycin, which is one of the "last line of defense" antibiotics [6, 7]. In February of 2016, resistance to colisitin, which is another last resort antibiotic, was for the first time found to be plasmid-borne and readily transmissible between different Gram-negative bacteria; a scenario that suggests that colisitin resistance is now poised to spread rapidly [8]. The Center for Disease Control estimates that more than two million antibiotic-resistant infections occur in the USA annually, resulting in approximately 23 000 deaths [4]. The rising threat that antibiotic resistance poses to public health prompted President Obama to issue an executive order in September of 2014, which called for the formulation of a national action plan to combat antibiotic-resistant bacteria [9]. That action plan, which was published in March of 2015, detailed guidelines to slow resistance development and advance research efforts, with one of its major goals being the development of new antibiotics and other therapeutics [10].

Anti-virulence therapies have been proposed as an alternative to traditional antibiotics, and such agents focus on inhibiting bacterial pathogenicity rather than growth [11–13]. Inhibition of cell adhesion, quorum sensing, biofilm production, or reactive oxygen (ROS) and nitrogen (RNS) species detoxification all constitute anti-virulence strategies that have the potential to limit or eliminate bacterial infections without targeting vital cell functions, as traditional antibiotics do [11–14]. This has several compelling advantages that include focusing selective pressure to the site

Bacterial Resistance to Antibiotics – From Molecules to Man, First Edition.
Edited by Boyan B. Bonev and Nicholas M. Brown.
© 2020 John Wiley & Sons Ltd. Published 2020 by John Wiley & Sons Ltd.

of infection, which is projected to slow resistance development [11], and reducing collateral damage to commensal bacteria, which protects against antibiotic-induced illnesses such as *Clostridium difficile* infections [15]. Confinement of selective pressure to within a host will avoid resistance development in *ex vivo* locations, such as water, soil, and sewage. When one considers that antibiotics have been detected at alarming concentrations in tap water and agricultural products (both animal- and plant-derived), for example in China [16], and that resistant bacteria can be selected for at antibiotic concentrations far below their minimum inhibitory concentrations (MIC) [17], the importance of host-restricted antibacterial activity is evident. Further, targeting pathogenic processes, rather than essential functions, will lead to healthier microbiomes following anti-infective treatment and lower incidences of infections caused by *C. difficile*, which in 2011 accounted for approximately 500 000 infections and 29 000 deaths in the USA alone [18]. To realize the potential of anti-virulence therapies to re-stock the antibacterial medicine cabinet, deep understanding of virulence factors and therapeutically relevant means to modulate them are needed.

Potentiating ROS represents a type of anti-virulence strategy that could be broadly applicable, given the large number of pathogens that employ antioxidant metabolites or enzymes, as well as repair systems for oxidative damage, as virulence factors (*e.g.,* *Enterococcus faecalis* [19], *Helicobacter pylori* [20], *M. tuberculosis* [21, 22], *Salmonella enterica* serovar Typhimurium [23], *S. aureus* [24, 25], *Streptococcus pyogenes* [26]). Sensitization of bacteria to oxidative stress can be achieved by decreasing the abilities of bacteria to detoxify ROS, increasing the amount of ROS bacteria have to cope with (*e.g.,* enhancing endogenous ROS production), increasing the capacities of ROS to damage biomolecules, or hindering the abilities of bacteria to repair oxidative damage. Given these potential routes of intervention, understanding the interactions between bacterial ROS production and detoxification, as well as oxidative damage and repair, will aid in realizing these types of therapies. In this chapter, we review the scientific rationale for developing ROS-potentiating anti-virulence therapies, describe several promising agents that use this mode of anti-infective action, and highlight the contributions that computational modeling could provide to increase understanding of bacterial ROS metabolism.

10.2 Weapons of Phagosomes

Phagocytic cells of the innate immune response expose infecting bacteria to numerous stresses, including acidified pH, RNS, ROS, proteases, antimicrobial peptides, and nutrient deprivation (Figure 10.1) [27, 28]. The phagosomal pH is decreased to ~4.5 by V-ATPases, which hydrolyze ATP within the cytoplasm to pump protons into the phagosome [27]. Nitric oxide (NO•) is generated by inducible nitric oxide synthase (iNOS) within the cytoplasm, and it readily diffuses into the phagosome, where it can form other RNS [27]. Antimicrobial peptides are generally cationic and amphipathic, and act by making the microbial cell membrane permeable [29], whereas proteases are present to degrade bacterial proteins [30]. Additionally, microbes within the phagosome are deprived of nutrients such as iron, oxygen, and glucose through various mechanisms [28].

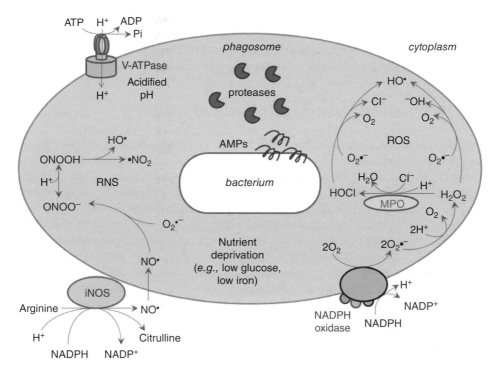

Figure 10.1 **Antimicrobial mechanisms of phagocytes.** During maturation, the pH in the phagosome is decreased to ~4.5 *via* V-ATPase activity, which consumes ATP to transport protons across the membrane. NO• is produced in the cytoplasm by iNOS, and it readily diffuses into the phagosome to form other reactive intermediates (*e.g.*, peroxynitrite, ONOO⁻; peroxynitrous acid, ONOOH; nitrogen dioxide, •NO₂). Additionally, O₂ is converted to superoxide ($O_2\bullet^-$) by NADPH oxidase and it can then be dismutated to hydrogen peroxide (H_2O_2). Hypochlorous acid (HOCl) can be formed from H_2O_2 and Cl^- by the action of myeloperoxidase (MPO), whereas hydroxyl radicals (HO•) can be generated *via* Haber-Weiss chemistry or reaction of HOCl with $O_2\bullet^-$. Pathogens are also exposed to proteases and antimicrobial peptides (AMPs), while they are starved of important nutrients, including iron and glucose. We note that this is a simplified portrayal of the biochemistry in phagosomes, and for illustrative purposes some reactions are not balanced in terms of elements and charge.

After engulfing pathogens, phagocytic cells of the innate immune response (*e.g.*, macrophages, neutrophils) recruit the components of NADPH oxidase to the phagosome to generate an "oxidative burst" of $O_2\bullet^-$ [31]. The $O_2\bullet^-$ generated can be dismutated to H_2O_2 and react further to form other ROS such as HOCl *via* myeloperoxidase and HO• by reaction with superoxide, which can be catalyzed by iron [32, 33]. H_2O_2 readily diffuses into bacterial cells [34], where it can damage proteins [35–38], react with antioxidant metabolites such as α-keto acids [39, 40], be detoxified by catalases and hydroperoxide reductases [41, 42], or be reduced by Fe^{2+} to form HO•, which reacts with most biomolecules (including DNA) at diffusion-limited rates [33]. The importance of ROS in the innate immune response is highlighted clinically by patients with chronic granulomatous disease (CGD), who have defects in genes that code for the subunits of NADPH oxidase, have recurrent infections, and a shortened life expectancy [43, 44]. To circumvent killing by immune-generated ROS, some pathogens, including *Bacillus anthracis, Chlamydia trachomatis, Francisella tularensis,* and *Coxiella burnetii*, inhibit assembly or activity of NADPH oxidase [45–48]. Alternatively, many pathogens employ ROS detoxification and repair systems, most notably peroxidases (with various electron donors, such as NADH, thiol, and glutathione) and catalases, but also pathogen-specific antioxidants, such as pigment molecules [19–26].

10.3 ROS Defense Systems as Virulence Factors

ROS defense systems are broadly important to bacterial pathogenesis in species ranging from *M. tuberculosis* and *S. enterica* to *S. aureus* and *E. faecalis* [19–26]. For example, *S. aureus* encodes catalase KatA and alkyl hydroperoxide reductase AhpC, which detoxify H_2O_2, and are required for nasal colonization of mice by *S. aureus* [24]. Furthermore, *S. aureus* synthesizes a carotenoid pigment, staphyloxanthin, which confers resistance to oxidants and NADPH oxidase-dependent killing by human neutrophils and whole blood [25]. Staphyloxanthin biosynthesis is also required for subcutaneous infection of mouse models, and the effect is eliminated in mice with a defect in NADPH oxidase $(gp91^{Phox-/-})$ [25]. *Salmonella typhimurium* encodes at least five H_2O_2 detoxification systems, three catalases and two alkyl hydroperoxide reductases, and deletion of all five enzymes led to increased NADPH oxidase-dependent killing by murine macrophages and attenuated virulence in a mouse model [23]. The attenuated virulence in strains lacking ROS defenses suggests that inhibiting those systems could serve as a potent anti-virulence strategy.

10.4 Therapeutic Potential of Oxidant Potentiating Anti-Virulence Therapies

Several compounds have been identified to sensitize pathogens to oxidative stress and immune clearance. NTBC (2-[2-nitro-4-trifluoromethylbenzoyl]-1,3-cyclohexanedione), an FDA-approved treatment for type 1 tyrosemia, was shown to also inhibit production of pyomelanin, a pigment with iron acquisition and antioxidant properties found in a clinical isolate of *Pseudomonas aeruginosa* [49]. Notably, NTBC treatment resulted in enhanced sensitivity of pyomelanin-expressing *P. aeruginosa* to oxidative stress [49]. A cholesterol biosynthesis inhibitor, BPH-652 (tripotassium;4-[3-phenoxyphenyl]-1-phosphonatobutane-1-sulfonate), was shown to block synthesis of staphyloxanthin, the antioxidant pigment discussed earlier, sensitizing *S. aureus* to killing by H_2O_2 and human blood, and enhancing clearance in a mouse infection model [50]. Since that influential study, additional compounds to inhibit staphyloxanthin synthesis have been identified [51, 52].

These studies exemplify the potential of targeting oxidative stress resistance as an anti-infective strategy. Neither pyomelanin nor staphyloxanthin are essential for growth of *P. aeruginosa* [49] or *S. aureus* [25, 50], respectively, but enhance their resistance to antimicrobials of innate immune cells. Selective pressure against these inhibitors would therefore be isolated to the site of infection, and limit the development of resistance to these agents [11]. However, these virulence factors are pathogen-specific, which limits the spectrum of bacteria that therapies based on their inhibition could treat. Catalases and hydroperoxide reductases are more widely distributed [53], but more difficult to target. For example, catalase can be inhibited by compounds that bind to its heme active site, such as cyanide, azide, hydroxylamine, aminotriazole, and mercaptoethanol [54], but all are toxic to humans and therefore not therapeutically relevant. Additionally, human catalase plays a critical role in protecting host cells from oxidative stress, so inadvertent inhibition of this host enzyme would likely result in undesirable side effects. A deeper understanding of bacterial ROS defenses may identify more druggable targets, and due to the complexity of ROS biochemical reaction networks, computational modeling has proven useful for the interpretation and prediction of how ROS metabolism changes in response to perturbations.

10.5 Modeling of Bacterial ROS Metabolism

The outcomes of oxidative stress are governed by complex networks, where both exogenous and endogenous sources supply ROS that is consumed by either detoxification systems (*e.g.*, catalases, superoxide dismutases) or reactions that damage biomolecules (*e.g.*, protein carbonylation, iron-sulfur cluster oxidation) that then need to be repaired or resynthesized to restore function, which concomitantly regenerates oxidative targets [33, 41, 55]. The connectivity of these reaction networks suggests that alterations to system components could produce cascades of events that lead to distal consequences. Computational models are useful tools for analyzing such systems and identifying scenarios where the knowledge base is adequate or where uncertainty exists. In particular, systems-level models have been used to increase understanding of both bacterial ROS production and consumption [41, 56, 57], and several of these accomplishments are discussed in the sections that follow.

10.6 Modeling Bacterial ROS Production

In addition to host-derived ROS (Figure 10.1) [27], pathogens are exposed to ROS they generate on their own, and this usually occurs from the inadvertent collision of reduced electron carriers (*e.g.*, reduced flavins, quinols, metal centers) with O_2 prior to their intended electron acceptor [58, 59]; although exceptions exist [60]. In *Escherichia coli*, whole-cell $O_2 \bullet^-$ and H_2O_2 production rates have been quantified under steady growth [42, 61]; however, the quantitative contributions of individual enzymes to cellular production rates remain for the most part elusive [56, 62, 63]. Given this uncertainty, as well as the paucity of kinetic information available for unintended ROS production from individual enzymes, ensemble modeling of steady-state metabolism has been used to assess the effect of different variables (*e.g.*, gene deletions [56], futile cycles [64]) on bacterial-derived ROS. Specifically, genome-scale metabolic models (GSMMs) have been used to predict the impact of perturbations to cellular ROS production [56, 64]. GSMMs have been used extensively to tackle a wide variety of biological questions, ranging from increasing production of desired metabolites to understanding the exchange of metabolites in interspecies interactions [65]. Currently, GSMMs have been constructed for 101 bacteria, 42 eukaryota, and 6 archaea [66] and software is readily accessible for flux calculations (*e.g.*, the COBRA Toolbox [67]). Flux balance analysis (FBA) utilizes these models to predict metabolic flux given a set of constraints (*i.e.*, reaction stoichiometry, Equation 10.2; bounds on reaction fluxes, Equation 10.3) [68].

Optimizing for a specified cellular objective function (*e.g.*, biomass, Equation 10.1) allows for identification of optimal flux distributions within the solution space, without the need for extensive knowledge regarding kinetic parameters and metabolite or enzyme concentrations [68]. This is shown below mathematically, where $\mathbf{c}^T$ is the transposed vector that weights the contribution of each reaction to the objective function, $\mathbf{v}$ is a vector of reaction fluxes, $\mathbf{S}$ is the metabolite × reaction stoichiometric matrix, and $\mathbf{lb}$ and $\mathbf{ub}$ represent lower and upper bounds on reaction fluxes, respectively.

$$Maximize/minimize \ \left(\mathbf{c}^T\mathbf{v}\right) \tag{10.1}$$

$$\text{subject to } \mathbf{Sv} = \mathbf{0} \tag{10.2}$$

$$\mathbf{lb} \le \mathbf{v} \le \mathbf{ub} \tag{10.3}$$

Since the requirements of GSMMs and FBA were well matched with the knowledge base of ROS production (*e.g.*, whole-cell ROS production levels, little kinetic

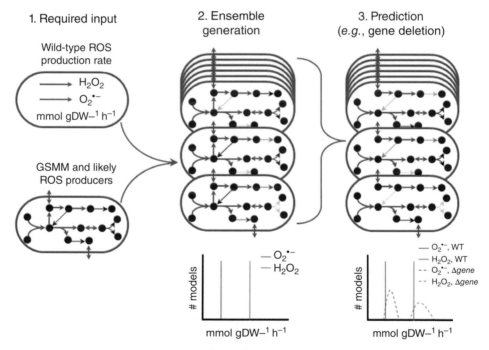

Figure 10.2 Steady-state modeling of bacterial ROS production. Model construction requires measurement of wild-type ROS production rates (*i.e.*, H_2O_2 and $O_2 \cdot^-$), a GSMM of the organism of interest, and knowledge of which enzymes are capable of generating ROS (*i.e.*, contain flavin, quinone, or transition metal cofactors). Indeterminacy regarding the proportion of ROS produced by each potential generator is accounted for by generating an ensemble of models that all produce wild-type levels of H_2O_2 and $O_2 \cdot^-$, but quantitatively attribute ROS production differently among the potential generators. Predictions can then be made about genetic perturbations, such as gene deletions or overexpressions. At this stage, the ensemble will predict a range of production rates depending on which reactions contribute substantially to generation.

information), an FBA approach, which is summarized in Figure 10.2, was developed to simulate whole-cell ROS ($O_2\bullet^-$ and H_2O_2) production in *E. coli* [56]. Notably, to address the uncertainty in the quantitative contributions of the different ROS generators, an ensemble modeling approach was used. Each model within the ensemble was constrained to match experimentally measured H_2O_2 and $O_2\bullet^-$ production rates for *E. coli*, but each varied in the contributions of potential ROS sources [56]. Ensemble simulations could then be analyzed to uncover trends in ROS production that were robust to the specific sources of ROS. For example, this approach qualitatively predicted changes in ROS production in approximately 90% of the single gene deletion mutants tested, and increased endogenous ROS production was found to correlate experimentally with enhanced oxidant sensitivity [56]. Notably, mutations that introduced inefficiencies in ATP production or usage increased sensitivity to oxidants [56]. This inspired subsequent work implementing futile cycles, which are sets of reactions with no net effect other than ATP hydrolysis, as a flexible method to introduce ATP inefficiency into the cell [64]. Futile cycling increased H_2O_2 production in *E. coli*, as predicted by the steady-state ROS model, and potentiated killing by H_2O_2 [64], which suggested that enhancing metabolic futile cycling could constitute an interesting therapeutic approach.

10.7 Modeling Bacterial ROS Consumption

In contrast to ROS production, the major pathways for ROS consumption in some bacteria, such as *E. coli*, have been identified and kinetically characterized [69–72]. This enables dynamic models to be developed to understand how those systems coordinate bacterial responses to transient oxidative stress, such as that provided by oxidative bursts in phagosomes [27]. Deterministic ordinary differential equation models (Equations 10.4 and 10.5) simulate species concentrations (X) over time, given reaction stoichiometry (S) and reactions rates (v) calculated from current concentrations (X) and kinetic parameters (p). In contrast to FBA, this modeling technique requires knowledge of initial species concentrations, reaction mechanisms, and kinetic parameters, where unknown values must be trained on experimental data. Often, multiple parameter sets can sufficiently describe the training data while making different forward predictions, which results from poorly constrained parameters and has been termed model "sloppiness" [41, 73–82]. To account for this, ensembles of parameter sets that all describe training data equally well can be identified, allowing assessment of forward prediction robustness [41, 73–82].

$$\frac{dX}{dt} = Sv \tag{10.4}$$

$$v = f(X, p) \tag{10.5}$$

Due to its importance as a signaling molecule, there have been notable efforts to capture the dynamics of H_2O_2 metabolism in mammalian systems [83–85]. These models have included detoxification by antioxidants (*e.g.*, thioredoxin, glutathione) and enzymes (*e.g.*, catalase, glutathione peroxidase), as well as repair of oxidized protein thiols. Inspired by the successes of such models in mammalian systems, a compartmentalized kinetic model of the H_2O_2 response network in *E. coli* was developed, which included H_2O_2-dependent transcriptional regulation; detoxification by antioxidant metabolites and enzymes; generation of $O_2\bullet^-$ and $HO\bullet$; and damage of metabolites and enzymes by ROS [41]. Model construction and capabilities are depicted in Figure 10.3. This model was used to quantitatively interrogate contributions from the major detoxification systems following challenge with varying boluses of H_2O_2, and to analyze the effect of a physiologically relevant environmental perturbation (*i.e.*, carbon starvation) on H_2O_2 detoxification [41]. Similar models for $NO\bullet$ have allowed for mechanistic dissection of system behavior [82, 86, 87], and we anticipate that the *E. coli* H_2O_2 model will be a similarly useful tool. In addition, it is worth noting that models can be continually updated to reconcile disagreements with experimental observations or any poorly constrained predictions, leading to ever-improving descriptive and predictive capabilities [41, 82].

10.8 Future Prospects for Computational Models of Bacterial Oxidative Stress

The models discussed here were constructed for a nonpathogenic, but well-characterized and genetically tractable strain of *E. coli*, K-12 MG1655. With such proof-of-concept studies now available, translation of these computational methodologies to pathogens is needed to realize the utility of these techniques for anti-virulence identification and development. GSMMs have been constructed for 101 bacterial species, including multiple pathogens (*e.g.*, *M. tuberculosis* [88], *P. aeruginosa* [89], *S. aureus* [90]). Each of these models can be adapted to incorporate ROS production by coupling generation of $O_2\bullet^-$ and H_2O_2 to its potential sources, and utilizing an ensemble approach to address uncertainty, as previously described [56]. Translating the dynamic model will entail a thorough analysis of all major consumption pathways in the target organism. Spontaneous, organism-independent reactions will be maintained across models, as well as the kinetics of enzymes with sufficient homology with the established *E. coli* model. New network components (*e.g.*, staphyloxanthin in *S. aureus* [25]) would need to be added, and any reactions not present in the target organism removed. Initial metabolite and enzyme concentrations will be updated if available, or trained on experimental data if uncertain. This generalized approach has proven effective in the translation of an analogous dynamic model for $NO\bullet$ from non-pathogenic *E. coli* to enterohemorrhagic *E. coli* (EHEC) [91].

In addition to translating the models to other organisms, more complete models that can account for combinatorial phagosomal stresses will provide a deeper understanding of the functionalities of bacterial ROS defenses as virulence systems *in vivo*. Combinations of phagosomal stresses have been shown to impact aspects of the H_2O_2 bacterial response network. For example, nutrient deprivation delays clearance of physiologically relevant levels H_2O_2 in *E. coli* [41], and $NO\bullet$ reversibly binds to and inhibits catalase activity [92]. Additionally, $NO\bullet$ has been shown to potentiate killing by high concentrations of H_2O_2 [93, 94]. *In vivo*, pathogens are exposed ROS and RNS, acidified pH, antimicrobial peptides, and proteases, all while being starved of nutrients [27, 28, 30]. Exploring individual stresses represents a necessary step in the understanding of bacterial responses to combinatorial stresses; however, unpredicted interactions

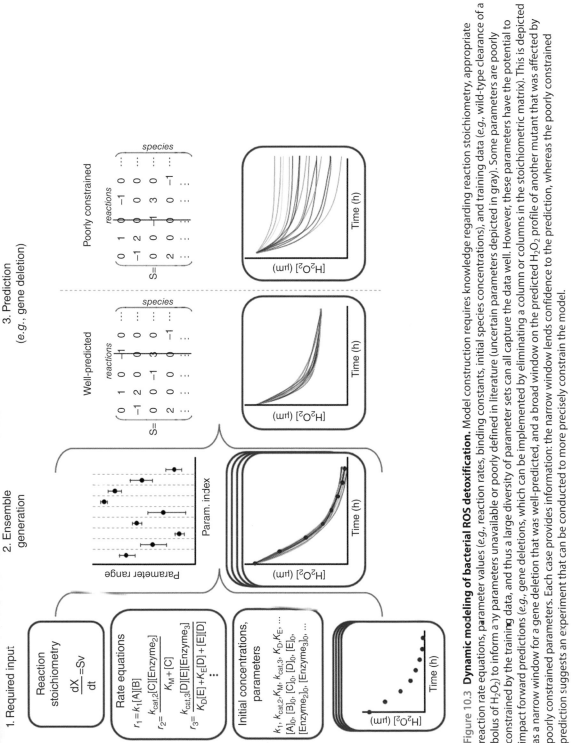

Figure 10.3 Dynamic modeling of bacterial ROS detoxification. Model construction requires knowledge regarding reaction stoichiometry, appropriate reaction rate equations, parameter values (*e.g.*, reaction rates, binding constants, initial species concentrations), and training data (*e.g.*, wild-type clearance of a bolus of H_2O_2) to inform a γ parameters unavailable or poorly defined in literature (uncertain parameters depicted in gray). Some parameters are poorly constrained by the training data, and thus a large diversity of parameter sets can all capture the data well. However, these parameters have the potential to impact forward predictions (*e.g.*, gene deletions, which can be implemented by eliminating a column or columns in the stoichiometric matrix). This is depicted as a narrow window for a gene deletion that was well-predicted, and a broad window on the predicted H_2O_2 profile of another mutant that was affected by poorly constrained parameters. Each case provides information: the narrow window lends confidence to the prediction, whereas the poorly constrained prediction suggests an experiment that can be conducted to more precisely constrain the model.

can manifest, which underscore the importance of considering conditions more reminiscent of phagosomes. Experimentally, multi-stress models that recapitulate key environmental features within phagosomes have been used, such as the non-replicating assay model developed for *M. tuberculosis*, which includes mild acidity (pH 5), RNS (0.5 mM sodium nitrite), microaerobic conditions (gas phase O_2 of 1%), and fatty acid as the sole carbon source (0.05% butyrate) [95]. Computationally, by contrast, more complex combinations of phagosomal stressors have been examined, such as the study by Winterbourn and colleagues where a model to analyze the biochemistry of several oxidative species in the phagosomal compartment (*e.g.*, $O_2\bullet^-$, H_2O_2, HOCl) was developed [96]. Naturally, an important next step will be to use experimentally tractable multistress conditions and computational models that can account for the complex biochemistry to analyze how bacteria respond to conditions that are more reflective of those they would encounter *in vivo*.

10.9 Summary and Conclusions

Antibiotic resistance is a growing public health crisis, and current methods of antibiotic discovery are unable to replenish a depleting reservoir of effective therapies [1–3]. This is largely due to the existence of selective pressure that is environmentally independent driving resistance development [17]. Anti-virulence therapies, which target the pathogen's ability to infect a host, rather than essential cellular functions, are projected to be less prone to resistance development because selective pressure will be limited to within the host [11–13]. ROS are critically important weapons used by the immune system against pathogens [27, 31, 43], and potentiating their activity could be leveraged for the development of anti-virulence therapies. For example, sensitizing pathogens by targeting the synthesis of antioxidant pigments is a promising and exciting option [49–52]. However, these pigments are largely organism-specific, so alternative therapies would have to be developed for each pathogen. In this review, we highlighted the capacity of computational techniques to increase understanding of bacterial oxidative stress in different scenarios (*e.g.*, genetic or environmental perturbations), which we anticipate will be useful for the identification of novel anti-virulence strategies. Specifically, recent advances in modeling of ROS production and detoxification have provided a framework to probe network perturbations that enhance sensitivity to oxidative stress [41, 56]. To leverage these tools to aid in the development of ROS-potentiating anti-infectives, they must be translated to pathogenic organisms and further developed so that they might capture the complexities of the phagosomal environment (*e.g.*, simultaneous ROS and RNS stress, acidified pH, nutrient deprivation). Once those milestones have been achieved, deeper quantitative insight of host and bacterial variables important for pathogenicity can be gained, which could reveal new anti-infective targets.

Acknowledgments

This work was supported by the National Science Foundation (CBET-1453325) and Princeton University. The content is solely the responsibility of the authors and does not necessarily represent the official views of the funding agencies.

References

1 Boucher, H.W., Talbot, G.H., Bradley, J.S. et al. (2009). Bad bugs, no drugs: no ESKAPE! An update from the infectious diseases society of America. *Clin. Infect. Dis.* 48: 1–12.

2 Payne, D.J., Gwynn, M.N., Holmes, D.J., and Pompliano, D.L. (2007). Drugs for bad bugs: confronting the challenges of antibacterial discovery. *Nat. Rev. Drug Discov.* 6: 29–40.

3 Taubes, G. (2008). The bacteria fight back. *Science* 321: 356–361.

4 Centers for Disease Control and Prevention. Antibiotic resistance threats in the United States, 2013. Centers for Disease Control and Prevention, Atlanta, GA, USA. Available at: https://www.cdc.gov/drugresistance/pdf/ar-threats-2013-508.pdf (2013).

5 Davies, J. and Davies, D. (2010). Origins and evolution of antibiotic resistance. *Microbiol. Mol. Biol. Rev.: MMBR* 74: 417–433.

6 Kali, A., Stephen, S., Umadevi, S. et al. (2013). Changing trends in resistance pattern of methicillin resistant *Staphylococcus aureus. J. Clin. Diagn. Res.: JCDR* 7: 1979–1982.

7 Saravolatz, L.D., Pawlak, J., and Johnson, L.B. (2012). In vitro susceptibilities and molecular analysis of vancomycin-intermediate and vancomycin-resistant *Staphylococcus aureus* isolates. *Clin. Infect. Dis.* 55: 582–586.

8 Liu, Y.Y., Wang, Y., Walsh, T.R. et al. (2016). Emergence of plasmid-mediated colistin resistance mechanism MCR-1 in animals and human beings in China: a microbiological and molecular biological study. *Lancet Infect. Dis.* 16: 161–168.

9 Obama, B. H. *Executive Order – Combating Antibiotic-Resistant Bacteria*, Exec. Order No. 13676, 3 C.F.R. 1-6 (2014). Available at https://obamawhitehouse.archives.gov/the-press-office/2014/09/18/executive-order-combating-antibiotic-resistant-bacteria (2014).

10 The White House. National Action Plan for Combating Antibiotic-Resistant Bacteria, https://www.cdc.gov/drugresistance/pdf/national_action_plan_for_combating_antibotic-resistant_bacteria.pdf (2015).

11 Allen, R.C., Popat, R., Diggle, S.P., and Brown, S.P. (2014). Targeting virulence: can we make evolution-proof drugs? *Nat. Rev. Microbiol.* 12: 300–308.

12 Escaich, S. (2008). Antivirulence as a new antibacterial approach for chemotherapy. *Curr. Opin. Chem. Biol.* 12: 400–408.

13 Rasko, D.A. and Sperandio, V. (2010). Anti-virulence strategies to combat bacteria-mediated disease. *Nat. Rev. Drug Discov.* 9: 117–128.

14 Robinson, J.L., Adolfsen, K.J., and Brynildsen, M.P. (2014). Deciphering nitric oxide stress in bacteria with quantitative modeling. *Curr. Opin. Microbiol.* 19: 16–24.

15 Hickson, M. (2011). Probiotics in the prevention of antibiotic-associated diarrhoea and *Clostridium difficile* infection. *Ther. Adv. Gastroenterol.* 4: 185–197.

16 Hao, R., Zhao, R., Qiu, S et al. (2015). Antibiotics crisis in China. *Science* 348: 1100–1101.

17 Gullberg, E., Cao, S., Berg, O.G. et al. (2011). Selection of resistant bacteria at very low antibiotic concentrations. *PLoS Pathog.* 7: e1002158.

18 Lessa, F.C., Winston, L.G., and McDonald, L.C. (2015). Burden of *Clostridium difficile* infection in the United States. *N. Engl. J. Med.* 372: 2369–2370.

19 La Carbona, S., Sauvageot, N., Giard, J.C. et al. (2007). Comparative study of the physiological roles of three peroxidases (NADH peroxidase, Alkyl hydroperoxide reductase and thiol peroxidase) in oxidative stress response, survival inside macrophages and virulence of *Enterococcus faecalis. Mol. Microbiol.* 66: 1148–1163.

20 Harris, A.G., Wilson, J.E., Danon, S.J. et al. (2003). Catalase (KatA) and KatA-associated protein (KapA) are essential to persistent colonization in the *Helicobacter pylori* SS1 mouse model. *Microbiology* 149: 665–672.

21 Li, Z., Kelley, C., Collins, F. et al. (1998). Expression of katG in *Mycobacterium tuberculosis* is associated with its growth and persistence in mice and Guinea pigs. *J. Infect. Dis.* 177: 1030–1035.

22 Manca, C., Paul, S., Barry, C.E. et al. (1999). *Mycobacterium tuberculosis* catalase and peroxidase activities and resistance to oxidative killing in human monocytes *in vitro*. *Infect. Immun.* 67: 74–79.

23 Hebrard, M., Viala, J.P.M., Meresse, S. et al. (2009). Redundant hydrogen peroxide scavengers contribute to *Salmonella* virulence and oxidative stress resistance. *J. Bacteriol.* 191: 4605–4614.

24 Cosgrove, K., Coutts, G., Jonsson, I.M. et al. (2007). Catalase (KatA) and alkyl hydroperoxide reductase (AhpC) have compensatory roles in peroxide stress resistance and are required for survival, persistence, and nasal colonization in *Staphylococcus aureus*. *J. Bacteriol.* 189: 1025–1035.

25 Liu, G.Y., Essex, A., Buchanan, J.T. et al. (2005). *Staphylococcus aureus* golden pigment impairs neutrophil killing and promotes virulence through its antioxidant activity. *J. Exp. Med.* 202: 209–215.

26 Brenot, A., King, K.Y., Janowiak, B. et al. (2004). Contribution of glutathione peroxidase to the virulence of *Streptococcus pyogenes*. *Infect. Immun.* 72: 408–413.

27 Flannagan, R.S., Cosio, G., and Grinstein, S. (2009). Antimicrobial mechanisms of phagocytes and bacterial evasion strategies. *Nat. Rev. Microbiol.* 7: 355–366.

28 Appelberg, R. (2006). Macrophage nutriprive antimicrobial mechanisms. *J. Leukoc. Biol.* 79: 1117–1128.

29 Ganz, T. (2003). The role of antimicrobial peptides in innate immunity. *Integr. Comp. Biol.* 43: 300–304.

30 Haas, A. (2007). The phagosome: compartment with a license to kill. *Traffic* 8: 311–330.

31 Diacovich, L. and Gorvel, J.P. (2010). Bacterial manipulation of innate immunity to promote infection. *Nat. Rev. Microbiol.* 8: 117–128.

32 Hampton, M.B., Kettle, A.J., and Winterbourn, C.C. (1998). Inside the neutrophil phagosome: oxidants, myeloperoxidase, and bacterial killing. *Blood* 92: 3007–3017.

33 Imlay, J.A. (2003). Pathways of oxidative damage. *Annu. Rev. Microbiol.* 57: 395–418.

34 Seaver, L.C. and Imlay, J.A. (2001). Hydrogen peroxide fluxes and compartmentalization inside growing *Escherichia coli*. *J. Bacteriol.* 183: 7182–7189.

35 Davies, M.J. (2005). The oxidative environment and protein damage. *Biochim. Biophys. Acta. Proteins Proteom.* 1703: 93–109.

36 Jang, S.J. and Imlay, J.A. (2007). Micromolar intracellular hydrogen peroxide disrupts metabolism by damaging iron-sulfur enzymes. *J. Biol. Chem.* 282: 929–937.

37 Luo, S. and Levine, R.L. (2009). Methionine in proteins defends against oxidative stress. *FASEB J.* 23: 464–472.

38 Stadtman, E.R. and Levine, R.L. (2000). Protein oxidation. *Ann. N. Y. Acad. Sci.* 899: 191–208.

39 Perera, A., Parkes, H.G., Herz, H. et al. (1997). High resolution 1H NMR investigations of the reactivities of alpha-keto acid anions with hydrogen peroxide. *Free Radic. Res.* 26: 145–157.

40 Vlessis, A.A., Bartos, D., and Trunkey, D. (1990). Importance of spontaneous alpha-ketoacid decarboxylation in experiments involving peroxide. *Biochem. Biophys. Res. Commun.* 170: 1281–1287.

41 Adolfsen, K.J. and Brynildsen, M.P. (2015). A kinetic platform to determine the fate of hydrogen peroxide in *Escherichia coli*. *PLoS Comput. Biol.* 11: e1004562.

42 Seaver, L.C. and Imlay, J.A. (2001). Alkyl hydroperoxide reductase is the primary scavenger of endogenous hydrogen peroxide in *Escherichia coli*. *J. Bacteriol.* 183: 7173–7181.

43 Fang, F.C. (2004). Antimicrobial reactive oxygen and nitrogen species: concepts and controversies. *Nat. Rev. Microbiol.* 2: 820–832.

44 Heyworth, P.G., Cross, A.R., and Curnutte, J.T. (2003). Chronic granulomatous disease. *Curr. Opin. Immunol.* 15: 578–584.

45 Crawford, M.A., Aylott, C.V., Bourdeau, R.W., and Bokoch, G.M. (2006). *Bacillus anthracis* toxins inhibit human neutrophil NADPH oxidase activity. *J. Immunol.* 176: 7557–7565.

46 Tauber, A.I., Pavlotsky, N., Lin, J.S., and Rice, P.A. (1989). Inhibition of human neutrophil NADPH oxidase by chlamydia serovars E, K, and L2. *Infect. Immun.* 57: 1108–1112.

47 Siemsen, D.W., Kirpotina, L.N., Jutila, M.A., and Quinn, M.T. (2009). Inhibition of the human neutrophil NADPH oxidase by *Coxiella burnetii*. *Microbes Infect.* 11: 671–679.

48 McCaffrey, R.L., Schwartz, J.T., Lindemann, S.R. et al. (2010). Multiple mechanisms of NADPH oxidase inhibition by type A and type B *Francisella tularensis*. *J. Leukoc. Biol.* 88: 791–805.

49 Ketelboeter, L.M., Potharla, V.Y., and Bardy, S.L. (2014). NTBC treatment of the Pyomelanogenic *Pseudomonas aeruginosa* clinical isolate PA1111 inhibits pigment production and increases sensitivity to oxidative stress. *Curr. Microbiol.* 69: 343–348.

50 Liu, C.I., Liu, G.Y., Song, Y. et al. (2008). A cholesterol biosynthesis inhibitor blocks *Staphylococcus aureus* virulence. *Science* 319: 1391–1394.

51 Leejae, S., Hasap, L., and Voravuthikunchai, S.P. (2013). Inhibition of staphyloxanthin biosynthesis in *Staphylococcus aureus* by rhodomyrtone, a novel antibiotic candidate. *J. Med. Microbiol.* 62: 421–428.

52 Song, Y., Liu, C.I., Lin, F.Y. et al. (2009). Inhibition of staphyloxanthin virulence factor biosynthesis in *Staphylococcus aureus*: *in vitro*, *in vivo* and crystallographic results. *J. Med. Chem.* 52: 3869–3880.

53 Scheer, M., Grote, A., Chang, A. et al. (2011). BRENDA, the enzyme information system in 2011. *Nucleic Acids Res.* 39: D670–D676.

54 Switala, J. and Loewen, P.C. (2002). Diversity of properties among catalases. *Arch. Biochem. Biophys.* 401: 145–154.

55 Jang, S. and Imlay, J.A. (2010). Hydrogen peroxide inactivates the *Escherichia coli* Isc iron-Sulphur assembly system, and OxyR induces the Suf system to compensate. *Mol. Microbiol.* 78: 1448–1467.

56 Brynildsen, M.P., Winkler, J.A., Spina, C.S. et al. (2013) Potentiating antibacterial activity by predictably enhancing endogenous microbial ROS production. *Nat. Biotechnol.* 31: 160–165.

57 Pillay, C.S., Hofmeyr, J.H.S., and Rohwer, J.M. (2011). The logic of kinetic regulation in the thioredoxin system. *BMC Syst. Biol.* 5: 15.

58 Messner, K.R. and Imlay, J.A. (1999). The identification of primary sites of superoxide and hydrogen peroxide formation in the aerobic respiratory chain and sulfite reductase complex of *Escherichia coli*. *J. Biol. Chem.* 274: 10119–10128.

59 Massey, V. (1994). Activation of molecular oxygen by flavins and flavoproteins. *J. Biol. Chem.* 269: 22459–22462.

60 Selva, L., Viana, D., Regev-Yochay, G. et al. (2009). Killing niche competitors by remote-control bacteriophage induction. *Proc. Natl. Acad. Sci. U. S. A.* 106: 1234–1238.

61 Imlay, J.A. and Fridovich, I. (1991). Assay of metabolic superoxide production in *Escherichia coli. J. Biol. Chem.* 266: 6957–6965.

62 Seaver, L.C. and Imlay, J.A. (2004). Are respiratory enzymes the primary sources of intracellular hydrogen peroxide? *J. Biol. Chem.* 279: 48742–48750.

63 Korshunov, S. and Imlay, J.A. (2010). Two sources of endogenous hydrogen peroxide in *Escherichia coli. Mol. Microbiol.* 75: 1389–1401.

64 Adolfsen, K.J. and Brynildsen, M.P. (2015). Futile cycling increases sensitivity toward oxidative stress in *Escherichia coli. Metab. Eng.* 29: 26–35.

65 McCloskey, D., Palsson, B.O., and Feist, A.M. (2013). Basic and applied uses of genome-scale metabolic network reconstructions of *Escherichia coli. Mol. Syst. Biol.* 9: 661.

66 Feist, A.M., Herrgard, M.J., Thiele, I. et al. (2009). Reconstruction of biochemical networks in microorganisms. *Nat. Rev. Microbiol.* 7: 129–143.

67 Becker, S.A., Feist, A.M., Mo, M.L. et al. (2007). Quantitative prediction of cellular metabolism with constraint-based models: the COBRA toolbox. *Nat. Protoc.* 2: 727–738.

68 Orth, J.D., Thiele, I., and Palsson, B.O. (2010). What is flux balance analysis? *Nat. Biotechnol.* 28: 245–248.

69 Gray, B. and Carmichael, A.J. (1992). Kinetics of superoxide scavenging by dismutase enzymes and manganese mimics determined by electron spin resonance. *Biochem. J.* 281 (Pt 3): 795–802.

70 Singh, R., Wiseman, B., Deemagarn, T. et al. (2008). Comparative study of catalase-peroxidases (KatGs). *Arch. Biochem. Biophys.* 471: 207–214.

71 Obinger, C., Maj, M., Nicholls, P., and Loewen, P. (1997). Activity, peroxide compound formation, and heme d synthesis in *Escherichia coli* HPII catalase. *Arch. Biochem. Biophys.* 342: 58–67.

72 Parsonage, D., Karplus, P.A., and Poole, L.B. (2008). Substrate specificity and redox potential of AhpC, a bacterial peroxiredoxin. *Proc. Natl. Acad. Sci. U. S. A.* 105: 8209–8214.

73 Gutenkunst, R.N., Waterfall, J.J., Casey, F.P. et al. (2007). Universally sloppy parameter sensitivities in systems biology models. *PLoS Comput. Biol.* 3: 1871–1878.

74 Zamora-Sillero, E., Hafner, M., Ibig, A. et al. (2011). Efficient characterization of high-dimensional parameter spaces for systems biology. *BMC Syst. Biol.* 5: 142.

75 Wang, L. and Hatzimanikatis, V. (2006). Metabolic engineering under uncertainty. I: framework development. *Metab. Eng.* 8: 133–141.

76 Wang, L. and Hatzimanikatis, V. (2006). Metabolic engineering under uncertainty – II: analysis of yeast metabolism. *Metab. Eng.* 8: 142–159.

77 Tran, L.M., Rizk, M.L., and Liao, J.C. (2008). Ensemble modeling of metabolic networks. *Biophys. J.* 95: 5606–5617.

78 Miskovic, L. and Hatzimanikatis, V. (2011). Modeling of uncertainties in biochemical reactions. *Biotechnol. Bioeng.* 108: 413–423.

79 Jia, G., Stephanopoulos, G., and Gunawan, R. (2012). Ensemble kinetic modeling of metabolic networks from dynamic metabolic profiles. *Metabolites* 2: 891–912.

80 Robinson, J.L. and Brynildsen, M.P. (2016). Ensemble modeling enables quantitative exploration of bacterial nitric oxide stress networks. In: *Stress and Environmental*

Regulation of Gene Expression and Adaptation in Bacteria (ed. F.J. de Bruijn), 1009–1014. Wiley-Blackwell.

81 Khodayari, A., Zomorrodi, A.R., Liao, J.C., and Maranas, C.D. (2014). A kinetic model of *Escherichia coli* core metabolism satisfying multiple sets of mutant flux data. *Metab. Eng.* 25: 50–62.

82 Robinson, J.L. and Brynildsen, M.P. (2016). Discovery and dissection of metabolic oscillations in the microaerobic nitric oxide response network of *Escherichia coli*. *Proc. Natl. Acad. Sci. U. S. A.* 113 (12): E1757–E1766.

83 Adimora, N.J., Jones, D.P., and Kemp, M.L. (2010). A model of redox kinetics implicates the thiol proteome in cellular hydrogen peroxide responses. *Antioxid. Redox Signal.* 13: 731–743.

84 Makino, N., Sasaki, K., Hashida, K., and Sakakura, Y. (2004). A metabolic model describing the H2O2 elimination by mammalian cells including H2O2 permeation through cytoplasmic and peroxisomal membranes: comparison with experimental data. *Biochim. Biophys. Acta* 1673: 149–159.

85 Brito, P.M. and Antunes, F. (2014). Estimation of kinetic parameters related to biochemical interactions between hydrogen peroxide and signal transduction proteins. *Front. Chem.* 2: 82.

86 Robinson, J.L. and Brynildsen, M.P. (2015). An ensemble-guided approach identifies ClpP as a major regulator of transcript levels in nitric oxide-stressed *Escherichia coli*. *Metab. Eng.* 31: 22–34.

87 Robinson, J.L., Miller, R.V., and Brynildsen, M.P. (2014). Model-driven identification of dosing regimens that maximize the antimicrobial activity of nitric oxide. *Metab. Eng. Commun.* 1: 12–18.

88 Fang, X., Wallqvist, A., and Reifman, J. (2010). Development and analysis of an in vivo-compatible metabolic network of *Mycobacterium tuberculosis*. *BMC Syst. Biol.* 4: 160.

89 Oberhardt, M.A., Puchalka, J., Fryer, K.E. et al. (2008). Genome-scale metabolic network analysis of the opportunistic pathogen *Pseudomonas aeruginosa* PAO1. *J. Bacteriol.* 190: 2790–2803.

90 Lee, D.S., Burd, H., Liu, J. et al. (2009). Comparative genome-scale metabolic reconstruction and flux balance analysis of multiple *Staphylococcus aureus* genomes identify novel antimicrobial drug targets. *J. Bacteriol.* 191: 4015–4024.

91 Robinson, J.L. and Brynildsen, M.P. (2016). Construction and experimental validation of a quantitative kinetic model of nitric oxide stress in Enterohemorrhagic *Escherichia coli* O157:H7. *Bioengineering* 3: 9.

92 Brown, G.C. (1995). Reversible binding and inhibition of catalase by nitric oxide. *Eur. J. Biochem.* 232: 188–191.

93 Yadav, R., Sumuni, Y., Abramson, A. et al. (2014). Pro-oxidative synergic bactericidal effect of NO: kinetics and inhibition by nitroxides. *Free Radic. Biol. Med.* 67: 248–254.

94 Pacelli, R., Wink, D.A., Cook, J.A. et al. (1995). Nitric oxide potentiates hydrogen peroxide-induced killing of *Escherichia coli*. *J. Exp. Med.* 182: 1469–1479.

95 Gold, B., Warrier, T., and Nathan, C. (2015). A multi-stress model for high throughput screening against non-replicating *Mycobacterium tuberculosis*. *Methods Mol. Biol.* 1285: 293–315.

96 Winterbourn, C.C., Hampton, M.B., Livesey, J.H., and Kettle, A.J. (2006). Modeling the reactions of superoxide and myeloperoxidase in the neutrophil phagosome: implications for microbial killing. *J. Biol. Chem.* 281: 39860–39869.

Index

Note: Page numbers in *italics* refer to figures and those in **bold** refer to tables.

Bacterial Resistance to Antibiotics – From Molecules to Man, First Edition.
Edited by Boyan B. Bonev and Nicholas M. Brown.
© 2020 John Wiley & Sons Ltd. Published 2020 by John Wiley & Sons Ltd.